Cellular Architecture and Dynamics in Female Meiosis

Binyam Mogessie
Editor

Cellular Architecture and Dynamics in Female Meiosis

Springer

Editor
Binyam Mogessie
Department of Molecular, Cellular, and Developmental Biology
Yale University
New Haven, CT, USA

ISBN 978-3-031-97172-3 ISBN 978-3-031-97173-0 (eBook)
https://doi.org/10.1007/978-3-031-97173-0

This Springer imprint is published by the registered company Springer Nature Switzerland AG
The registered company address is: Gewerbestrasse 11, 6330 Cham, Switzerland

Preface

It is with great enthusiasm that I present *Cellular Architecture and Dynamics in Female Meiosis*, a comprehensive volume that brings together a diverse and insightful collection of chapters exploring the intricate cellular and molecular mechanisms that govern female meiosis. This book aims to provide an integrated view of the dynamic processes underpinning chromosome behavior, spindle assembly, and cell cycle control in oocytes across a range of model systems, from worms to echinoderms to humans.

Female meiosis is a highly specialized and asymmetric cell division essential for reproductive success and species continuity. The precise orchestration of chromosome segregation, spindle dynamics, and cytoskeletal organization during oocyte maturation has profound implications for fertility, developmental potential, and the prevention of aneuploidy. In recent years, advances in live cell imaging, molecular genetics, and model organism research have dramatically expanded our understanding of these processes, yet many fundamental questions remain. This volume captures the current state of the field, highlighting emerging concepts, technological innovations, and outstanding challenges.

I would like to express my deepest appreciation to all contributing authors for their outstanding scholarship and dedication. Their expertise and collaborative spirit have made this volume possible. I am equally grateful to the peer reviewers whose thoughtful evaluations and constructive feedback have elevated the clarity and impact of each chapter.

It is my hope that this book will serve not only as a valuable reference for established investigators but also as an inspiration for trainees and new entrants to the field of reproductive cell biology. As our understanding of female meiosis continues to evolve, the discoveries and discussions presented here will undoubtedly stimulate further inquiry into this fascinating and essential biological process.

New Haven, CT, USA Binyam Mogessie

Contents

About the Editor

Binyam Mogessie studied Biochemistry and Cell Biology as an undergraduate at Jacobs University Bremen, Germany, and earned his Ph.D. in Cell Biology from the Institute of Cancer Research, University of London. He conducted postdoctoral research at the MRC Laboratory of Molecular Biology, Cambridge, and the Max Planck Institute, Göttingen, where he discovered novel functions of F-actin in chromosome segregation. In 2019, he established his independent research group at the University of Bristol as a Wellcome Trust and Royal Society Sir Henry Dale Fellow and HFSP Young Investigator. In 2022, he joined Yale University as an Assistant Professor of Molecular, Cellular, and Developmental Biology and OBGYN. His lab investigates cytoskeletal organization and chromosome segregation mechanisms in mammalian oocytes, advancing fundamental understanding of female reproductive cell biology.

Cytoskeletal Organization and Dynamics in Female Meiosis and Early Embryogenesis

Andrea Rix, Federica Giannini, Helena van Eck, and Binyam Mogessie

Abstract Insights into the interaction between actin and microtubule cytoskeletons during female meiosis and early embryogenesis have reshaped traditional perspectives on cell division. Once viewed as functionally exclusive entities, these cytoskeletal structures are now recognized for cooperatively driving vital processes like chromosome congression, proper segregation, and nuclear positioning in oocytes and zygotes. This review discusses recent findings that reveal how actin filaments aid in chromosome capture and alignment and influence microtubule-driven forces during segregation. In oocytes, spindle-associated actin helps stabilize kinetochore-microtubule connections and mitigates age-related loss of cohesion, thereby reducing the risk of chromosome segregation errors that exacerbate with maternal age. In zygotes, synchronized actin polymerization and microtubule motor activities facilitate pronuclear migration and parental genome unification. Innovative experimental methods, such as live-cell 4D microscopy and targeted protein degradation techniques, allow for the mechanistic exploration of cytoskeletal functions with remarkable spatial and temporal precision.

Keywords Actin · Microtubules · Meiosis · Oocyte · Aneuploidy · Cytoskeleton

Actin and microtubules have historically been considered as functionally non-interacting components of the cytoskeleton. However, coordination between these two cytoskeletal polymers in space and time is emerging as a common feature of living systems [1]. Surprising findings range from new roles of microtubules in muscle physiology [2–4] to a function of actin in spindle organization and

Andrea Rix, Federica Giannini and Helena van Eck contributed equally.

A. Rix · F. Giannini · H. van Eck · B. Mogessie (✉)
Department of Molecular, Cellular, and Developmental Biology, Yale University, New Haven, CT, USA
e-mail: binyam.mogessie@yale.edu

B. Mogessie (ed.), *Cellular Architecture and Dynamics in Female Meiosis*,
https://doi.org/10.1007/978-3-031-97173-0_1

chromosome dynamics [5–9], processes previously thought to be driven solely by either cytoskeletal system. As new discoveries continue to reveal completely unexpected functions of actin-microtubule crosstalk, understanding the mechanistic basis of these interactions and how they underpin cellular processes has become a fundamental open question in biology.

Mammalian oocytes and embryos offer exciting opportunities to interrogate functional crosstalk between actin and microtubules. Oocyte-to-embryo transition is underpinned by meiosis, a highly specialized form of cell division that generates eggs and is completed upon fertilization, leading to the formation of a single-cell embryo (zygote) [10]. Recent advances in long-term live-cell microscopy have enabled visualization of oocyte meiosis from nuclear envelope disassembly and spindle formation to chromosome segregation at high spatiotemporal resolution [11] (Fig. 1). Among discoveries driven by these technological advances is the finding that chromosome segregation, a highly conserved microtubule-based process, uniquely requires actin-microtubule crosstalk in oocytes [6]. Complete, high-resolution live imaging of pronuclear formation and migration similarly revealed that nuclear positioning in mammalian zygotes is motorized by coordination between actin and microtubule networks [10–12]. Integration of these quantitative microscopy assays with innovative targeted protein degradation tools is expected to provide answers to longstanding mysteries in female meiosis and early embryogenesis, such as the molecular origins of egg aneuploidies associated with female reproductive aging.

In this chapter, we highlight recent conceptual and technical advances in our understanding of cytoskeletal processes that support accurate chromosome

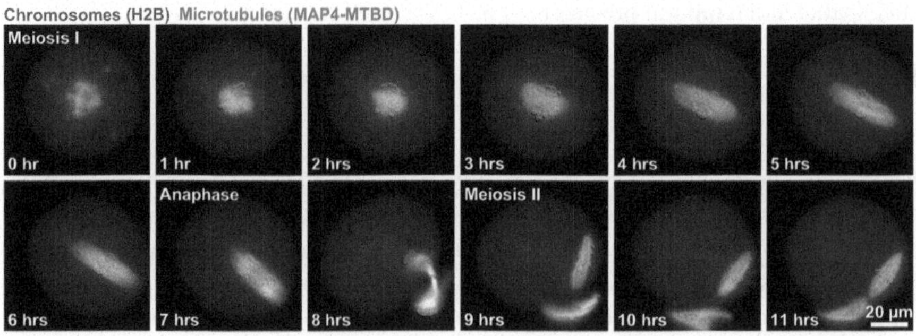

Fig. 1 High-resolution live imaging of meiotic spindle assembly and chromosome segregation in mouse oocytes. Laser scanning confocal microscopes with highly sensitive detectors enable gentle imaging of meiosis in live oocytes. In this experiment, mouse oocytes were microinjected with mRNAs encoding H2B-mScarlet (to label chromosomes) and mNeonGreen-MAP4-MTBD (to label microtubules). Imaging was performed using a Zeiss LSM 900 microscope equipped with a 40× C-Apochromat 1.2 NA water-immersion objective. Z-stacks spanning 81 µm were captured every 6 min, with confocal slices acquired at 1.5-µm intervals. Time lapse images are presented at 1-h intervals to highlight key meiotic events, including acentriolar microtubule organization center (aMTOC) convergence, nuclear envelope breakdown, spindle assembly and bipolarization, asymmetric spindle position, anaphase I chromosome segregation, and metaphase II arrest in the fertilizable egg

organization, transport, and segregation in mammalian oocytes and early embryos. We also underscore innovations that are required to push forward the state of the art in understanding cellular architecture and dynamics during female meiosis.

Emerging Roles of Cytoskeletal Crosstalk in Oocytes and Zygotes

A growing body of work has revealed that coordination between actin and microtubule networks is central to early development. Most notably, advanced microscopy-driven mechanistic studies showed that these cytoskeletal interactions govern chromosome organization and segregation in oocytes [13], as well as pronuclear migration in newly formed zygotes, processes generally thought to rely exclusively on microtubules [14].

Actin-Mediated Chromosome Congression in Large Eggs

Oocytes are unusually large cells, spanning ~80–130 μm in mammalian species [15, 16]. With such excess cytoplasmic volume come fundamental logistical challenges such as capturing and organizing chromosomes that are scattered after nuclear envelope disassembly. This hurdle is more pronounced in starfish oocytes, even larger cells (~180 μm) [17] where microtubules are insufficient to capture and congress chromosomes [18]. A contractile network of actin filaments was found to facilitate capture by transporting chromosomes closer to microtubule ends. However, this actin fishnet was found to collect scattered chromosomes through steric interactions rather than direct connection between DNA and actin filaments. Until recently, how actin-mediated chromosome congression is coordinated with timely formation of kinetochore–microtubule attachments in this system remained unknown. Mechanistic experiments revealed that Arp2/3-mediated branched patches of actin are quickly assembled on chromosomes after nuclear envelope breakdown [9]. Surprisingly, this is a Ran-GTP-dependent process, analogous to chromatin-mediated microtubule nucleation events that support meiotic and mitotic spindle assembly [19, 20]. These actin patches sterically block microtubule access to kinetochores until chromosomal transport is completed by the actin fishnet. Experimentally disassembling these actin patches immediately after nuclear envelope breakdown results in the premature capture of chromosomes by microtubules, which interferes with their accurate transport and causes chromosome loss. Although the underlying mechanisms of actin patch disassembly and how this is coordinated with the formation of stable kinetochore-microtubule attachments are still obscure, this seminal discovery highlights the importance of functional actin-microtubule interactions in female meiosis.

Actin-Based Chromosome Segregation in Oocytes

The role of actin and its associated myosin motors in cytokinesis has been studied in great mechanistic details across different models and continues to be a topic of extensive investigation [21–24]. However, despite early interest in actin and myosin involvement in spindle form and function dating back to more than five decades [25], a direct role of actin in chromosome segregation had until recently remained a mystery. The first evidence for actin-based chromosome segregation mechanism in a eukaryotic cell came from our study of mammalian oocyte meiosis. Using high-resolution quantitative microscopy assays, we found that meiotic spindles in oocytes of several species including humans contain prominent actin filaments that are aligned with microtubules [6]. Similarly to filaments that make up the oocyte cytoplasmic actin network and are required for vesicle transport [26, 27] and spindle positioning [28, 29], this spindle actin structure is assembled by the unbranched actin nucleation protein Formin-2. Disrupting spindle actin function genetically or chemically blocks the formation of robust, stable connections between kinetochores and microtubules that are critical for accurate chromosome segregation. Consequently, oocytes depleted of spindle actin are unable to faithfully separate chromosomes and display several lagging chromosomes during anaphase, common features of cancer cells [30–32]. These findings have challenged textbook knowledge of cell division principles because microtubules were previously thought to be the only cytoskeletal components directly involved in chromosome segregation. How spindle actin is assembled and exerts its function at the chromosome-microtubule interface is unknown and among fundamental open questions in the field. Addressing this conceptual gap in female meiosis requires overcoming technical limitations that have hampered mechanistic studies in reproductive cell biology areas (discussed later in this chapter). Interestingly, increasing the actin content of meiotic spindles is equally detrimental to chromosome segregation in oocytes [6]. Quantitative microscopy analyses of kinetochore-microtubule attachments revealed that excess spindle actin over-stabilizes these connections. Since initial kinetochore-microtubule attachments formed early in meiosis I are predominantly erroneous [33] and need to be destabilized by the error correction machinery [34], it is conceivable that premature stabilization of incorrect connections could predispose oocytes to severe chromosome segregation errors. Unraveling the regulatory basis for maintaining finely balanced spindle actin content in oocytes therefore necessitates identifying mechanisms that prevent excessive actin assembly inside meiotic spindles.

Meiotic Chromosome Cohesion in Aging Oocytes

Until their programmed separation during anaphase, oocyte chromosomes are held together by the cohesin complex, a highly conserved molecular glue mechanism consisting of the meiosis-specific cohesin subunit REC8 [35–38]. In most mammalian species including humans, cohesion proteins loaded onto chromosomes before birth must serve this gluing function throughout the oocyte's long lifespan [35]. Consistent with increasing oocyte aneuploidy rates during aging, REC8 and other cohesion proteins are gradually depleted from chromosome arms and centromeric regions of DNA [39–41]. Premature chromosome separation events arising from cohesion dysfunction are thus thought to predispose oocytes to chromosome segregation errors harmful to embryo development [35]. However, in depth mechanistic studies are needed to resolve decades of conjecture on how such gradual cohesion depletion triggers a near-exponential rise in oocyte aneuploidy at advanced reproductive ages. Because both ovarian and systemic aging are multifactorial processes [42, 43], it is highly plausible that synergistic dysfunction of other meiotic features and age-related cohesion depletion underlie aneuploidy in aged oocytes. Here, our discovery of a unique, actin-based chromosome segregation mechanism [6], whose disruption in aged oocytes accelerates premature chromosome separation [5], has provided new avenues to identify cytoskeletal origins of oocyte aneuploidy. In a recent study, we directly examined the role of age-related spindle actin dysfunction in inducing oocyte aneuploidy. By combining advanced microscopy assays [11] with rapid protein degradation tools [44], we innovated a unique experimental system in which REC8 cohesin levels can be experimentally manipulated to induce aging-like premature chromosome separation events [5]. Disrupting actin in this meiotic chromosome cohesion reduction system revealed that spindle-associated actin filaments prevent complete separation of weakly linked chromosomes when canonical cohesion mechanisms fail during female reproductive aging. Our findings collectively suggest that spindle actin exerts this protective function by regulating microtubule-based chromosome pulling forces (Fig. 2). How this works mechanistically and the molecular basis of meiotic spindle-specific actin disruption in aged oocytes require further investigation. Importantly, spindle actin loss in young eggs is sufficient to induce aging-like untimely chromosome separation events [5]. Age-related dysfunction of actin-based chromosome segregation mechanisms thus constitutes a new, cohesion depletion-independent cause of oocyte aneuploidy in reproductively older females.

Actin-Microtubule Coordination in Zygotic Pronuclear Migration

After fertilization, the large cytoplasmic volume of the zygote remains a logistical challenge, particularly for packaging the parental genomes into pronuclei and transporting them to the cell center [10]. In metazoans, this pronuclear migration is typically driven by microtubules and associated motor proteins [14, 45, 46]. Earlier studies focusing on later stages of this process revealed that the actin cytoskeleton also plays a key role in centering pronuclei in mouse zygotes [47]. With the development of the first high spatiotemporal resolution microscopy assay capable of capturing the entire course of mouse zygotic development—from pronuclear envelope

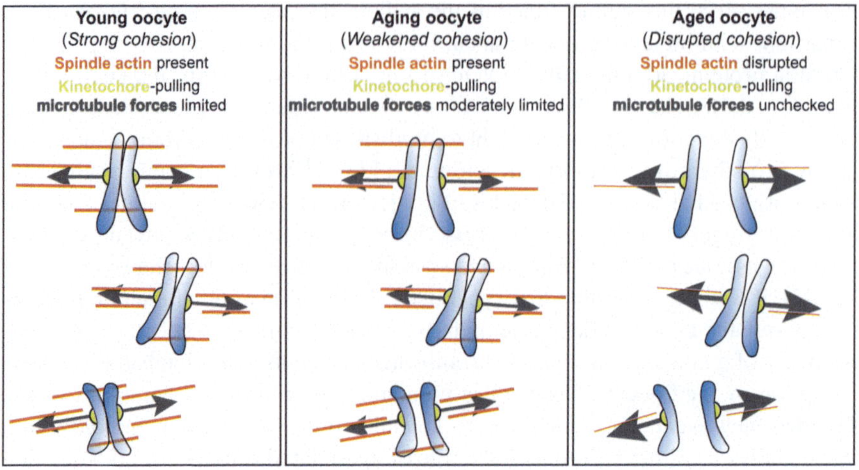

Fig. 2 Age-related disruption of spindle actin compromises sister chromatid cohesion in oocytes. This schematic illustrates the progressive effects of female reproductive aging on spindle actin integrity and sister chromatid cohesion in oocytes. In young oocytes (left panel), robust spindle actin (orange) is present, supporting strong cohesion between sister chromatids (blue) and limiting microtubule (gray) pulling forces at the kinetochores (yellow), thus maintaining proper chromosome association. In aging oocytes (middle panel), cohesion weakens while spindle actin remains present, leading to moderately increased kinetochore-mediated microtubule pulling forces and a greater risk of premature chromatid separation. In aged oocytes (right panel), spindle actin is disrupted, resulting in unregulated microtubule forces that prematurely separate sister chromatids due to severely compromised cohesion. These age-associated cytoskeletal changes contribute to chromosome segregation errors in female meiosis, increasing the risk of aneuploidy in oocytes from older individuals

formation to the first mitotic division—new evidence has emerged supporting a cooperative role between actin and microtubules in pronuclear migration [12].

Quantitative 3D analysis of nuclear volume and movement revealed that pronuclear migration occurs in two distinct phases. First, pronuclei are rapidly launched from the cell cortex by a burst of actin polymerization mediated by Spire proteins, nucleators that are required for actin network assembly and spindle migration in oocytes [48, 49]. The extent of this actin assembly depends on how close the pronuclei form to the cell surface: male pronuclei, which typically form closer to the cortex, trigger more robust actin polymerization than female pronuclei. When female pronuclei are experimentally positioned near the cortex, they can induce similarly strong actin polymerization responses. In the second, slower phase, pronuclei migrate toward the cell center via a network of microtubules, actin filaments, and dynein—a minus end-directed motor known to power pronuclear movement in other systems [45, 46]. While this study has advanced our understanding of actin-microtubule coordination, several questions remain. Specifically, the role of actin in the second migration phase is unclear, as actin disruption was found to also impair microtubule dynamics [12]. Additionally, dynein-dependent inward movement requires a net orientation of microtubule minus ends toward the center—yet such organization has not been observed, suggesting other, unknown cytoskeletal processes may be involved. Uncovering how this microtubule network assembles de novo after fertilization will be key. Moving forward, combining high-resolution live imaging assays developed here with innovative targeted protein loss-of-function studies will be essential for addressing these fundamental questions in reproductive cell biology.

Discovery-Enabling Technologies in Female Meiosis Research

Cytoskeletal organization and dynamics are fundamental to cellular structure and function in female meiosis [50] and beyond. Advancing mechanistic insights into these processes in mammalian oocytes faces two key challenges. Firstly, reproductive cell biology still lacks effective methods to visualize and measure endogenous protein dynamics under physiological conditions. Secondly, there are limited tools to control protein function with high spatial and temporal precision during complex events like oocyte meiosis. Emerging technologies such as advanced microscopy and tools for targeted protein degradation offer exciting opportunities to explore cytoskeletal architecture and cell division mechanisms in oocytes with unprecedented detail.

Quantitative 4D Microscopy

Mammalian oocytes are highly sensitive to light-induced damage and often fail to mature properly when exposed to standard live imaging techniques. Unlike most cultured cells, oocytes divide along a plane randomly oriented within the large cytoplasmic volume. Therefore, to maximize the chances of capturing spindle assembly and chromosome segregation during live imaging, the largest possible cell volume must be continuously visualized over the ~7 h period of oocyte maturation. Consequently, live imaging of mammalian oocyte meiosis involves a compromise between spatial and temporal resolution to minimize phototoxicity. Recent technological advances in microscopy have enabled detailed description of meiotic processes including cytoskeletal organization and function. In particular, the use of highly sensitive photon detectors (e.g., GaAsP, HyD, and Airyscan) combined with motorized piezo Z-stages now enables high-resolution live imaging of cytoskeletal dynamics and chromosome segregation in oocytes, reducing the need for high laser power and allowing fast, stable vertical stage movement [11]. The enhanced resolution makes it possible to move beyond subjective image analysis toward objective, quantitative approaches, using 3D volume reconstruction and tracking software to measure key meiotic processes such as spindle assembly, bipolarization, and chromosome dynamics [5, 6, 11, 12].

Beyond instrumentation, studying cellular dynamics in mammalian oocytes is complicated by a lack of probes that do not interfere with endogenous protein behavior. This is particularly challenging for the actin cytoskeleton, where commonly used actin probes (e.g., UtrCH [6, 25], actin chromobody [51]) can stabilize actin filaments. Additionally, proteins that are stably associated with cellular structures in developing oocytes, like CENP-A [52] and REC8 [53], are not amenable to overexpression of fluorescent variants for studying their dynamics. A promising solution is nanobodies, single-domain antibodies typically derived from Camelidae species [54–56]. We previously demonstrated that a fluorescence-tagged actin nanobody with a nuclear localization signal can visualize nuclear actin filaments in oocytes and generate stable nuclear actin structures [51]. However, the biophysical properties of this nanobody and its precise effect on actin filament behavior are not well understood, limiting its use in studying actin dynamics. Nanobodies targeting specific proteins can now be discovered using synthetic libraries and then optimized for high specificity and affinity through ribosome display technologies and biochemical approaches [57–60]. Furthermore, advancements in antibody engineering now allow conversion of well-established monoclonal antibodies into live imaging-compatible fluorescent intracellular antibodies [61]. These approaches offer transformative potential for high-resolution live imaging of endogenous proteins in mammalian oocytes without disrupting physiological function.

Targeted Protein Degradation

Despite major advances enabled by innovations in quantitative microscopy, the field is now at a juncture where new experimental tools are needed to gain deeper mechanistic understanding of how oocyte meiosis operates at the molecular level, and why this process becomes disrupted during female reproductive aging. A key requirement for this is the experimental ability to interrogate protein function in oocytes. Slow protein turnover and a generally transcriptionally silent oocyte genome [62, 63] impede genome editing and RNA interference efforts to study protein function in this system. Meiotic proteins are also often functionally redundant or play key roles in the especially prolonged mammalian oocyte growth and maturation process. Similarly, several cytoskeletal proteins are multifunctional and participate in various stages of oocyte meiosis (spindle assembly, spindle bipolarization, chromosome alignment, chromosome segregation). This severely constrains the use of knockout animal models to gain holistic understanding of protein function during oocyte meiosis.

The pioneering of TRIM-Away, a rapid protein degradation method, has addressed this key challenge by enabling targeted proteasomal degradation of endogenous proteins using heavy chain antibodies [44]. This powerful technique also allows for the disruption of protein function in specific meiotic processes, as the antibodies that mediate degradation can be introduced into cells at any desired stage of oocyte meiosis. As anticipated, its implementation has catalyzed a series of advances in the field, spanning from oocyte centromere biology [64, 65] and DNA repair [66] to early embryo nuclear and protein dynamics [12, 67–69]. Importantly, our recent discovery of actin's function in limiting age-related egg aneuploidy was made possible by using TRIM-Away in young oocytes to mimic chromosome cohesion defects associated with reproductive aging [5]. Manipulating protein function at a specific stage of a cellular process has remained a challenge across cell biology areas. Filling this technical gap, TRIM-Away has rapidly been implemented to interrogate fundamental biological processes in a variety of models including cultured mammalian cells [70], human embryos [71], fish embryos [72], insect cells [73], frog oocytes [74, 75], and frog egg extracts [76].

Despite the powerful new experimental capabilities that TRIM-Away offers for mechanistic studies of oocyte meiosis, several challenges continue to limit its broader use. First, the method depends on highly specific heavy chain antibodies, which are not reliably available from commercial sources. Second, delivering antibodies into cells requires microinjection or electroporation, techniques that are not easily scalable for widespread applications. Beyond these technical hurdles, performing TRIM-Away rescue experiments presents additional challenges. Specifically, ruling out off-target effects requires the generation of degradation-resistant protein variants that retain full functionality. This is a non-trivial task, as it demands prior knowledge of the antibody's binding site on the target protein. Finally, TRIM-Away is neither inducible nor reversible by design, limiting its potential for manipulating endogenous protein function with precise temporal control. In the original study that introduced TRIM-Away, we demonstrated that it could be coupled with nanobodies to degrade endogenous proteins [44]. Building on this, the synthetic nanobody discovery platforms discussed above offer significant potential to address the limitations associated with heavy chain antibody use. Because nanobodies can be expressed as fusion proteins, their integration into TRIM-Away experiments could further enable spatially targeted and more temporally controlled protein degradation assays.

Targeted protein degradation has emerged as a promising strategy for clinical interventions in cancer and other physiological disorders [77–80]. In particular, proteolysis-targeting chimeras (PROTACs), heterobifunctional molecules that induce the proteasomal degradation of target proteins, are being actively investigated for cancer therapeutics [79, 80]. Importantly, because PROTACs harness a natural cellular process present in healthy cells, they have also become valuable tools for investigating protein function across various biological pathways [79, 81]. For example, the PROTAC dTAG-13 enables the degradation of cellular proteins fused to a small fragment of FKBP12 [82, 83] (Fig. 3). Since FKBP12 itself plays essential roles in cellular processes such as cell cycle regulation [84], dTAG-13 was specifically designed to selectively recognize and degrade the engineered FKBP12^{F36V} mutant [83]. We have developed a new experimental system that leverages this PROTAC system to investigate the molecular origins of age-related egg aneuploidy in females. In this system, we generated transgenic mice in which REC8 is endogenously fused to FKBP12^{F36V}. Treatment of REC8-FKBP12^{F36V} oocytes from reproductively older females with dTAG-13 enables rapid, aging-like

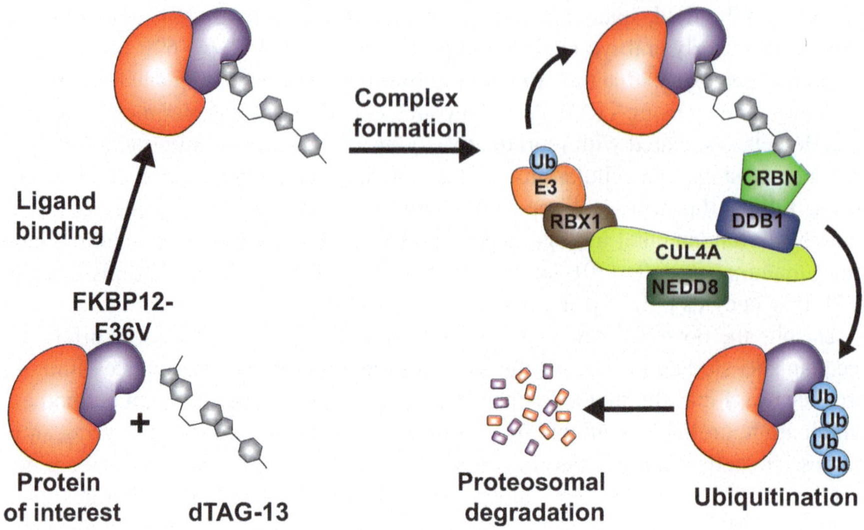

Fig. 3 PROTAC-mediated targeted degradation of FKBP12^{F36V}-tagged proteins for functional studies in cells. This schematic depicts the mechanism of induced protein degradation using a PROTAC (proteolysis-targeting chimera) strategy. A protein of interest is fused to an FKBP12^{F36V} tag, which selectively binds the small-molecule ligand dTAG-13. Upon ligand binding, the tagged protein forms a complex with the dTAG-13 molecule, which simultaneously recruits the CRBN E3 ubiquitin ligase complex. This recruitment leads to polyubiquitination of the FKBP12^{F36V}-tagged protein, marking it for proteasomal degradation. This system enables rapid, selective, and inducible degradation of target proteins, facilitating controlled studies of protein function and cellular phenotypes following acute protein loss

degradation of REC8 cohesin, leading to premature sister chromatid separation events characteristic of advanced maternal age [85]. Furthermore, combining PROTAC-mediated cohesin disruption with TRIM-Away revealed that aging-like depletion of the histone H3 variant CENP-A further accelerates premature chromatid separation and scattering [85]. Together, these findings highlight the experimental potential of PROTAC-based approaches to uncover previously inaccessible mechanisms underlying the high incidence of egg aneuploidy in reproductively older females.

Concluding Remarks

Cytoskeletal organization is essential for oocyte physiology and successful embryogenesis [50]. While our focus here has been on the functional interplay between the actin and microtubule cytoskeletons, it is likely that other cytoskeletal components also play crucial roles in female meiosis. Intermediate filaments [86], which are central to maintaining cellular structure and known to interact with actin and microtubules in other systems, remain understudied in oocytes. Additionally, septins [87], a polymeric system that is poorly understood in oocyte cytoskeletal function, represent promising candidates for supporting female meiosis. Ongoing innovations in advanced microscopy techniques and targeted protein degradation tools hold great potential for accelerating a comprehensive understanding of cytoskeletal organization and crosstalk mechanisms essential to female meiosis.

Acknowledgments This work was funded by a Wellcome Trust and Royal Society Sir Henry Dale Fellowship 213470/Z/18/Z (BM) and a National Institutes of Health grant R35GM146725 (BM). For the purpose of open access, the corresponding author has applied a CC BY public copyright license to any Author Accepted Manuscript version arising from this submission.

Author Contributions Conceptualization: AR, FG, HvE, BM.

Funding acquisition: BM.
Project administration: BM.
Writing—original draft: AR, FG, HvE.
Writing—review & editing: AR, BM.

Competing Interests Authors declare that they have no competing interests.

References

1. Dogterom M, Koenderink GH (2019) Actin-microtubule crosstalk in cell biology. Nat Rev Mol Cell Biol 20:38–54. https://doi.org/10.1038/s41580-018-0067-1
2. Mogessie B, Roth D, Rahil Z, Straube A (2015) A novel isoform of MAP4 organises the paraxial microtubule array required for muscle cell differentiation. elife 4:e05697. https://doi.org/10.7554/eLife.05697
3. Scarborough EA et al (2021) Microtubules orchestrate local translation to enable cardiac growth. Nat Commun 12:1547. https://doi.org/10.1038/s41467-021-21685-4
4. Dhanyasi N, VijayRaghavan K, Shilo BZ, Schejter ED (2021) Microtubules provide guidance cues for myofibril and sarcomere assembly and growth. Dev Dyn 250:60–73. https://doi.org/10.1002/dvdy.227
5. Dunkley S, Mogessie B (2023) Actin limits egg aneuploidies associated with female reproductive aging. Sci Adv 9:eadc9161. https://doi.org/10.1126/sciadv.adc9161
6. Mogessie B, Schuh M (2017) Actin protects mammalian eggs against chromosome segregation errors. Science 357:eaal1647. https://doi.org/10.1126/science.aal1647
7. Weber KL, Sokac AM, Berg JS, Cheney RE, Bement WM (2004) A microtubule-binding myosin required for nuclear anchoring and spindle assembly. Nature 431:325–329. https://doi.org/10.1038/nature02834
8. Kita AM et al (2019) Spindle-F-actin interactions in mitotic spindles in an intact vertebrate epithelium. Mol Biol Cell 30:1645–1654. https://doi.org/10.1091/mbc.E19-02-0126
9. Burdyniuk M, Callegari A, Mori M, Nedelec F, Lenart P (2018) F-actin nucleated on chromosomes coordinates their capture by microtubules in oocyte meiosis. J Cell Biol 217:2661–2674. https://doi.org/10.1083/jcb.201802080
10. Mogessie B, Scheffler K, Schuh M (2018) Assembly and positioning of the oocyte meiotic spindle. Annu Rev Cell Dev Biol 34:381–403. https://doi.org/10.1146/annurev-cellbio-100616-060553

11. Mogessie B (2020) Visualization and functional analysis of spindle actin and chromosome segregation in mammalian oocytes. Methods Mol Biol 2101:267–295. https://doi.org/10.1007/978-1-0716-0219-5_17
12. Scheffler K et al (2021) Two mechanisms drive pronuclear migration in mouse zygotes. Nat Commun 12:841. https://doi.org/10.1038/s41467-021-21020-x
13. Schneider I, Lenart P (2017) Chromosome segregation: is the spindle all about microtubules? Curr Biol 27:R1168–R1170. https://doi.org/10.1016/j.cub.2017.09.022
14. Zhang Y, Kawashima T (2025) Cytoskeletal dynamics of gamete nuclear migration in flowering plants, animals, and yeast. Curr Top Dev Biol 162:33–53. https://doi.org/10.1016/bs.ctdb.2024.10.010
15. Kyogoku H, Kitajima TS (2023) The large cytoplasmic volume of oocyte. J Reprod Dev 69:1–9. https://doi.org/10.1262/jrd.2022-101
16. Kitasaka H, Konuma Y, Tokoro M, Fukunaga N, Asada Y (2022) Oocyte cytoplasmic diameter of >/=130 mum can be used to determine human giant oocytes. F&S Sci 3:10–17. https://doi.org/10.1016/j.xfss.2021.11.004
17. Terasaki M et al (2001) A new model for nuclear envelope breakdown. Mol Biol Cell 12:503–510. https://doi.org/10.1091/mbc.12.2.503
18. Lenart P et al (2005) A contractile nuclear actin network drives chromosome congression in oocytes. Nature 436:812–818. https://doi.org/10.1038/nature03810
19. Kalab P, Heald R (2008) The RanGTP gradient—A GPS for the mitotic spindle. J Cell Sci 121:1577–1586. https://doi.org/10.1242/jcs.005959
20. Dumont J et al (2007) A centriole- and RanGTP-independent spindle assembly pathway in meiosis I of vertebrate oocytes. J Cell Biol 176:295–305. https://doi.org/10.1083/jcb.200605199
21. Pollard TD (2017) Nine unanswered questions about cytokinesis. J Cell Biol 216:3007–3016. https://doi.org/10.1083/jcb.201612068
22. Pollard TD (2010) Mechanics of cytokinesis in eukaryotes. Curr Opin Cell Biol 22:50–56. https://doi.org/10.1016/j.ceb.2009.11.010
23. Pollard TD, O'Shaughnessy B (2019) Molecular mechanism of cytokinesis. Annu Rev Biochem 88:661–689. https://doi.org/10.1146/annurev-biochem-062917-012530
24. Satterwhite LL, Pollard TD (1992) Cytokinesis. Curr Opin Cell Biol 4:43–52. https://doi.org/10.1016/0955-0674(92)90057-j
25. Sandquist JC, Kita AM, Bement WM (2011) And the dead shall rise: actin and myosin return to the spindle. Dev Cell 21:410–419. https://doi.org/10.1016/j.devcel.2011.07.018
26. Holubcova Z, Howard G, Schuh M (2013) Vesicles modulate an actin network for asymmetric spindle positioning. Nat Cell Biol 15:937–947. https://doi.org/10.1038/ncb2802
27. Schuh M (2011) An actin-dependent mechanism for long-range vesicle transport. Nat Cell Biol 13:1431–1436. https://doi.org/10.1038/ncb2353
28. Schuh M, Ellenberg J (2008) A new model for asymmetric spindle positioning in mouse oocytes. Curr Biol 18:1986–1992. https://doi.org/10.1016/j.cub.2008.11.022
29. Dumont J et al (2007) Formin-2 is required for spindle migration and for the late steps of cytokinesis in mouse oocytes. Dev Biol 301:254–265. https://doi.org/10.1016/j.ydbio.2006.08.044
30. Ganem NJ, Godinho SA, Pellman D (2009) A mechanism linking extra centrosomes to chromosomal instability. Nature 460:278–282. https://doi.org/10.1038/nature08136
31. Gregan J, Polakova S, Zhang L, Tolic-Norrelykke IM, Cimini D (2011) Merotelic kinetochore attachment: causes and effects. Trends Cell Biol 21:374–381. https://doi.org/10.1016/j.tcb.2011.01.003
32. Silkworth WT, Nardi IK, Scholl LM, Cimini D (2009) Multipolar spindle pole coalescence is a major source of kinetochore mis-attachment and chromosome mis-segregation in cancer cells. PLoS One 4:e6564. https://doi.org/10.1371/journal.pone.0006564
33. Kitajima TS, Ohsugi M, Ellenberg J (2011) Complete kinetochore tracking reveals error-prone homologous chromosome biorientation in mammalian oocytes. Cell 146:568–581. https://doi.org/10.1016/j.cell.2011.07.031

34. Zahm JA, Harrison SC (2024) A communication hub for phosphoregulation of kinetochore-microtubule attachment. Curr Biol 34:2308–2318.e2306. https://doi.org/10.1016/j.cub.2024.04.067
35. Herbert M, Kalleas D, Cooney D, Lamb M, Lister L (2015) Meiosis and maternal aging: insights from aneuploid oocytes and trisomy births. Cold Spring Harb Perspect Biol 7:a017970. https://doi.org/10.1101/cshperspect.a017970
36. Garcia-Cruz R et al (2010) Dynamics of cohesin proteins REC8, STAG3, SMC1 beta and SMC3 are consistent with a role in sister chromatid cohesion during meiosis in human oocytes. Hum Reprod 25:2316–2327. https://doi.org/10.1093/humrep/deq180
37. Ishiguro KI (2019) The cohesin complex in mammalian meiosis. Genes Cells 24:6–30. https://doi.org/10.1111/gtc.12652
38. Lee JY, Orr-Weaver TL (2001) The molecular basis of sister-chromatid cohesion. Annu Rev Cell Dev Biol 17:753–777. https://doi.org/10.1146/annurev.cellbio.17.1.753
39. Lister LM et al (2010) Age-related meiotic segregation errors in mammalian oocytes are preceded by depletion of cohesin and Sgo2. Curr Biol 20:1511–1521. https://doi.org/10.1016/j.cub.2010.08.023
40. Chiang T, Duncan FE, Schindler K, Schultz RM, Lampson MA (2010) Evidence that weakened centromere cohesion is a leading cause of age-related aneuploidy in oocytes. Curr Biol 20:1522–1528. https://doi.org/10.1016/j.cub.2010.06.069
41. Mihalas BP et al (2024) Age-dependent loss of cohesion protection in human oocytes. Curr Biol 34:117–131.e115. https://doi.org/10.1016/j.cub.2023.11.061
42. Wang X, Wang L, Xiang W (2023) Mechanisms of ovarian aging in women: a review. J Ovarian Res 16:67. https://doi.org/10.1186/s13048-023-01151-z
43. Kroemer G et al (2025) From geroscience to precision geromedicine: understanding and managing aging. Cell 188:2043–2062. https://doi.org/10.1016/j.cell.2025.03.011
44. Clift D et al (2017) A method for the acute and rapid degradation of endogenous proteins. Cell 171:1692–1706.e1618. https://doi.org/10.1016/j.cell.2017.10.033
45. Meaders JL, Burgess DR (2020) Microtubule-based mechanisms of pronuclear positioning. Cells 9:505. https://doi.org/10.3390/cells9020505
46. Reinsch S, Karsenti E (1997) Movement of nuclei along microtubules in Xenopus egg extracts. Curr Biol 7:211–214. https://doi.org/10.1016/s0960-9822(97)70092-7
47. Chaigne A et al (2016) F-actin mechanics control spindle centring in the mouse zygote. Nat Commun 7:10253. https://doi.org/10.1038/ncomms10253
48. Pfender S, Kuznetsov V, Pleiser S, Kerkhoff E, Schuh M (2011) Spire-type actin nucleators cooperate with Formin-2 to drive asymmetric oocyte division. Curr Biol 21:955–960. https://doi.org/10.1016/j.cub.2011.04.029
49. Montaville P et al (2014) Spire and Formin 2 synergize and antagonize in regulating actin assembly in meiosis by a ping-pong mechanism. PLoS Biol 12:e1001795. https://doi.org/10.1371/journal.pbio.1001795
50. Dunkley S, Scheffler K, Mogessie B (2022) Cytoskeletal form and function in mammalian oocytes and zygotes. Curr Opin Cell Biol 75:102073. https://doi.org/10.1016/j.ceb.2022.02.007
51. Scheffler K, Giannini F, Lemonnier T, Mogessie B (2022) The prophase oocyte nucleus is a homeostatic G-actin buffer. J Cell Sci 135:jcs259807. https://doi.org/10.1242/jcs.259807
52. Smoak EM, Stein P, Schultz RM, Lampson MA, Black BE (2016) Long-term retention of CENP-A nucleosomes in mammalian oocytes underpins Transgenerational inheritance of centromere identity. Curr Biol 26:1110–1116. https://doi.org/10.1016/j.cub.2016.02.061
53. Tachibana-Konwalski K et al (2010) Rec8-containing cohesin maintains bivalents without turnover during the growing phase of mouse oocytes. Genes Dev 24:2505–2516. https://doi.org/10.1101/gad.605910
54. Dong JX et al (2019) A toolbox of nanobodies developed and validated for use as intrabodies and nanoscale immunolabels in mammalian brain neurons. elife 8:e48750. https://doi.org/10.7554/eLife.48750

55. Barakat S, Berksoz M, Zahedimaram P, Piepoli S, Erman B (2022) Nanobodies as molecular imaging probes. Free Radic Biol Med 182:260–275. https://doi.org/10.1016/j.freeradbiomed.2022.02.031
56. Hamers-Casterman C et al (1993) Naturally occurring antibodies devoid of light chains. Nature 363:446–448. https://doi.org/10.1038/363446a0
57. Hanes J, Schaffitzel C, Knappik A, Pluckthun A (2000) Picomolar affinity antibodies from a fully synthetic naive library selected and evolved by ribosome display. Nat Biotechnol 18:1287–1292. https://doi.org/10.1038/82407
58. Schaffitzel C, Hanes J, Jermutus L, Pluckthun A (1999) Ribosome display: an in vitro method for selection and evolution of antibodies from libraries. J Immunol Methods 231:119–135. https://doi.org/10.1016/s0022-1759(99)00149-0
59. Dreier B, Pluckthun A (2011) Ribosome display: a technology for selecting and evolving proteins from large libraries. Methods Mol Biol 687:283–306. https://doi.org/10.1007/978-1-60761-944-4_21
60. Dreier B, Pluckthun A (2018) Rapid selection of high-affinity antibody scFv fragments using ribosome display. Methods Mol Biol 1827:235–268. https://doi.org/10.1007/978-1-4939-8648-4_13
61. Galindo G et al (2025) AI-assisted protein design to rapidly convert antibody sequences to intrabodies targeting diverse peptides and histone modifications. bioRxiv. https://doi.org/10.1101/2025.02.06.636921
62. De La Fuente R, Eppig JJ (2001) Transcriptional activity of the mouse oocyte genome: companion granulosa cells modulate transcription and chromatin remodeling. Dev Biol 229:224–236. https://doi.org/10.1006/dbio.2000.9947
63. Zuccotti M, Bellone M, Longo F, Redi CA, Garagna S (2011) Fully-mature antral mouse oocytes are transcriptionally silent but their heterochromatin maintains a transcriptional permissive histone acetylation profile. J Assist Reprod Genet 28:1193–1196. https://doi.org/10.1007/s10815-011-9562-4
64. Dudka D et al (2025) Satellite DNA shapes dictate pericentromere packaging in female meiosis. Nature 638:814–822. https://doi.org/10.1038/s41586-024-08374-0
65. El Yakoubi W, Akera T (2023) Condensin dysfunction is a reproductive isolating barrier in mice. Nature 623:347–355. https://doi.org/10.1038/s41586-023-06700-6
66. Leem J, Kim JS, Oh JS (2023) Oocytes can repair DNA damage during meiosis via a microtubule-dependent recruitment of CIP2A-MDC1-TOPBP1 complex from spindle pole to chromosomes. Nucleic Acids Res 51:4899–4913. https://doi.org/10.1093/nar/gkad213
67. Israel S et al (2023) Intracellular fraction of zona pellucida protein 3 is required for the oocyte-to-embryo transition in mice. Mol Hum Reprod 29:gaad038. https://doi.org/10.1093/molehr/gaad038
68. Israel S, Drexler HCA, Fuellen G, Boiani M (2021) The COP9 signalosome subunit 3 is necessary for early embryo survival by way of a stable protein deposit in mouse oocytes. Mol Hum Reprod 27:gaab048. https://doi.org/10.1093/molehr/gaab048
69. Zhou CJ et al (2019) Loss of CENPF leads to developmental failure in mouse embryos. Cell Cycle 18:2784–2799. https://doi.org/10.1080/15384101.2019.1661173
70. Fant CB et al (2020) TFIID enables RNA polymerase II promoter-proximal pausing. Mol Cell 78:785–793.e788. https://doi.org/10.1016/j.molcel.2020.03.008
71. Zhang T et al (2025) Maternal transcription factor OTX2 directly induces SETD1A and promotes embryonic genome activation in human pre-implantation embryos. Sci China Life Sci. https://doi.org/10.1007/s11427-024-2875-3
72. Chen X et al (2019) Degradation of endogenous proteins and generation of a null-like phenotype in zebrafish using Trim-away technology. Genome Biol 20:19. https://doi.org/10.1186/s13059-019-1624-4
73. Sun X et al (2022) Generation of resistance to Nosema bombycis (Dissociodihaplophasida: Nosematidae) by degrading NbSWP12 using the ubiquitin-proteasome pathway in Sf9-III cells. J Econ Entomol 115:2068–2074. https://doi.org/10.1093/jee/toac145

74. Heim A, Niedermeier ML, Stengel F, Mayer TU (2022) The translation regulator Zar1l controls timing of meiosis in Xenopus oocytes. Development 149:dev200900. https://doi.org/10.1242/dev.200900
75. Schunk R et al (2025) A phosphate-binding pocket in cyclin B3 is essential for XErp1/Emi2 degradation in meiosis I. EMBO Rep 26:768–790. https://doi.org/10.1038/s44319-024-00347-8
76. Mevissen TET, Prasad AV, Walter JC (2023) TRIM21-dependent target protein ubiquitination mediates cell-free TRIM-away. Cell Rep 42:112125. https://doi.org/10.1016/j.celrep.2023.112125
77. Churcher I (2018) Protac-induced protein degradation in drug discovery: breaking the rules or just making new ones? J Med Chem 61:444–452. https://doi.org/10.1021/acs.jmedchem.7b01272
78. Lai AC, Crews CM (2017) Induced protein degradation: an emerging drug discovery paradigm. Nat Rev Drug Discov 16:101–114. https://doi.org/10.1038/nrd.2016.211
79. Burslem GM, Crews CM (2020) Proteolysis-targeting chimeras as therapeutics and tools for biological discovery. Cell 181:102–114. https://doi.org/10.1016/j.cell.2019.11.031
80. Bekes M, Langley DR, Crews CM (2022) PROTAC targeted protein degraders: the past is prologue. Nat Rev Drug Discov 21:181–200. https://doi.org/10.1038/s41573-021-00371-6
81. Sun X et al (2019) PROTACs: great opportunities for academia and industry. Signal Transduct Target Ther 4:64. https://doi.org/10.1038/s41392-019-0101-6
82. Nabet B et al (2020) Rapid and direct control of target protein levels with VHL-recruiting dTAG molecules. Nat Commun 11:4687. https://doi.org/10.1038/s41467-020-18377-w
83. Nabet B et al (2018) The dTAG system for immediate and target-specific protein degradation. Nat Chem Biol 14:431–441. https://doi.org/10.1038/s41589-018-0021-8
84. Aghdasi B et al (2001) FKBP12, the 12-kDa FK506-binding protein, is a physiologic regulator of the cell cycle. Proc Natl Acad Sci U S A 98:2425–2430. https://doi.org/10.1073/pnas.041614198
85. Leem J et al (2025) A versatile cohesion manipulation system reveals CENP-A dysfunction accelerates female reproductive age-related egg aneuploidy. bioRxiv. https://doi.org/10.1101/2025.02.27.640570
86. Pradeau-Phelut L, Etienne-Manneville S (2024) Cytoskeletal crosstalk: a focus on intermediate filaments. Curr Opin Cell Biol 87:102325. https://doi.org/10.1016/j.ceb.2024.102325
87. Spiliotis ET, Nakos K (2021) Cellular functions of actin- and microtubule-associated septins. Curr Biol 31:R651–R666. https://doi.org/10.1016/j.cub.2021.03.064

Conserved Functions of Mos-MAPK in Oocyte Meiosis

Ivan Avilov and Peter Lenart

Abstract The Mos kinase is exclusively expressed in oocytes and is degraded upon fertilization. When present, Mos constitutively activates the ERK/MAPK pathway. This atypical activation of ERK/MAPK during meiotic divisions has been documented across animal phyla, from the non-bilaterian cnidarians to mammals. However, the inherent diversity of reproductive strategies and the consequent variability in timing and morphology of meiotic divisions complicates the identification of conserved molecular mechanisms under the control of Mos-MAPK. In this review, we synthesize data from different species in an attempt to identify conserved downstream functional modules. We first provide an overview of the cellular functions necessary for execution of the specialized meiotic divisions of oocytes, and detail those that have been found to be under the control of Mos-MAPK. We then discuss how Mos-MAPK may be a universal regulator of the meiotic cell cycle by controlling translation of cyclin B and other factors. Secondly, we discuss how ensuring the asymmetry of meiotic divisions may be another conserved function regulated by Mos-MAPK acting on the microtubule spindle and on the cytokinetic machinery.

Keywords Oocyte meiosis · Oocyte maturation · Mos-MAPK · Cell division · Regulation of translation

---

I. Avilov · P. Lenart (✉)
Research Group Cytoskeletal Dynamics in Oocytes, Max Planck Institute for Multidisciplinary Sciences, Göttingen, Germany
e-mail: peter.lenart@mpinat.mpg.de

B. Mogessie (ed.), *Cellular Architecture and Dynamics in Female Meiosis*,
https://doi.org/10.1007/978-3-031-97173-0_2

Conservation and Diversity in Oocyte Meiosis

Oocyte Meiosis Is a Specialized form of Cell Division Producing the Fertilizable Egg

The female gamete, the oocyte, is an exceptionally large cell that stores nutrients to support embryo development. Oocytes also feature a very large nucleus, commonly referred to as the germinal vesicle (GV). After fertilization by the sperm, the zygote uses proteins, mRNAs, lipids, and other nutrients stored in the oocyte's cytoplasm and nucleus to head-start embryo development. Male and female gametes are highly specialized in adaptation to this "division of labor" between small and motile sperm cells produced in large numbers having to find the few, large and immotile eggs. Consequently, the processes of spermatogenesis and oogenesis differ substantially in their morphology and timing.

In oocytes, the meiotic prophase, during which homologous chromosomes pair and recombine, is temporally separated from the actual meiotic divisions. This often involves a selection process whereby the mother organism only "invests" into growing those oocytes to full size, in which meiotic recombination has been successfully completed. The entire process of oocyte growth occurs then later in the meiotic prophase, which is generally referred to as the prophase arrest state. In this review, we will focus on the subsequent steps of the actual oocyte divisions, commonly referred to as oocyte maturation. For an extensive discussion of oogenesis, including its earlier steps, we refer the reader to a recent excellent review by Jessus and coworkers [1].

Conserved Features of Oocyte Meiosis: Reductional, Asymmetric, and Coordinated with Fertilization

Undoubtedly, the most important function of meiotic divisions—equally for sperm and egg—is the reduction of the genomic content, i.e., haploidization. This is achieved by running two division cycles without DNA replication in-between: during meiosis the oocyte undergoes two divisions (M-phases), meiosis I (MI) and meiosis II (MII) without an intervening S-phase. Biochemically, it is well understood that these M-phases are driven by the conserved cell cycle module culminating in the activation of the Cdk1-Cyclin B complex (also known as MPF, i.e., M-phase or maturation promoting factor [2]). Indeed, MPF as a maturation promoting activity was first identified in *Xenopus* and starfish oocytes [3, 4], followed by yeast genetics and biochemical work in early embryos to define the molecular identity of Cdk1-Cyclin B [5, 6]. The key to meiotic function here is that Cdk1-Cyclin B is reactivated after meiosis I in such a way that the oocyte does not enter an interphase state and does not undergo an S-phase, as it would be the case in mitosis. At the cellular level, this typically correlates with the absence of nuclear reformation between meiosis I and II (although there appears to be some diversity between species, and only in a few cases has this process been characterized in detail [7]). Strikingly, after fertilization, just a few hours later in the same egg cytoplasm, rapid mitotic cell cycles take place with alternating M- and S-phases, and with the nuclear envelope disassembling and re-forming in every cycle. Taken together, one critical meiosis-specific feature of the cell cycle is the reactivation of Cdk1-Cyclin B immediately after meiosis I to drive meiosis II without an S-phase.

The second key meiotic feature, specific to female gametes, is that the division is highly asymmetric—first recognized by Theodor Boveri [8]. The purpose of this is to preserve all accumulated nutrients in a single large egg, rather than producing four equal daughter cells [9]. Therefore, the extra genomic content is expelled into small and unviable polar bodies [10]: half of the genome, a complete set of chromosomes is segregated into the first polar body in meiosis I. Chromatids of replicated chromosomes are then segregated into the second polar body in meiosis II. It is important to recognize that polar body divisions are fundamentally distinct in their function from other forms of asymmetric division. Asymmetric divisions produce two viable daughter cells with different sizes and typically also different fates. The function of polar body divisions is instead to produce the smallest possible "container" to remove the extra genomic content while minimizing cytoplasmic volume to prevent the further development of polar bodies, and thereby control embryo number. In other words, in the case of polar bodies, it is an important part of the function that polar bodies do not survive. If they were, they would compete with the egg for fertilization and development. In rare cases this occurs in humans, whereby polar body fertilization can lead to parasitic twin formation [11]. In the mouse, the formation of two mature eggs within the same *zona pellucida* can lead to blastomere mixing of the two embryos [12].

Thirdly, oocyte meiosis has to be coordinated with other events of reproduction. It is a general, widely conserved feature that maturation is timed to start just before fertilization is expected to occur. Thus, the induction of the meiotic divisions is directly linked to the hormonal stimuli regulating mating and reproduction [1]. Only fully-grown oocytes are receptive to this hormonal stimulus, which ensures that only oocytes enter meiosis that are able to support embryo development. In most species, there is then a second synchronization event to halt meiotic progression awaiting fertilization or spawning/ovulation. In vertebrates, this arrest is at the metaphase of meiosis II and is referred to as CSF arrest (cytostatic factor arrest) and is released by fertilization. In other species, oocytes may arrest at the metaphase of meiosis I or at the G1 stage after both meiotic divisions are completed [13].

Diversity in Meiosis Results from Adaptation to Varied Reproductive Strategies

While the three features listed are essential for oocyte meiosis across animal species, there are major species-to-species differences in the way they are executed; in their morphology, and in their timing. This should not come as a surprise, as animals show an enormous diversity in reproductive strategies. Oocyte meiosis needs to continuously "keep up" with these changes, i.e., cellular mechanisms need to evolve that ensure the production and survival of the gametes in diverse environments and adapted morphologies. As a prominent example, oocytes and eggs have very different sizes ranging from tens of micrometers to centimeters in diameter [7]. There are large differences even between closely related species driven by the amount of nutrients they need to store until the embryo is able to feed on its own—which in turn is highly dependent on the lifestyle of the species [14]. How does cellular architecture, such as cytoskeletal systems adapt to cope with these enormous size differences?

As an example, we now understand the basic laws of how microtubule structures are built, what determines their shape, and what limits their size [15]. This predicts that in cells smaller than approximately 140 micrometers in diameter, astral microtubules of the spindle are able to reach to the cell cortex [16]. Therefore, microtubules directly interacting with the cell cortex can position the spindle in a smaller egg cell, in *C. elegans*, for example [17]. However, the same model also predicts that the majority of species with oocytes larger than 140 μm must have evolved different mechanisms for spindle positioning—which remain largely uncharacterized. A similar scaling problem exists early in cell division: the efficiency of chromosome capture is dependent on the distance between chromosomes and microtubule nucleation centers (centrosomes or acentrosomal microtubule organizing centers) [18]. Across species and cell types, the capture range of microtubule asters appears to be rather universal and is approximately 30 micrometers [19]. In oocytes, chromosomes scattered in the large oocyte nucleus (GV) need to be collected from much larger distances of up to several hundreds of micrometers. This requires additional mechanisms to facilitative chromosome capture, including specialized microtubule structures in *Xenopus* [20], and actin-driven mechanisms that have been characterized in starfish and more recently in mammalian species [21, 22].

Besides the variable physical size of oocytes, the timing of meiosis also shows substantial diversity: whether fertilization occurs internally or externally, whether on the land or in a marine environment, there are substantial differences in when and where oocytes need to be released from the ovary and in what stage and for how long should they be waiting for fertilization [1]. Therefore, while the fact that maturation needs to be coordinated with fertilization is universal, the actual schedule of meiotic progression is widely variable among animal species. For example, the time from hormone stimulus to maturation varies from many hours (e.g., frogs or mammalian species) to just about 10–20 min in marine species such as starfish or the jellyfish *Clytia* [1]. When and for how long oocytes then arrest awaiting fertilization is also highly variable. In vertebrate oocytes, the MII arrest is released by fertilization which may be correlated with the fact that vertebrate species typically mate, so spawning/ovulation is reliably followed by fertilization [13]. Starfish oocytes instead sense the event of spawning likely by responding to the change in pH upon release from the ovary to the outside environment, where then fertilization occurs [23].

Taken together, meiotic divisions produce the fertilizable egg that necessitates specialized functions such as reduction of the genomic content, the high asymmetry of divisions to produce a single, large egg cell, and the coordination of maturation with fertilization and other events of reproduction. These features are common to animal species. At the same time, there is a huge diversity in the size and geometry of oocytes and the timing of divisions adapted to the widely diverse reproductive strategies of animal species. This raises the exciting question of to what extent are the underlying molecular mechanisms evolutionarily conserved, and in which cases diversification led to the evolution of new solutions to carry out the function in a different geometry or with a different timing.

Mos-MAPK Is a Conserved Regulator of Oocyte Meiosis

Mos Is a Constitutive Activator of the ERK/MAPK Cascade Exclusively Expressed in Oocytes

Mos is a constitutive activator of the extracellular signal-regulated kinase/mitogen-activated protein kinase (ERK/MAPK) cascade (Fig. 1). Specifically, Mos is a MAPKKK, a kinase that directly phosphorylates MEK1, the MAPK-kinase [24]. In somatic cells, the ERK/MAPK pathway is regulated by extracellular signals, such as growth factors, and is thereby one of the main pathways controlling proliferation, survival, and differentiation. Therefore, if Mos is erroneously expressed in somatic cells, it acts as a potent oncogene through overactivation of ERK/MAPK that originally led to its identification as the transforming gene of the Moloney murine sarcoma virus [25, 26].

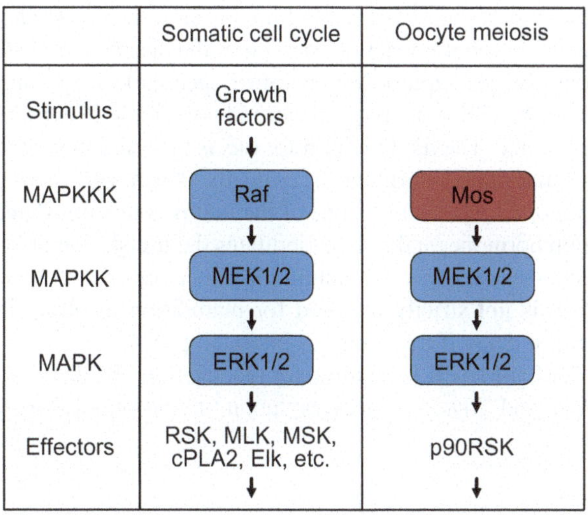

Fig. 1 Mos constitutively activates the ERK-MAPK pathway in oocyte meiosis. *MAPK* mitogen-activated protein kinase, *MAPKK* MAPK-kinase, *MAPKKK* MAPK-kinase-kinase, *ERK* extracellular signal-regulated kinase, *MEK* MAPK/ERK-kinase

Mos is physiologically not expressed in somatic cells; the Mos protein is present exclusively in meiotic oocytes and is degraded upon fertilization. This expression pattern appears to be strictly conserved across animal species, with very few exceptions (for example, in mice, Mos is also expressed in developing spermatids [27]). Mos is found in oocytes of all vertebrate species investigated, it is present in tunicates [28], and it has been extensively characterized in the echinoderm starfish [23, 29–31], as well as in the cnidarian jellyfish, *Clytia hemisphaerica* [32]. Mos is also found in *Drosophila*, however, it appears to play a less important role [33], and in *C. elegans* the Mos gene is absent [34, 35]. To the contrary, it has been shown that Mos-MAPK is involved in CSF arrest in the sawfly [36]. Overall, this indicates that the Mos-MAPK pathway is an ancient essential regulator of oocyte meiosis across eumetazoa that appears to have secondarily lost its importance in some protostome species.

The Mos-MAPK Pathway Mediates Key Meiotic Functions

What are the functions of this atypical activation of ERK/MAPK in meiosis? Broadly, Mos-MAPK appears to be involved in two types of functions: control of cell cycle progression and cytoskeletal regulation. For the former, Mos-MAPK is known to have a role in (*i*) initiating meiotic maturation in response to the maturation hormone; (*ii*) in reactivation of cdk1-cyclin B right after meiosis I and thereby mediating the skipping S-phase between MI and MII; and finally, (*iii*) Mos-MAPK also plays a critical role in the arrests awaiting fertilization. As for cytoskeletal regulation, Mos-MAPK's functions generally appear to ensure the asymmetry of polar body divisions by regulating both the microtubule spindle and the actin cytoskeleton.

Mos was first characterized in *Xenopus* oocytes and was found to be essential for meiotic maturation [37]—although more recent studies demonstrated that with a delay maturation can occur even in the absence of Mos [38]. However, as it turned out, this function of Mos is rather specific to frogs, and in the majority of species, Mos-MAPK is not required for inducing meiotic entry in response of the maturation hormone. The likely reason for this is that only in a minority of species the response to maturation hormone is critically dependent on translation of new protein. In *Xenopus*, Mos itself is one of the first proteins translated in response to the maturation hormone, and it then facilitates the translation of key cell cycle regulators [39]. In most other species, maturation either does not require translation at all, or translation is not strictly required for maturation resulting in no observable effect or a minor delay if translation is inhibited [40–43].

Mos-MAPK is required for reactivation of Cdk1-Cyclin B after meiosis I in starfish and *Xenopus* oocytes: upon inhibition of Mos-MAPK, oocytes exit after meiosis I, re-form the nucleus and initiate S-phase [29, 30, 44]. Whether Cdk1-Cyclin B reactivation is under the control of Mos-MAPK is less clear in other species. In mouse oocytes, results are somewhat varied as in one of the knockout lines Mos does not seem to be required for Cdk1-Cyclin B reactivation [45], whereas in another knockout line reformation of the nucleus after meiosis I is observed in a substantial proportion of the oocytes [46]. In the jellyfish, *Clytia* meiosis II appears to occur even when Mos-MAPK is inhibited [32].

There is an even greater degree of diversity in the regulation of meiotic arrest points. On the one hand, it is remarkable that Mos-MAPK is required for the CSF-type arrest in all species investigated so far. Indeed, the most prominent and most universal phenotype of Mos-MAPK inhibition is the override of CSF-like arrest and induction of parthenogenetic development [47, 48]. This clearly indicates conservation. On the other hand, the timing of the arrest, whether the arrest occurs in meiosis I, meiosis II, or the G1 phase following meiosis, is variable, and as far as we know, there is significant diversity in the underlying molecular mechanisms as well.

Mos-MAPK's functions in the regulation of the asymmetric divisions appear similarly conserved and diverse at the same time. Across species, Mos-MAPK inhibition results in enlarged spindles and leads to the formation of abnormally large or aborted polar bodies. To the reverse, in the case of the tunicate *Phallusia mammillata,* Mos overexpression in the fertilized egg was shown to be sufficient to turn the first mitotic divisions highly asymmetric and meiotic like [28]. On the other hand, the size of the oocytes, the morphology of spindles, the way spindles are positioned varies widely between species. Therefore, the effects and phenotypes of Mos-MAPK inhibition also show substantial diversity in different species, as we will detail below.

We summarized published functions of Mos-MAPK in different species in Table 1. In the subsequent sections, we will discuss possible evolutionarily conserved molecular mechanisms under the control of the Mos-MAPK pathway that mediate this wide range of meiotic functions. As detailed above, the conservation of the Mos protein, its expression pattern, the almost universally observed meiotic activation of ERK/MAPK, as well as the conservation of functions controlled by Mos-MAPK all argue for conservation of the underlying molecular mechanisms. At the same time, this conserved core must have diversified substantially, lost some, and gained additional functions in adaptation to the diverse reproductive strategies observed across species.

We will discuss here two hypotheses: (*i*) May translational regulation of stored mRNAs, specifically those encoding Cyclin B and other key cell cycle proteins, explain a large portion of Mos-MAPK's functions in cell cycle regulation? (*ii*) May there be conserved molecular modules regulating the meiotic spindle and cytokinesis under the control of Mos-MAPK to "switch" between the equal mitotic and asymmetric meiotic divisions?

Table 1 Conserved functions of Mos-MAPK observed across animal species

Mos-MAPK functions	Species
Hormone-induced entry into meiosis	*Xenopus* [37]
Driving meiosis II/S-phase skipping	Mouse [49]
	Xenopus [50]
	Starfish [30]
Arrest awaiting fertilization/CSF arrest	Mouse [47, 48]
	Xenopus [37]
	Tunicates [28]
	Starfish [30]
	Sawfly *Athalia rosae* [36]
	Clam [51]
	Jellyfish [32]
Downregulation of spindle poles/astral microtubules	Somatic cells [52]
	Mouse [45, 49]
	Xenopus [53]
	Tunicates [28]
	Starfish [31]
	Urechis [54]
	Jellyfish [32]
Spindle translocation to the cortex	Mouse [45, 49]
	Xenopus [53]
	Jellyfish [32]
Reduction of polar body size	Human [55]
	Mouse [45, 49]
	Tunicates [28]
	Starfish [30]
	Jellyfish [32]

Regulation of Translation by the Mos-MAPK Pathway

Cytoplasmic Polyadenylation Regulates Translation in Oocyte Meiosis

Fully-grown oocytes store large amounts of maternal mRNA while being transcriptionally inactive. This renders translation of maternal mRNAs the predominant mechanism to control protein amounts in oocytes. Indeed, meiotic progression is guided by the translation of mRNAs of cell cycle regulatory proteins and their targeted degradation [56]. Before maturation, the majority of stored maternal mRNAs are kept "dormant," that is, translationally inactive by multiple mechanisms and generally have relatively short (<20 bases) poly(A) tails [57, 58]. Among other mechanisms, extension of this tail by the poly(A) polymerase enzyme (PAP) renders the mRNAs competent for translation [59]. Polyadenylation of maternal mRNAs depends on their 3′ untranslated regions (UTRs). 3'-UTRs carry short conserved sequences recognized by specific RNA-interacting proteins, that in turn recruit PAP [60]. Among these, best studied are the highly conserved polyadenylation signal hexanucleotide (AAUAAA, a.k.a. HEX) recognized by cleavage and polyadenylation specificity factor (CPSF). Second is the meiosis-specific U-rich cytoplasmic polyadenylation element (CPE), which is regulated by the CPE-binding protein CPEB [61].

CPEB plays a dual role in the storage of maternal mRNAs by both protecting them from degradation and preventing their translation [62]. According to detailed studies in *Xenopus*, CPEB binds two adjacent (<100 bases apart) CPE elements and forms a dimer that assembles a translation repression complex [63]. This controls the polyadenylation of the CPE-containing mRNAs in several ways [62]. In *Xenopus*, a complex consisting of Maskin (a.k.a. TACC3), Symplekin, the poly(A)-specific nuclease PARN, and PAP assembles on CPEB [64]. It has been proposed that Maskin binds the 5'cap-binding protein, eIF4E, and thereby interferes with the ability of eIF4E to bind to the mRNA cap and to initiate translation. Others identified another complex recruited to CPEB composed of DDX2, PAT1, LSM14A and the eIF4E-binding protein, 4E-T (a.k.a. EIF4ENIF1) [65]. In this complex, CPEB represses translation by interacting with 4E-T and eIF4E1b, a germline specific paralog of eIF4E that is unable to support translation as it cannot interact with eIF4G. Interestingly, sea urchin 4E-T is a specific translation inhibitor of the cyclin B mRNA that may suggest conservation of this mechanism [66].

While the un-phosphorylated form CPEB inhibits translation by the mechanisms outlined above, phosphorylation switches CPEB into a translational activator. Upon phosphorylation, CPEB's affinity to CPSF is increased which results in the activation of the cytoplasmic poly(A) polymerase (GLD2), and displacement of PARN [64].

In Xenopus, Two Distinct Waves of Polyadenylation Define Phases of Meiosis

In *Xenopus*, two major polyadenylation waves have been distinguished, each driving the translation of a specific subset of meiotic regulators [60, 67]. While there are substantial species-specific variations, this serves as a good example of how translational regulation can guide multiple events and ensure their temporal order.

The maturation hormone, progesterone induces the *early* polyadenylation wave, which drives entry to meiosis. In *Xenopus,* these *early* translated proteins include cyclins B2 and B5, Mos, and the APC/C inhibitor Emi1, all of which are required for the meiosis entry. The *early* wave involves transcripts that contain the early-acting Musashi binding element (MBE) in their 3'-UTR [68] and contain an *early* CPE configuration of two adjacent CPEs. These CPE elements are populated by excess CPEB that is phosphorylated in a Cdk1-independent manner. Initial reports showed that upon stimulation by progesterone, CPEB is phosphorylated by the

Aurora A kinase [69, 70]. However, there is evidence that CPEB at this stage may be phosphorylated by other kinases, namely xGEF and xRINGO [71].

As mentioned above, the strict dependence of meiotic maturation on translation is rather specific to *Xenopus*. Consistently, the *early* wave shows species-specific features. In *Xenopus,* Mos translation strongly depends on the early 3'UTR binding protein Musashi [72, 73], which is not the case for other species. Another specific feature appears to be an opposing feedback loop that is mediated by C3H4, a zinc-finger protein involved in the destabilization of AU-rich element (ARE)-containing mRNAs. In meiosis I, this results in de-adenylation of Emi1 mRNA, the subsequent degradation of Emi1 protein, which in turn is required to allow exit from metaphase I [74, 75].

The second, *late* wave of polyadenylation occurs at metaphase I and is responsible for the meiosis-specific skipping of the S-phase, by driving the reactivation of Cdk1-Cyclin B through the translation of cyclin B1 and B4 [67]. The *late* wave is characterized by cdk1-mediated CPEB phosphorylation, resulting in partial degradation of CPEB. Furthermore, the *late* mRNAs feature a CPE overlapping with the polyadenylation signal (HEX). In the *late* wave, CPEB destruction opens these HEX-overlapping CPEs, creating a binding site for CPSF. In addition, while the *early* CPEs are predominantly regulated by CPEB1, this is gradually replaced by CPEB4 for *late* mRNAs [76].

Thus in *Xenopus, early* cyclins B5 and B2 with an *early* CPE configuration are translated in an MBE-dependent manner and are driving hormone-induced meiotic entry. The *late* cyclins B1 and B4 with a *late* CPE configuration drive cdk1-cyclinB reactivation between meiosis I and meiosis II [68]. In contrast, in mouse oocytes, hormone-induced translation is not required for maturation. Correspondingly, cyclin B1 transcripts contain either short 3'UTRs without a CPE, likely responsible for baseline, unregulated cyclin B1 synthesis. The other form of cyclin B1 mRNAs contains a long 3'UTR with multiple CPE elements, which appear to be translated in a Mos-MAPK-dependent manner [77].

Temporal Control of Cyclin B Translation Is a Key Regulator of the Meiotic Cell Cycle

Thus, it is clear that translational regulation is critical for meiotic progression. From many studies in multiple species, a consistent picture emerges that among stored transcripts, translation of Cyclin B is one of the main drivers of meiotic events. On the one hand, this is expected because the meiotic M-phases, as mitotic M-phase, are driven by activation of Cdk1-Cyclin B. As oocytes are typically transcriptionally inactive, Cyclin B needs to be produced from stored mRNAs. On the other hand, this raises the question of how does meiosis-specific regulation of cyclin B synthesis underlie meiosis-specific features of the cell cycle.

In *Xenopus*, Cyclin B is synthesized early and is required for hormone-induced meiotic entry, whereas in other species activation of stored Cdk1-Cyclin B complexes appear to be mostly sufficient to drive this first step of maturation. However, the maintenance of high Cdk1-Cyclin B activity through meiosis I to II transition is critical for skipping S-phase between the two meiotic divisions, and is likely to involve new Cyclin B synthesis in most if not all species. In *Xenopus*, it is the above detailed *late* polyadenylation wave that is driving thc continued synthesis of new Cyclin B protein after meiosis I. The excess Cyclin B protein then appears to override degradation by the anaphase-promoting complex (APC/C) in anaphase I, and thereby prevents the total Cdk1-Cyclin B activity to drop below the threshold to initiate interphase [78, 79]. In many species, the CSF-type cytostatic arrest halts cell cycle progression in metaphase of the first or the second meiotic division. This means that high cdk1-cyclin B activity is maintained during CSF arrest, which also requires a shifted balance between production and degradation of Cyclin B. Taken together, translational regulation of Cyclin B levels appears to be one core module of the meiotic cell cycle involved in induction of meiotic maturation, driving the second meiotic division and CSF-type cell cycle arrests.

Evidence for Regulation of Cyclin B Translation by Mos-MAPK

There is a strong correlation between the above functions mediated by Cyclin B translation and those under the control of Mos-MAPK. In *Xenopus* oocytes, Mos-MAPK is essential for all three main functions: Mos is required for meiotic maturation, reactivation of Cdk1-Cyclin B after meiosis I, and Mos inhibition also overrides the CSF arrest [37, 80, 81]. While the essential role of Mos-MAPK in meiotic maturation has been challenged by more recent studies, these are consistent with the picture that Mos-MAPK is a potent enhancer of Cdk1-Cyclin B activation—by facilitating Cyclin B synthesis—and therefore required for timely maturation [38]. In mouse oocytes, observations are somewhat varied that may necessitate further clarification of Mos-MAPK function in this species. Mos depletion by morpholinos, data from two different knock-out strains, and inhibition of the Mos-MAPK pathway by small-molecule inhibitors resulted in slightly different phenotypes [45, 46, 49]. Nevertheless, there is consensus that in mice Mos is not required for induction of maturation, but it may be needed for reactivation of Cdk1-Cyclin B [46], and it is certainly required for maintenance of CSF arrest [47, 48]. In starfish oocytes, Mos is not required for meiotic maturation, which in starfish is completely mediated by a post-translational signaling pathway [30]. However, Mos is clearly essential for the reactivation of Cdk1-Cyclin B and driving meiosis II in starfish, as well as for the maintenance of the CSF-like arrest in G1 [30]. In the jellyfish *Clytia*, Mos-MAPK is also required for maintenance of the CSF arrest [32]. Taken together, wherever data is available, there is a close correlation between functions dependent on Mos-MAPK activity and those requiring translation of Cyclin B.

How much evidence is there for a direct molecular link between these processes? Without a known exception, CPEB emerges as a widely conserved regulator of meiotic translation. Indeed, in *Xenopus* oocytes, MAPK directly phosphorylates CPEB before NEBD, thereby participating in the *early* polyadenylation wave required for meiosis entry [71]. After NEBD, the polyadenylation of *late* Cyclin B1 is also Mos-dependent [82]. In mouse oocytes, the *late* Cyclin B1 translation requires not only CPEB phosphorylation but also CPEB degradation which in mice has been shown to depend on Mos-MAPK [77, 83]. Interestingly, the Mos protein itself has been shown in several species to be under the control of the same translational regulation [77, 84]. This suggests that Mos-MAPK may act in a positive feedback loop to facilitate Cyclin B translation. In starfish oocytes, meiosis I proceeds normally even when translation is completely inhibited [85, 86]. Strikingly, the very few proteins that show changes in protein levels include CPEB with some of its regulators and, prominently, Cyclin B [86, 87]. Although in the starfish species *A. aranciacus* and *M. glacialis* no evidence has been found for Mos-MAPK-dependent CPEB phosphorylation [87], studies in another starfish species, *A. pectinifera* indicate that translation of Cyclin B does depend on CPEB phosphorylation [23, 88].

Taken together, evidence from multiple species indicates that Cyclin B translation is facilitated by Mos-MAPK activity, in particular between meiosis I and II. This mechanism appears to be controlled by CPEB phosphorylation, likely directly downstream of the Mos-MAPK pathway. CPEB phosphorylation then leads to polyadenylation of Cyclin B mRNA, and subsequent translation of the Cyclin B protein. The identification of the specific CPEB kinase and the sites of phosphorylation remain important future tasks, ideally addressed in multiple species to allow evolutionary comparisons.

The Functions of Mos-MAPK in the Regulation of Asymmetric Meiotic Divisions

The Organization of the Microtubule Spindle

The molecular mechanisms underlying the organization of the microtubule spindle are beginning to be rather well understood, detailed in recent excellent reviews [89]. While historically centrosomes were considered as essential constituents of spindle assembly in animal species, it is now clear that a bipolar spindle can also assemble without centrosomes. As a prominent example, in most oocytes, centrioles are eliminated during the meiotic prophase, nevertheless oocytes are able to form a bipolar spindle and faithfully segregate chromosomes. We now also understand the mechanisms of acentrosomal spindle formation that have been studied extensively both in vertebrate and invertebrate animals, reviewed in [90, 91]. One possibility is that in absence of centrioles and thus proper centrosomes, aMTOCs (acentriolar microtubule organizing centers) functionally replace centrosomes. Alternatively, chromatin-dependent, Ran-GTP-mediated microtubule nucleation mechanisms appear to take the major role [92]. Indeed, there seems to be a large variability between species in the extent to which aMTOCs versus chromatin-mediated microtubule nucleation contributes to meiotic spindle assembly. For example, among mammalian species, the mouse meiosis I spindle is dominated by aMTOCs, and correspondingly it is only weakly affected by inhibition of the Ran pathway [93]. By contrast, human oocyte spindles appear to almost fully depend on chromatin and Ran-GTP-dependent microtubule nucleation [94].

The aMTOCs can be seen as organelles largely equivalent to centrosomes. Apart from the fact that they are not nucleated by centrioles, they contain a similar set of pericentriolar material proteins. These include the pericentriolar material components pericentrin, CEP120, CEP192, and NEDD1, as well as the microtubule nucleator γ-tubulin. Intriguingly and as shown recently in mouse and other mammalian oocytes, components of aMTOCs are organized at spindle poles into a liquid-like spindle domain (LISD), a phase-separated condensate of key microtubule regulators [95].

As for the chromosomal nucleation pathway, Ran-GTP produced by its chromatin-bound exchange factor RCC1 is responsible for the activation of several spindle assembly factors such as HURP, NuMa, and TPX2 [9, 96, 97]. HURP is a microtubule-associated protein that stabilizes microtubules in the proximity of mitotic chromosomes and helps sort them into a bipolar structure [96]. The NUMA protein anchors microtubule minus ends at the spindle poles [97]. Another, and best characterized Ran-dependent factor is TPX2, which has several roles in spindle assembly. Firstly, it is a microtubule nucleation factor [98], secondly, it phosphorylates Maskin/TACC3 and thereby regulates spindle pole organization [99], and thirdly it binds AURKA, and enhances its activity [99–101].

Meiotic Spindles Have Much-Reduced Asters and Interpolar Microtubules

Theodor Boveri already recognized the fundamental difference between the asymmetric meiotic divisions and subsequent embryonic mitoses and contributed this difference to the absence of centrosomes in oocytes [8]. Furthermore, he proposed that the absence of centrosomes in meiosis is crucial to prevent parthenogenesis [91]. However, we now know that while in most species centrioles are indeed eliminated from oocytes, there are also many variations on this theme. In rodents, oocytes and sperm both lack centrioles and thus meiotic as well as the first mitotic divisions are acentrosomal [102, 103]. The size and morphology of meiotic and mitotic divisions are nevertheless strikingly different. Conversely, in some mollusks, as well as in echinoderms meiotic spindles do contain centrioles, and meiotic and mitotic spindle morphologies are again very distinct [104–109]. Thus, independent of whether spindles have centrioles or not, meiotic spindles and mitotic spindles are consistently very different from one another, demonstrated by sampling several species across animal phylogeny [105] (Fig. 2).

The most prominent difference is that in mitosis spindles feature microtubule asters emanating from the centrosomes that interact with and establish a mechanical link to the cell cortex. This is required for the orientation of the division plane, and it also contributes to the fidelity of chromosome segregation [90]. In contrast, meiotic spindles have shorter and less abundant astral microtubules [105]. These short meiotic asters are key to meiotic function: they allow the spindle to localize close to the cortex, and thereby making an essential contribution to reduce the volume of polar bodies. Consistently, if the size of the asters is artificially increased, for example, by stabilization of microtubules by taxol or by other similar treatments, oocytes form larger polar bodies [110].

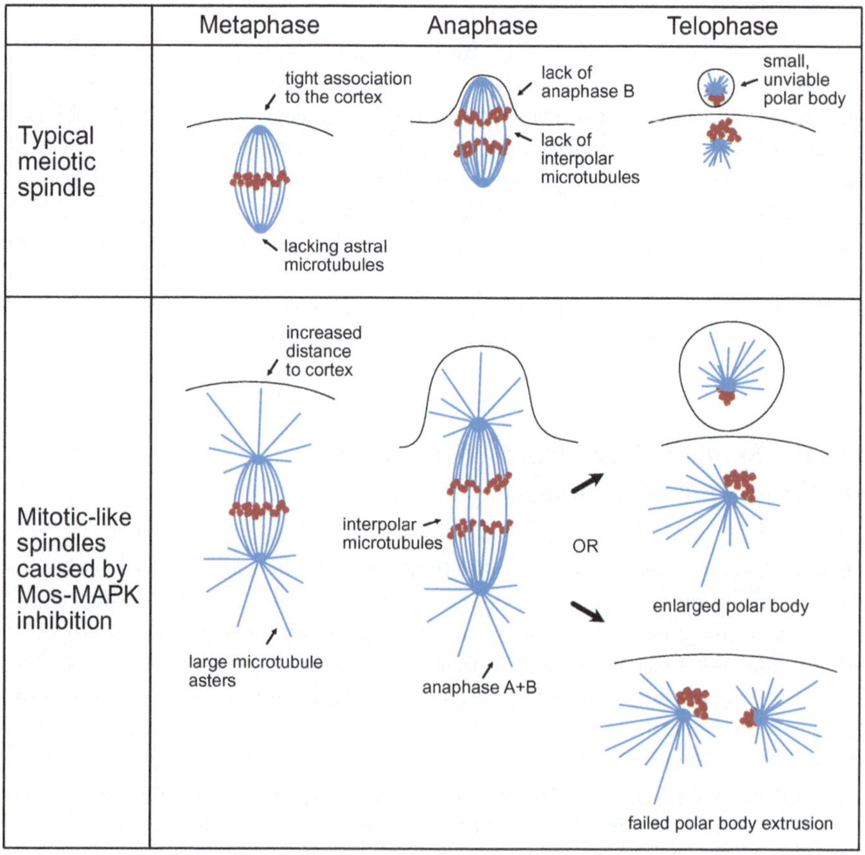

Fig. 2 Schematics comparing meiotic spindle architecture to mitotic-like architecture upon Mos-MAPK inhibition

Possibly linked to the reduced number and length of centrosome-nucleated microtubules, another long-observed difference between mitotic and meiotic spindles is the "discontinuity" of the microtubule array at the spindle midzone [105]. In mitotic spindles long microtubules extending from each centrosome to the midzone and beyond are clearly observed, and referred to as interpolar microtubules [111]. By contrast, electron tomography studies in *C. elegans* oocytes have found few if any microtubules that would reach from the poles to the midzone or the chromosomes [112]. Meiotic spindles appear to be constituted almost exclusively from a short, tiled array of microtubules not only in *C. elegans*, but in *Drosophila* oocytes and *Xenopus* oocyte extracts as well [113–115].

In mitotic spindles, the interpolar microtubules cross-link to one another and form antiparallel bundles that stabilize the mitotic spindle [116]. After the separation of sister chromatids in anaphase, these antiparallel bundles constitute the central spindle that promotes the initiation of cytokinesis [111]. In meiotic spindles these interpolar microtubules seem to be largely absent. As a consequence, meiotic spindles appear to rely almost exclusively on the shortening of the kinetochore fibers for pulling chromosomes apart, referred to as anaphase A [117, 118]. In mitosis, in parallel to anaphase A, the overlapping interpolar microtubule arrays are pushed apart by kinesin motors leading to the separation of the spindle poles, referred to as anaphase B. Thus, while in mitotic anaphase, both anaphase A and B are observed, in oocyte meiosis anaphase is dominated by anaphase A. This serves as another key mechanisms to minimize the size of the polar body.

Evidence for Regulation of Meiotic Spindle Architecture by Mos-MAPK

Practically at the same time around 1996, three independent groups reported that Mos-MAPK inhibition results in severe spindle abnormalities in mouse meiosis [45, 46, 49], which later were confirmed in other species [28, 31, 32, 53]. Overall, Mos-MAPK inhibition made spindles appear more "mitotic-like," larger and featuring large aster-like structures. On the other hand, Mos overexpression in somatic cells or embryos resulted in a reversed "meiotic-like" spindle phenotype with short asters that fail to position relative to the cortex [28, 52].

As an important note, and as explained above, in several species Mos-MAPK activity is required to drive the second meiotic division. Therefore, when Mos-MAPK is inhibited, meiosis II is skipped, and the first meiotic division is followed by the first mitosis, after which embryos continue parthenogenetic development [30, 49]. This caused some confusion in the literature, because what has been considered by some as an abnormal meiosis II spindle may in fact be a rather normal mitotic spindle. This is further complicated by the fact that these parthenogenetic mitotic spindles are not completely normal, because depending on the exact way parthenogenetic development proceeds in a given species, these spindles may contain the half or the double number of chromosomes. However, in most species Mos-MAPK inhibition already affects the morphology of the first meiotic spindle, which is independent of Mos-MAPK's role in meiotic cell cycle regulation, and may thus be better suited for such comparisons.

Looking into the details, Mos-MAPK inhibition appears to downregulate spindle poles and the growth of astral microtubules independently of whether the oocyte has a centrosomal or acentrosomal spindle. In the species with acentrosomal spindles, such as mouse, *Xenopus* and the jellyfish *Clytia* Mos-MAPK inhibition results in the appearance of astral-like microtubules that are almost completely absent in untreated meiotic spindles [32, 45, 46, 49, 53]. In species that retain centrosomes in oocyte meiosis, such as *Urechis* and starfish, Mos-MAPK inhibition has an even more obvious effect to activate centrosomes in meiosis [31, 54]. By downregulation of spindle poles, Mos-MAPK facilitates positioning the spindle close to the oocyte cortex. Consistently, Mos-MAPK inhibition in mouse and jellyfish impairs the normal off-center localization of the spindle, leading to formation of enlarged polar bodies [32, 45, 49]. In addition, Mos-MAPK activity has been shown to be required for the formation of proper attachments to kinetochores, and in mouse, *Xenopus* and the *Clytia* jellyfish Mos-MAPK inhibition results in chromosome misalignment leading to aneuploidy [32, 45, 46, 49, 53]. Finally, Mos-MAPK also appears to play an important role later in meiotic divisions. As detailed above, meiotic spindles typically segregate chromosomes by shortening of the kinetochore fibers, i.e., anaphase A. Anaphase B, the separation of poles, typically does not occur in meiosis. This has only been looked at in much fewer species, but in mouse oocytes spindles in Mos-MAPK inhibited oocytes keep elongating during anaphase [119]. Whether this is a direct Mos effect or just a mechanism to compensate for the off-centric spindle positioning, remains to be investigated.

Taken together, these observations indicate conserved roles of Mos-MAPK in regulating spindle architecture acting as a switch that changes the spindle to the meiotic morphology adapted to the highly asymmetric meiotic divisions, to produce the smallest possible polar body. What do we know about the molecules that may underlie these changes in spindle morphology controlled by Mos-MAPK? First, there is an early evidence showing that Mos may directly interact with and phosphorylate tubulin [120–122] as well as microtubule regulators. In particular, in mouse oocytes, MISS [123] and DOC1R [124] have been identified as regulators specifically under the control of Mos-MAPK. While these regulators identified so far serve as important examples, systematic identification of downstream targets will be necessary to mechanistically understand how Mos-MAPK regulates spindle architecture.

The Lack of Astral Microtubules Necessitates Meiosis-Specific Mechanisms of Spindle Assembly

Short asters are critical in meiosis to reduce the size of the polar body, but at the same time, they are then unable to perform their original function in spindle positioning. This necessitates meiosis-specific mechanisms that function independently of astral microtubules. For example, in mammals, the meiotic spindle is translocated to the cortex in an actin-dependent manner [93, 119] and in starfish, the chromosomes are translocated by contraction of a disassembling actin network to the site of spindle assembly [125].

It is likely that due to the reduced number of interpolar microtubules, the meiotic spindle also requires additional mechanisms to support its mechanical stability, especially in anaphase. Intriguingly, the spindle in mouse oocytes [126, 127] as well as in human, pig, and sheep oocytes [128] is permeated by prominent actin filaments (spindle actin), required for the accurate segregation of chromosomes in both meiosis I and II [128]. Notably, actin has also been reported in meiotic spindles in a variety of non-mammalian species [129], including *Xenopus* oocytes [130], and spermatocytes of insects [131].

Regulation of Polar Body Extrusion

The geometry of polar body extrusion differs from cytokinesis in somatic cell division. While in somatic cells principally the spindle midzone defines the position of the cleavage furrow, during polar body formation an extrusion, and out-bulging forms in parallel with the closure of the cleavage furrow. In fact, typically the polar body bud containing the spindle protrudes from the oocyte cortex first, which is then followed by closing, by "pinching off" the bud neck during polar body cytokinesis. Both of these processes, bulging and closure involve the regulation of the actin cortex. However, as it emerges from recent studies, the regulatory mechanisms are distinct. While closure, the "pinching off" of the polar body is mediated by the same spindle-induced RhoA module as somatic cytokinesis, bulging out of the polar body appears to rely on another small GTPase regulating the actin cortex, Cdc42 [132].

The "preparatory step" for polar body formation is called cortical polarization which takes place in prometaphase. Best characterized in mouse oocytes, in this process the cortex at the animal pole undergoes a reorganization manifesting in a decreased density of cortical granules [133] and removal of microvilli forming the so-called actin cap [134]. While there are substantial differences between species, similar processes have been observed from ascidians to mammals [135–138].

In mouse oocytes, the pathway leading to the polarization of the cortex is well understood. The formation of the actin cap coincides with spindle migration and has been shown to depend on the proximity of chromosomes to the cortex that are able to induce cortical polarization even in the absence of microtubules [134, 139, 140].

Indeed, Ran-GTP produced on chromatin is involved [139]: where chromosomes are in close proximity to the cortex, Ran-GTP activates two other small GTPases Rac [9] and Cdc42 [141, 142]. Rac recruits and activates N-WASP, which in turn activates the Arp2/3 complex to nucleate a branched network of actin filaments [140, 143, 144]. Cdc42 is a plasma membrane-bound GTPase that in addition to Rac is also contributing to N-WASP localization at the polarized cortex [145–147] and therefore is critical for cortical polarization [143, 145]. Cdc42 facilitates the formation of the plasma membrane protrusion by pulling the spindle outwards the cortex. Additionally, Cdc42 and Rac define the inner boundary of the contractile ring by antagonizing the RhoA activity as shown in *Xenopus* [132, 148] and in echinoderms [149]. A recent publication also revealed that a distinct pathway downstream of Cdc42 acting through the Myotonic dystrophy kinase-related Cdc42-binding kinase (MRCK) controls Myosin-2 in the actin cap and promotes polar body formation [150].

Interestingly, Mos-MAPK appears to be involved in this process of cortical polarization. In the mouse, Mos-MAPK has been shown to mediate actin cap formation by the Arp2/3 complex activator WAVE2 [151]. Furthermore, Mos-MAPK inhibition as well as genetic Mos knock-out perturbs chromatin-dependent actin cap formation [139, 152]. Although Ran does not affect the global MAPK activity, it has been proposed that Ran may restrict Mos-MAPK activity to chromosomes [139]. Thereby, the activities of Mos-MAPK and Ran may establish a positive feedback loop: chromosomes at the cortex induce cortical polarization, which in turn pushes chromosomes closer to the cortex and promotes actin cap maintenance [140, 146]. In addition, myosin light chain kinase (MLCK) has been shown to be phosphorylated by Mos-MAPK [153]. MLCK and downstream myosins are involved in the movement of the spindle to the cortex and in the closure of the contractile furrow [127]. This is supported by the fact that in Mos knock-out mice or when Mos-MAPK is inhibited, spindle migration does not occur [45, 46, 154]. Besides mice, it has been shown in starfish oocytes that Mos-MAPK regulates the Diaphanous formin which is responsible for the cleavage furrow closure [31].

Taken together, Mos-MAPK controls multiple steps of meiotic divisions that all appear to serve the purpose of rendering these divisions asymmetric in order to produce the smallest possible polar bodies, and a single, large mature egg. Therefore, Mos-MAPK changes spindle morphology in a way that the spindle can move as close to the cortex as possible, and segregate chromosomes in the most "compact" manner without further separation of the spindle poles. This appears to also involve specific effects of Mos-MAPK on the actin cytoskeleton and the cell cortex to mediate the specialized polar body cytokinesis.

Acknowledgments We thank all members of the Lenart laboratory at MPI-NAT for feedback and discussions. We would like to thank Evelyn Houliston (IMEV, Villefranche-sur-Mer, France) for providing comments. I.A. is a member and has been partially funded by the IMPRS Molecular Biology Programme. The laboratory of P.L. is funded by the Max Planck Society.

Competing Interests The authors declare no competing financial interests.

References

1. Jessus C, Munro C, Houliston E (2020) Managing the oocyte meiotic arrest-lessons from frogs and jellyfish. Cells 9:1150. https://doi.org/10.3390/cells9051150
2. Dorée M, Hunt T (2002) From Cdc2 to Cdk1: when did the cell cycle kinase join its cyclin partner? J Cell Sci 115:2461–2464. https://doi.org/10.1242/jcs.115.12.2461
3. Kishimoto T, Kanatani H (1976) Cytoplasmic factor responsible for germinal vesicle breakdown and meiotic maturation in starfish oocyte. Nature 260:321–322. https://doi.org/10.1038/260321a0
4. Masui Y, Markert CL (1971) Cytoplasmic control of nuclear behavior during meiotic maturation of frog oocytes. J Exp Zool 177:129–145. https://doi.org/10.1002/jez.1401770202
5. Evans T, Rosenthal ET, Youngblom J, Distel D, Hunt T (1983) Cyclin: a protein specified by maternal mRNA in sea urchin eggs that is destroyed at each cleavage division. Cell 33:389–396. https://doi.org/10.1016/0092-8674(83)90420-8
6. Nurse P, Bissett Y (1981) Gene required in G1 for commitment to cell cycle and in G2 for control of mitosis in fission yeast. Nature 292:558–560. https://doi.org/10.1038/292558a0
7. Lénárt P, Ellenberg J (2003) Nuclear envelope dynamics in oocytes: from germinal vesicle breakdown to mitosis. Curr Opin Cell Biol 15:88–95. https://doi.org/10.1016/s0955-0674(02)00011-x
8. Boveri T (1887) Zellen-Studien. G. Fischer, Jena
9. Brunet S, Verlhac MH (2011) Positioning to get out of meiosis: the asymmetry of division. Hum Reprod Update 17:68–75. https://doi.org/10.1093/humupd/dmq044
10. Maddox AS, Azoury J, Dumont J (2012) Polar body cytokinesis. Cytoskeleton (Hoboken) 69:855–868. https://doi.org/10.1002/cm.21064
11. Bieber FR, Nance WE, Morton CC, Brown JA, Redwine FO, Jordan RL, Mohanakumar T (1981) Genetic studies of an acardiac monster: evidence of polar body twinning in man. Science 213:775–777. https://doi.org/10.1126/science.7196086
12. Otsuki J, Nagai Y, Lopata A, Chiba K, Yasmin L, Sankai T (2012) Symmetrical division of mouse oocytes during meiotic maturation can lead to the development of twin embryos that amalgamate to form a chimeric hermaphrodite. Hum Reprod 27:380–387. https://doi.org/10.1093/humrep/der408
13. Dupré A, Haccard O, Jessus C (2010) Mos in the oocyte: how to use MAPK independently of growth factors and transcription to control meiotic divisions. J Signal Transduct 2011:e350412. https://doi.org/10.1155/2011/350412
14. Moran AL, McAlister JS (2009) Egg size as a life history character of marine invertebrates: is it all it's cracked up to be? Biol Bull 216:226–242. https://doi.org/10.1086/BBLv216n3p226
15. Mitchison TJ (2020) Cell biology: size scaling of mitotic spindles. Curr Biol 30:R1476–R1478. https://doi.org/10.1016/j.cub.2020.09.047
16. Wühr M, Chen Y, Dumont S, Groen AC, Needleman DJ, Salic A, Mitchison TJ (2008) Evidence for an upper limit to mitotic spindle length. Curr Biol 18:1256–1261. https://doi.org/10.1016/j.cub.2008.07.092
17. Grill SW, Howard J, Schäffer E, Stelzer EHK, Hyman AA (2003) The distribution of active force generators controls mitotic spindle position. Science 301:518–521. https://doi.org/10.1126/science.1086560
18. Rieder CL, Alexander SP (1990) Kinetochores are transported poleward along a single astral microtubule during chromosome attachment to the spindle in newt lung cells. J Cell Biol 110:81–95. https://doi.org/10.1083/jcb.110.1.81
19. Wollman R, Cytrynbaum EN, Jones JT, Meyer T, Scholey JM, Mogilner A (2005) Efficient chromosome capture requires a bias in the "search-and-capture" process during mitotic-spindle assembly. Curr Biol 15:828–832. https://doi.org/10.1016/j.cub.2005.03.019
20. Gard DL (1992) Microtubule organization during maturation of Xenopus oocytes: assembly and rotation of the meiotic spindles. Dev Biol 151:516–530. https://doi.org/10.1016/0012-1606(92)90190-r

21. Harasimov K, Uraji J, Mönnich EU, Holubcová Z, Elder K, Blayney M, Schuh M (2023) Actin-driven chromosome clustering facilitates fast and complete chromosome capture in mammalian oocytes. Nat Cell Biol 25:439–452. https://doi.org/10.1038/s41556-022-01082-9
22. Lénárt P, Bacher CP, Daigle N, Hand AR, Eils R, Terasaki M, Ellenberg J (2005) A contractile nuclear actin network drives chromosome congression in oocytes. Nature 436:812–818. https://doi.org/10.1038/nature03810
23. Harada K, Oita E, Chiba K (2003) Metaphase I arrest of starfish oocytes induced via the MAP kinase pathway is released by an increase of intracellular pH. Development 130:4581–4586. https://doi.org/10.1242/dev.00649
24. Zhang W, Liu HT (2002) MAPK signal pathways in the regulation of cell proliferation in mammalian cells. Cell Res 12:9–18. https://doi.org/10.1038/sj.cr.7290105
25. Maxwell SA, Arlinghaus RB (1985) Serine kinase activity associated with Maloney murine sarcoma virus-124-encoded p37mos. Virology 143:321–333. https://doi.org/10.1016/0042-6822(85)90119-9
26. Oskarsson M, McClements WL, Blair DG, Maizel JV, Vande Woude GF (1980) Properties of a normal mouse cell DNA sequence (sarc) homologous to the src sequence of Moloney sarcoma virus. Science 207:1222–1224. https://doi.org/10.1126/science.6243788
27. Goldman DS, Kiessling AA, Millette CF, Cooper GM (1987) Expression of c-Mos RNA in germ cells of male and female mice. Proc Natl Acad Sci U S A 84:4509–4513. https://doi.org/10.1073/pnas.84.13.4509
28. Dumollard R, Levasseur M, Hebras C, Huitorel P, Carroll M, Chambon J-P, McDougall A (2011) Mos limits the number of meiotic divisions in urochordate eggs. Development 138:885–895. https://doi.org/10.1242/dev.057133
29. Mori M, Hara M, Tachibana K, Kishimoto T (2006) p90Rsk is required for G1 phase arrest in unfertilized starfish eggs. Development 133:1823–1830. https://doi.org/10.1242/dev.02348
30. Tachibana K, Tanaka D, Isobe T, Kishimoto T (2000) c-Mos forces the mitotic cell cycle to undergo meiosis II to produce haploid gametes. Proc Natl Acad Sci U S A 97:14301–14306. https://doi.org/10.1073/pnas.97.26.14301
31. Ucar H, Tachibana K, Kishimoto T (2013) The Mos-MAPK pathway regulates diaphanous-related formin activity to drive cleavage furrow closure during polar body extrusion in starfish oocytes. J Cell Sci 126:5153–5165. https://doi.org/10.1242/jcs.130476
32. Amiel A, Leclère L, Robert L, Chevalier S, Houliston E (2009) Conserved functions for Mos in eumetazoan oocyte maturation revealed by studies in a cnidarian. Curr Biol 19:305–311. https://doi.org/10.1016/j.cub.2008.12.054
33. Ivanovska I, Lee E, Kwan KM, Fenger DD, Orr-Weaver TL (2004) The Drosophila MOS ortholog is not essential for meiosis. Curr Biol 14:75–80. https://doi.org/10.1016/j.cub.2003.12.031
34. Church DL, Guan KL, Lambie EJ (1995) Three genes of the MAP kinase cascade, mek-2, mpk-1/sur-1 and let-60 ras, are required for meiotic cell cycle progression in Caenorhabditis elegans. Development 121:2525–2535. https://doi.org/10.1242/dev.121.8.2525
35. Page BD, Guedes S, Waring D, Priess JR (2001) The C. elegans E2F- and DP-related proteins are required for embryonic asymmetry and negatively regulate Ras/MAPK signaling. Mol Cell 7:451–460. https://doi.org/10.1016/s1097-2765(01)00193-9
36. Yamamoto DS, Tachibana K, Sumitani M, Lee JM, Hatakeyama M (2008) Involvement of Mos-MEK-MAPK pathway in cytostatic factor (CSF) arrest in eggs of the parthenogenetic insect, Athalia rosae. Mech Dev 125(11–12):996–1008. https://doi.org/10.1016/j.mod.2008.08.004
37. Sagata N, Oskarsson M, Copeland T, Brumbaugh J, Woude GFV (1988) Function of c-mos proto-oncogene product in meiotic maturation in Xenopus oocytes. Nature 335:519–525. https://doi.org/10.1038/335519a0
38. Haccard O, Jessus C (2006) Oocyte maturation, Mos and cyclins—A matter of synthesis: two functionally redundant ways to induce meiotic maturation. Cell Cycle 5:1152–1159. https://doi.org/10.4161/cc.5.11.2800

39. Howard EL, Charlesworth A, Welk J, MacNicol AM (1999) The mitogen-activated protein kinase signaling pathway stimulates mos mRNA cytoplasmic polyadenylation during Xenopus oocyte maturation. Mol Cell Biol 19:1990–1999. https://doi.org/10.1128/MCB.19.3.1990
40. Galas S, Barakat H, Dorée M, Picard A (1993) A nuclear factor required for specific translation of cyclin B may control the timing of first meiotic cleavage in starfish oocytes. Mol Biol Cell 4:1295–1306. https://doi.org/10.1091/mbc.4.12.1295
41. Kuroda T, Naito K, Sugiura K, Yamashita M, Takakura I, Tojo H (2004) Analysis of the roles of cyclin B1 and cyclin B2 in porcine oocyte maturation by inhibiting synthesis with antisense RNA injection. Biol Reprod 70:154–159. https://doi.org/10.1095/biolreprod.103.021519
42. Ledan E, Polanski Z, Terret ME, Maro B (2001) Meiotic maturation of the mouse oocyte requires an equilibrium between cyclin B synthesis and degradation. Dev Biol 232:400–413. https://doi.org/10.1006/dbio.2001.0188
43. Lévesque JT, Sirard MA (1996) Resumption of meiosis is initiated by the accumulation of cyclin B in bovine oocytes. Biol Reprod 55:1427–1436. https://doi.org/10.1095/biolreprod55.6.1427
44. Gross SD, Schwab MS, Taieb FE, Lewellyn AL, Qian Y-W, Maller JL (2000) The critical role of the MAP kinase pathway in meiosis II in Xenopus oocytes is mediated by p90Rsk. Curr Biol 10:430–438. https://doi.org/10.1016/S0960-9822(00)00425-5
45. Verlhac MH, Kubiak JZ, Weber M, Géraud G, Colledge WH, Evans MJ, Maro B (1996) Mos is required for MAP kinase activation and is involved in microtubule organization during meiotic maturation in the mouse. Development 122:815–822. https://doi.org/10.1242/dev.122.3.815
46. Choi T, Fukasawa K, Zhou R, Tessarollo L, Borror K, Resau J, Vande Woude GF (1996) The Mos/mitogen-activated protein kinase (MAPK) pathway regulates the size and degradation of the first polar body in maturing mouse oocytes. Proc Natl Acad Sci U S A 93:7032–7035. https://doi.org/10.1073/pnas.93.14.7032
47. Colledge WH, Carlton MBL, Udy GB, Evans MJ (1994) Disruption of c-mos causes parthenogenetic development of unfertilized mouse eggs. Nature 370:65–68. https://doi.org/10.1038/370065a0
48. Hashimoto N, Watanabe N, Furuta Y, Tamemoto H, Sagata N, Yokoyama M, Okazaki K, Nagayoshi M, Takeda N, Ikawa Y (1994) Parthenogenetic activation of oocytes in c-mos-deficient mice. Nature 370:68–71. https://doi.org/10.1038/370068a0
49. Araki K, Naito K, Haraguchi S, Suzuki R, Yokoyama M, Inoue M, Aizawa S, Toyoda Y, Sato E (1996) Meiotic abnormalities of c-mos knockout mouse oocytes: activation after first meiosis or entrance into third meiotic metaphase. Biol Reprod 55:1315–1324. https://doi.org/10.1095/biolreprod55.6.1315
50. Furuno N, Nishizawa M, Okazaki K, Tanaka H, Iwashita J, Nakajo N, Ogawa Y, Sagata N (1994) Suppression of DNA replication via Mos function during meiotic divisions in Xenopus oocytes. EMBO J 13:2399–2410. https://doi.org/10.1002/j.1460-2075.1994.tb06524.x
51. Shibuya EK, Boulton TG, Cobb MH, Ruderman JV (1992) Activation of p42 MAP kinase and the release of oocytes from cell cycle arrest. EMBO J 11:3963–3975. https://doi.org/10.1002/j.1460-2075.1992.tb05490.x
52. Fukasawa K, Vande Woude GF (1995) Mos overexpression in Swiss 3T3 cells induces meiotic-like alterations of the mitotic spindle. Proc Natl Acad Sci U S A 92:3430–3434. https://doi.org/10.1073/pnas.92.8.3430
53. Bodart J-FL, Baert FY, Sellier C, Duesbery NS, Flament S, Vilain J-P (2005) Differential roles of p39Mos-Xp42Mpk1 cascade proteins on Raf1 phosphorylation and spindle morphogenesis in Xenopus oocytes. Dev Biol 283:373–383. https://doi.org/10.1016/j.ydbio.2005.04.031
54. Gould MC, Stephano JL (1999) MAP kinase, meiosis, and sperm centrosome suppression in Urechis caupo. Dev Biol 216:348–358. https://doi.org/10.1006/dbio.1999.9468

55. Jiao G, Lian H, Xing J, Chen L, Du Z, Liu X (2022) MOS mutation causes female infertility with large polar body oocytes. Gynecol Endocrinol 38:1158–1163. https://doi.org/10.1080/09513590.2022.2147158
56. Mendez R, Richter JD (2001) Translational control by CPEB: a means to the end. Nat Rev Mol Cell Biol 2:521–529. https://doi.org/10.1038/35080081
57. Barkoff A, Ballantyne S, Wickens M (1998) Meiotic maturation in Xenopus requires polyadenylation of multiple mRNAs. EMBO J 17:3168–3175. https://doi.org/10.1093/emboj/17.11.3168
58. Lorenzo-Orts L, Pauli A (2024) The molecular mechanisms underpinning maternal mRNA dormancy. Biochem Soc Trans 52:861–871. https://doi.org/10.1042/BST20231122
59. Eckmann CR, Rammelt C, Wahle E (2011) Control of poly(A) tail length. Wiley Interdiscip Rev RNA 2:348–361. https://doi.org/10.1002/wrna.56
60. Reyes JM, Ross PJ (2016) Cytoplasmic polyadenylation in mammalian oocyte maturation. Wiley Interdiscip Rev RNA 7:71–89. https://doi.org/10.1002/wrna.1316
61. Stebbins-Boaz B, Hake LE, Richter JD (1996) CPEB controls the cytoplasmic polyadenylation of cyclin, Cdk2 and c-mos mRNAs and is necessary for oocyte maturation in Xenopus. EMBO J 15:2582–2592
62. Conti M, Kunitomi C (2024) A genome-wide perspective of the maternal mRNA translation program during oocyte development. Semin Cell Dev Biol 154:88–98. https://doi.org/10.1016/j.semcdb.2023.03.003
63. Piqué M, López JM, Foissac S, Guigó R, Méndez R (2008) A combinatorial code for CPE-mediated translational control. Cell 132:434–448. https://doi.org/10.1016/j.cell.2007.12.038
64. Kim JH, Richter JD (2006) Opposing polymerase-deadenylase activities regulate cytoplasmic polyadenylation. Mol Cell 24:173–183. https://doi.org/10.1016/j.molcel.2006.08.016
65. Minshall N, Reiter MH, Weil D, Standart N (2007) CPEB interacts with an ovary-specific eIF4E and 4E-T in early Xenopus oocytes. J Biol Chem 282:37389–37401. https://doi.org/10.1074/jbc.M704629200
66. Chassé H, Mulner-Lorillon O, Boulben S, Glippa V, Morales J, Cormier P (2016) Cyclin B translation depends on mTOR activity after fertilization in sea urchin embryos. PLoS One 11:e0150318. https://doi.org/10.1371/journal.pone.0150318
67. Belloc E, Piqué M, Méndez R (2008) Sequential waves of polyadenylation and deadenylation define a translation circuit that drives meiotic progression. Biochem Soc Trans 36:665–670. https://doi.org/10.1042/BST0360665
68. Arumugam K, Wang Y, Hardy LL, MacNicol MC, MacNicol AM (2010) Enforcing temporal control of maternal mRNA translation during oocyte cell-cycle progression. EMBO J 29:387–397. https://doi.org/10.1038/emboj.2009.337
69. Mendez R, Hake LE, Andresson T, Littlepage LE, Ruderman JV, Richter JD (2000) Phosphorylation of CPE binding factor by Eg2 regulates translation of c-mos mRNA. Nature 404:302–307. https://doi.org/10.1038/35005126
70. Sarkissian M, Mendez R, Richter JD (2004) Progesterone and insulin stimulation of CPEB-dependent polyadenylation is regulated by Aurora A and glycogen synthase kinase-3. Genes Dev 18:48–61. https://doi.org/10.1101/gad.1136004
71. Keady BT, Kuo P, Martínez SE, Yuan L, Hake LE (2007) MAPK interacts with XGef and is required for CPEB activation during meiosis in Xenopus oocytes. J Cell Sci 120:1093–1103. https://doi.org/10.1242/jcs.03416
72. Charlesworth A, Wilczynska A, Thampi P, Cox LL, MacNicol AM (2006) Musashi regulates the temporal order of mRNA translation during Xenopus oocyte maturation. EMBO J 25:2792–2801. https://doi.org/10.1038/sj.emboj.7601159
73. Weill L, Belloc E, Castellazzi CL, Méndez R (2017) Musashi 1 regulates the timing and extent of meiotic mRNA translational activation by promoting the use of specific CPEs. Nat Struct Mol Biol 24:672–681. https://doi.org/10.1038/nsmb.3434
74. Belloc E, Méndez R (2008) A deadenylation negative feedback mechanism governs meiotic metaphase arrest. Nature 452:1017–1021. https://doi.org/10.1038/nature06809

75. Reimann JD, Freed E, Hsu JY, Kramer ER, Peters JM, Jackson PK (2001) Emi1 is a mitotic regulator that interacts with Cdc20 and inhibits the anaphase promoting complex. Cell 105:645–655. https://doi.org/10.1016/s0092-8674(01)00361-0
76. Igea A, Méndez R (2010) Meiosis requires a translational positive loop where CPEB1 ensues its replacement by CPEB4. EMBO J 29:2182–2193. https://doi.org/10.1038/emboj.2010.111
77. Cao L-R, Jiang J-C, Fan H-Y (2020) Positive feedback stimulation of Ccnb1 and Mos mRNA translation by MAPK cascade during mouse oocyte maturation. Front Cell Dev Biol 8:609430. https://doi.org/10.3389/fcell.2020.609430
78. Hochegger H, Klotzbücher A, Kirk J, Howell M, le Guellec K, Fletcher K, Duncan T, Sohail M, Hunt T (2001) New B-type cyclin synthesis is required between meiosis I and II during Xenopus oocyte maturation. Development 128:3795–3807. https://doi.org/10.1242/dev.128.19.3795
79. Iwabuchi M, Ohsumi K, Yamamoto TM, Sawada W, Kishimoto T (2000) Residual Cdc2 activity remaining at meiosis I exit is essential for meiotic M-M transition in Xenopus oocyte extracts. EMBO J 19:4513–4523. https://doi.org/10.1093/emboj/19.17.4513
80. Inoue D, Ohe M, Kanemori Y, Nobui T, Sagata N (2007) A direct link of the Mos-MAPK pathway to Erp1/Emi2 in meiotic arrest of Xenopus laevis eggs. Nature 446:1100–1104. https://doi.org/10.1038/nature05688
81. Palmer A, Gavin AC, Nebreda AR (1998) A link between MAP kinase and p34(cdc2)/cyclin B during oocyte maturation: p90(rsk) phosphorylates and inactivates the p34(cdc2) inhibitory kinase Myt1. EMBO J 17:5037–5047. https://doi.org/10.1093/emboj/17.17.5037
82. de Moor CH, Richter JD (1997) The Mos pathway regulates cytoplasmic polyadenylation in Xenopus oocytes. Mol Cell Biol 17:6419–6426. https://doi.org/10.1128/MCB.17.11.6419
83. Sha Q-Q, Dai X-X, Dang Y, Tang F, Liu J, Zhang Y-L, Fan H-Y (2017) A MAPK cascade couples maternal mRNA translation and degradation to meiotic cell cycle progression in mouse oocytes. Development 144:452–463. https://doi.org/10.1242/dev.144410
84. Gebauer F, Richter JD (1997) Synthesis and function of mos: the control switch of vertebrate oocyte meiosis. BioEssays 19:23–28. https://doi.org/10.1002/bies.950190106
85. Houk MS, Epel D (1974) Protein synthesis during hormonally induced meiotic maturation and fertilization in starfish oocytes. Dev Biol 40:298–310. https://doi.org/10.1016/0012-1606(74)90132-8
86. Swartz SZ, Nguyen HT, McEwan BC, Adamo ME, Cheeseman IM, Kettenbach AN (2021) Selective dephosphorylation by PP2A-B55 directs the meiosis I-meiosis II transition in oocytes. elife 10:e70588. https://doi.org/10.7554/eLife.70588
87. Lapasset L, Pradet-Balade B, Lozano J-C, Peaucellier G, Picard A (2005) Nuclear envelope breakdown may deliver an inhibitor of protein phosphatase 1 which triggers cyclin B translation in starfish oocytes. Dev Biol 285:200–210. https://doi.org/10.1016/j.ydbio.2005.06.016
88. Ochi H, Aoto S, Tachibana K, Hara M, Chiba K (2016) Block of CDK1-dependent polyadenosine elongation of cyclin B mRNA in metaphase-i-arrested starfish oocytes is released by intracellular pH elevation upon spawning. Mol Reprod Dev 83:79–87. https://doi.org/10.1002/mrd.22599
89. Valdez VA, Neahring L, Petry S, Dumont S (2023) Mechanisms underlying spindle assembly and robustness. Nat Rev Mol Cell Biol 24:523–542. https://doi.org/10.1038/s41580-023-00584-0
90. Blengini CS, Schindler K (2022) Acentriolar spindle assembly in mammalian female meiosis and the consequences of its perturbations on human reproduction†. Biol Reprod 106:253–263. https://doi.org/10.1093/biolre/ioab210
91. Severson AF, von Dassow G, Bowerman B (2016) Oocyte meiotic spindle assembly and function. Curr Top Dev Biol 116:65–98. https://doi.org/10.1016/bs.ctdb.2015.11.031
92. Heald R, Tournebize R, Blank T, Sandaltzopoulos R, Becker P, Hyman A, Karsenti E (1996) Self-organization of microtubules into bipolar spindles around artificial chromosomes in Xenopus egg extracts. Nature 382:420–425. https://doi.org/10.1038/382420a0

93. Schuh M, Ellenberg J (2007) Self-organization of MTOCs replaces centrosome function during acentrosomal spindle assembly in live mouse oocytes. Cell 130:484–498. https://doi.org/10.1016/j.cell.2007.06.025
94. Holubcová Z, Blayney M, Elder K, Schuh M (2015) Human oocytes. Error-prone chromosome-mediated spindle assembly favors chromosome segregation defects in human oocytes. Science 348:1143–1147. https://doi.org/10.1126/science.aaa9529
95. So C, Seres KB, Steyer AM, Mönnich E, Clift D, Pejkovska A, Möbius W, Schuh M (2019) A liquid-like spindle domain promotes acentrosomal spindle assembly in mammalian oocytes. Science 364:eaat9557. https://doi.org/10.1126/science.aat9557
96. Breuer M, Kolano A, Kwon M, Li C-C, Tsai T-F, Pellman D, Brunet S, Verlhac M-H (2010) HURP permits MTOC sorting for robust meiotic spindle bipolarity, similar to extra centrosome clustering in cancer cells. J Cell Biol 191:1251–1260. https://doi.org/10.1083/jcb.201005065
97. Kolano A, Brunet S, Silk AD, Cleveland DW, Verlhac M-H (2012) Error-prone mammalian female meiosis from silencing the spindle assembly checkpoint without normal interkinetochore tension. Proc Natl Acad Sci U S A 109:E1858–E1867. https://doi.org/10.1073/pnas.1204686109
98. Gruss OJ, Vernos I (2004) The mechanism of spindle assembly: functions of Ran and its target TPX2. J Cell Biol 166:949–955. https://doi.org/10.1083/jcb.200312112
99. Brunet S, Dumont J, Lee KW, Kinoshita K, Hikal P, Gruss OJ, Maro B, Verlhac M-H (2008) Meiotic regulation of TPX2 protein levels governs cell cycle progression in mouse oocytes. PLoS One 3:e3338. https://doi.org/10.1371/journal.pone.0003338
100. Eyers PA, Erikson E, Chen LG, Maller JL (2003) A novel mechanism for activation of the protein kinase Aurora A. Curr Biol 13:691–697. https://doi.org/10.1016/s0960-9822(03)00166-0
101. Tsai M-Y, Wiese C, Cao K, Martin O, Donovan P, Ruderman J, Prigent C, Zheng Y (2003) A Ran signalling pathway mediated by the mitotic kinase Aurora A in spindle assembly. Nat Cell Biol 5:242–248. https://doi.org/10.1038/ncb936
102. Manandhar G, Schatten H, Sutovsky P (2005) Centrosome reduction during gametogenesis and its significance1. Biol Reprod 72:2–13. https://doi.org/10.1095/biolreprod.104.031245
103. Szollosi D, Calarco P, Donahue RP (1972) Absence of centrioles in the first and second meiotic spindles of mouse oocytes. J Cell Sci 11:521–541. https://doi.org/10.1242/jcs.11.2.521
104. Borrego-Pinto J, Somogyi K, Karreman MA, König J, Müller-Reichert T, Bettencourt-Dias M, Gönczy P, Schwab Y, Lénárt P (2016) Distinct mechanisms eliminate mother and daughter centrioles in meiosis of starfish oocytes. J Cell Biol 212:815–827. https://doi.org/10.1083/jcb.201510083
105. Crowder ME, Strzelecka M, Wilbur JD, Good MC, von Dassow G, Heald R (2015) A comparative analysis of spindle morphometrics across metazoans. Curr Biol 25:1542–1550. https://doi.org/10.1016/j.cub.2015.04.036
106. Holland ND (1981) Electron microscopic study of development in a sea cucumber, Stichopus tremulus (Holothuroidea), from unfertilized egg through hatched blastula. Acta Zool 62:89–111. https://doi.org/10.1111/j.1463-6395.1981.tb00618.x
107. Nakashima S, Kato KH (2001) Centriole behavior during meiosis in oocytes of the sea urchin Hemicentrotus pulcherrimus. Dev Growth Differ 43:437–445. https://doi.org/10.1046/j.1440-169x.2001.00580.x
108. Sluder G, Miller FJ, Lewis K (1993) Centrosome inheritance in starfish zygotes. II: selective suppression of the maternal centrosome during meiosis. Dev Biol 155:58–67. https://doi.org/10.1006/dbio.1993.1006
109. Wu X, Palazzo RE (1999) Differential regulation of maternal vs. paternal centrosomes. Proc Natl Acad Sci U S A 96:1397–1402. https://doi.org/10.1073/pnas.96.4.1397
110. Kikuchi Y, Hamaguchi Y (2012) The effect of taxol microinjection on the microtubular structure in polar body formation of starfish oocytes. Cytoskeleton (Hoboken) 69:125–132. https://doi.org/10.1002/cm.21004

111. White EA, Glotzer M (2012) Centralspindlin: at the heart of cytokinesis. Cytoskeleton (Hoboken) 69:882–892. https://doi.org/10.1002/cm.21065
112. Srayko M, O'toole ET, Hyman AA, Müller-Reichert T (2006) Katanin disrupts the microtubule lattice and increases polymer number in C. elegans meiosis. Curr Biol 16:1944–1949. https://doi.org/10.1016/j.cub.2006.08.029
113. Burbank KS, Groen AC, Perlman ZE, Fisher DS, Mitchison TJ (2006) A new method reveals microtubule minus ends throughout the meiotic spindle. J Cell Biol 175:369–375. https://doi.org/10.1083/jcb.200511112
114. Sköld HN, Komma DJ, Endow SA (2005) Assembly pathway of the anastral Drosophilaoocyte meiosis I spindle. J Cell Sci 118:1745–1755. https://doi.org/10.1242/jcs.02304
115. Yang G, Houghtaling BR, Gaetz J, Liu JZ, Danuser G, Kapoor TM (2007) Architectural dynamics of the meiotic spindle revealed by single-fluorophore imaging. Nat Cell Biol 9:1233–1242. https://doi.org/10.1038/ncb1643
116. Matković J, Ghosh S, Ćosić M, Eibes S, Barišić M, Pavin N, Tolić IM (2022) Kinetochore- and chromosome-driven transition of microtubules into bundles promotes spindle assembly. Nat Commun 13:7307. https://doi.org/10.1038/s41467-022-34957-4
117. Cheeseman IM, Desai A (2008) Molecular architecture of the kinetochore-microtubule interface. Nat Rev Mol Cell Biol 9:33–46. https://doi.org/10.1038/nrm2310
118. Maiato H, DeLuca J, Salmon ED, Earnshaw WC (2004) The dynamic kinetochore-microtubule interface. J Cell Sci 117:5461–5477. https://doi.org/10.1242/jcs.01536
119. Verlhac MH, Lefebvre C, Guillaud P, Rassinier P, Maro B (2000) Asymmetric division in mouse oocytes: with or without Mos. Curr Biol 10:1303–1306. https://doi.org/10.1016/s0960-9822(00)00753-3
120. Bai W, Singh B, Yang Y, Ramagli LS, Nash M, Herzog NK, Arlinghaus RB (1992) The physical interactions between p37env-mos and tubulin structures. Oncogene 7:493–500
121. Zhou R, Daar I, Ferris DK, White G, Paules RS, Vande Woude G (1992) pp39mos is associated with p34cdc2 kinase in c-mosxe-transformed NIH 3T3 cells. Mol Cell Biol 12:3583–3589. https://doi.org/10.1128/mcb.12.8.3583-3589.1992
122. Zhou RP, Shen RL, Pinto da Silva P, Vande Woude GF (1991) In vitro and in vivo characterization of pp39mos association with tubulin. Cell Growth Differ 2:257–265
123. Lefebvre C, Terret ME, Djiane A, Rassinier P, Maro B, Verlhac M-H (2002) Meiotic spindle stability depends on MAPK-interacting and spindle-stabilizing protein (MISS), a new MAPK substrate. J Cell Biol 157:603–613. https://doi.org/10.1083/jcb.200202052
124. Terret ME, Lefebvre C, Djiane A, Rassinier P, Moreau J, Maro B, Verlhac M-H (2003) DOC1R: a MAP kinase substrate that control microtubule organization of metaphase II mouse oocytes. Development 130:5169–5177. https://doi.org/10.1242/dev.00731
125. Bun P, Dmitrieff S, Belmonte JM, Nédélec FJ, Lénárt P (2018) A disassembly-driven mechanism explains F-actin-mediated chromosome transport in starfish oocytes. elife 7:e31469. https://doi.org/10.7554/eLife.31469
126. Azoury J, Lee KW, Georget V, Rassinier P, Leader B, Verlhac M-H (2008) Spindle positioning in mouse oocytes relies on a dynamic meshwork of actin filaments. Curr Biol 18:1514–1519. https://doi.org/10.1016/j.cub.2008.08.044
127. Schuh M, Ellenberg J (2008) A new model for asymmetric spindle positioning in mouse oocytes. Curr Biol 18:1986–1992. https://doi.org/10.1016/j.cub.2008.11.022
128. Mogessie B, Schuh M (2017) Actin protects mammalian eggs against chromosome segregation errors. Science 357:eaal1647. https://doi.org/10.1126/science.aal1647
129. Sandquist JC, Kita AM, Bement WM (2011) And the dead shall rise: actin and myosin return to the spindle. Dev Cell 21:410–419. https://doi.org/10.1016/j.devcel.2011.07.018
130. Weber KL, Sokac AM, Berg JS, Cheney RE, Bement WM (2004) A microtubule-binding myosin required for nuclear anchoring and spindle assembly. Nature 431:325–329. https://doi.org/10.1038/nature02834

131. Silverman-Gavrila RV, Forer A (2000) Evidence that actin and myosin are involved in the poleward flux of tubulin in metaphase kinetochore microtubules of crane-fly spermatocytes. J Cell Sci 113(Pt 4):597–609. https://doi.org/10.1242/jcs.113.4.597
132. Zhang X, Ma C, Miller AL, Katbi HA, Bement WM, Liu XJ (2008) Polar body emission requires a rhoA contractile ring and Cdc42-mediated membrane protrusion. Dev Cell 15:386–400. https://doi.org/10.1016/j.devcel.2008.07.005
133. Ducibella T, Anderson E, Albertini DF, Aalberg J, Rangarajan S (1988) Quantitative studies of changes in cortical granule number and distribution in the mouse oocyte during meiotic maturation. Dev Biol 130:184–197. https://doi.org/10.1016/0012-1606(88)90425-3
134. Longo FJ, Chen DY (1984) Development of surface polarity in mouse eggs. Scan Electron Microsc Pt. 2:703–716
135. Gard DL, Cha BJ, Roeder AD (1995) F-actin is required for spindle anchoring and rotation in Xenopus oocytes: a re-examination of the effects of cytochalasin B on oocyte maturation. Zygote 3:17–26. https://doi.org/10.1017/s0967199400002331
136. Kim NH, Cho SK, Choi SH, Kim EY, Park SP, Lim JH (2000) The distribution and requirements of microtubules and microfilaments in bovine oocytes during in vitro maturation. Zygote 8:25–32. https://doi.org/10.1017/s0967199400000794
137. Sardet C, Speksnijder J, Terasaki M, Chang P (1992) Polarity of the ascidian egg cortex before fertilization. Development 115:221–237. https://doi.org/10.1242/dev.115.1.221
138. Sun QY, Lai L, Park KW, Kühholzer B, Prather RS, Schatten H (2001) Dynamic events are differently mediated by microfilaments, microtubules, and mitogen-activated protein kinase during porcine oocyte maturation and fertilization in vitro. Biol Reprod 64:879–889. https://doi.org/10.1095/biolreprod64.3.879
139. Deng M, Suraneni P, Schultz RM, Li R (2007) The Ran GTPase mediates chromatin signaling to control cortical polarity during polar body extrusion in mouse oocytes. Dev Cell 12:301–308. https://doi.org/10.1016/j.devcel.2006.11.008
140. Yi K, Rubinstein B, Unruh JR, Guo F, Slaughter BD, Li R (2013) Sequential actin-based pushing forces drive meiosis I chromosome migration and symmetry breaking in oocytes. J Cell Biol 200:567–576. https://doi.org/10.1083/jcb.201211068
141. Mei Q, Li H, Liu Y, Wang X, Xiang W (2022) Advances in the study of CDC42 in the female reproductive system. J Cell Mol Med 26:16–24. https://doi.org/10.1111/jcmm.17088
142. Mogessie B, Scheffler K, Schuh M (2018) Assembly and positioning of the oocyte meiotic spindle. Annu Rev Cell Dev Biol 34:381–403. https://doi.org/10.1146/annurev-cellbio-100616-060553
143. Dehapiot B, Halet G (2013) Ran GTPase promotes oocyte polarization by regulating ERM (Ezrin/Radixin/Moesin) inactivation. Cell Cycle 12:1672–1678. https://doi.org/10.4161/cc.24901
144. Rohatgi R, Ma L, Miki H, Lopez M, Kirchhausen T, Takenawa T, Kirschner MW (1999) The interaction between N-WASP and the Arp2/3 complex links Cdc42-dependent signals to actin assembly. Cell 97:221–231. https://doi.org/10.1016/s0092-8674(00)80732-1
145. Dehapiot B, Carrière V, Carroll J, Halet G (2013) Polarized Cdc42 activation promotes polar body protrusion and asymmetric division in mouse oocytes. Dev Biol 377:202–212. https://doi.org/10.1016/j.ydbio.2013.01.029
146. Yi K, Unruh JR, Deng M, Slaughter BD, Rubinstein B, Li R (2011) Dynamic maintenance of asymmetric meiotic spindle position through Arp2/3 complex-driven cytoplasmic streaming in mouse oocytes. Nat Cell Biol 13:1252–1258. https://doi.org/10.1038/ncb2320
147. Zhang J, Ma R, Li L, Wang L, Hou X, Han L, Ge J, Li M, Wang Q (2017) Intersectin 2 controls actin cap formation and meiotic division in mouse oocytes through the Cdc42 pathway. FASEB J 31:4277–4285. https://doi.org/10.1096/fj.201700179R
148. Ma C, Benink HA, Cheng D, Montplaisir V, Wang L, Xi Y, Zheng P-P, Bement WM, Liu XJ (2006) Cdc42 activation couples spindle positioning to first polar body formation in oocyte maturation. Curr Biol 16:214–220. https://doi.org/10.1016/j.cub.2005.11.067

149. Pal D, Ellis A, Sepúlveda-Ramírez SP, Salgado T, Terrazas I, Reyes G, De La Rosa R, Henson JH, Shuster CB (2020) Rac and Arp2/3-nucleated actin networks antagonize Rho during mitotic and meiotic cleavages. Front Cell Dev Biol 8:591141. https://doi.org/10.3389/fcell.2020.591141
150. Bourdais A, Dehapiot B, Halet G (2023) MRCK activates mouse oocyte myosin II for spindle rotation and male pronucleus centration. J Cell Biol 222:e202211029. https://doi.org/10.1083/jcb.202211029
151. Chaigne A, Campillo C, Gov NS, Voituriez R, Azoury J, Umaña-Diaz C, Almonacid M, Queguiner I, Nassoy P, Sykes C, Verlhac M-H, Terret M-E (2013) A soft cortex is essential for asymmetric spindle positioning in mouse oocytes. Nat Cell Biol 15:958–966. https://doi.org/10.1038/ncb2799
152. Deng M, Williams CJ, Schultz RM (2005) Role of MAP kinase and myosin light chain kinase in chromosome-induced development of mouse egg polarity. Dev Biol 278:358–366. https://doi.org/10.1016/j.ydbio.2004.11.013
153. Morrison DL, Sanghera JS, Stewart J, Sutherland C, Walsh MP, Pelech SL (1996) Phosphorylation and activation of smooth muscle myosin light chain kinase by MAP kinase and cyclin-dependent kinase-1. Biochem Cell Biol 74:549–557. https://doi.org/10.1139/o96-459
154. Phillips KP, Petrunewich MAF, Collins JL, Booth RA, Liu XJ, Baltz JM (2002) Inhibition of MEK or cdc2 kinase parthenogenetically activates mouse eggs and yields the same phenotypes as Mos−/− Parthenogenotes. Dev Biol 247:210–223. https://doi.org/10.1006/dbio.2002.0680

What Are the Fitness Costs of Centromere Drive?

Bo Pan and Takashi Akera

Abstract Centromere DNA and proteins are rapidly evolving despite their essential conserved functions in chromosome segregation. Centromere drive theory is based on the idea that centromere DNA are rapidly evolving to bias their transmission through female meiosis with associated fitness costs, which facilitate the evolution of the rest of the genome, especially the genes encoding centromere proteins to suppress the cheating and restore fitness. While the cell biological mechanisms of centromere drive and the rapid evolution of centromere proteins have been investigated and discussed, the exact nature of fitness costs associated with centromere drive remains largely unknown. In this chapter, we discuss different possibilities of fitness costs associated with centromere drive, especially focusing on how selfish centromeres can cause chromosome mis-segregation in mitosis and meiosis.

Keywords Centromere · Centromere drive · Meiotic drive · Female meiosis · Chromosome segregation

B. Pan
Cell and Developmental Biology Center, National Heart, Lungand, Blood Institute, National Institutes of Health, Bethesda, Maryland, USA

Department of Biochemistry and Molecular Biology, Uniformed Services University of the Health Sciences, Bethesda, Maryland, USA
e-mail: bo.pan.ctr@usuhs.edu

T. Akera (✉)
Cell and Developmental Biology Center, National Heart, Lungand, Blood Institute, National Institutes of Health, Bethesda, Maryland, USA
e-mail: takashi.akera@nih.gov

B. Mogessie (ed.), *Cellular Architecture and Dynamics in Female Meiosis*,
https://doi.org/10.1007/978-3-031-97173-0_3

Introduction

The centromere is the major site of the chromosome that mediates chromosome segregation by attaching to the spindle machinery [1–6]. Because centromere-directed chromosome segregation is a conserved feature of eukaryote cells, it is reasonable to assume that centromere DNA and the genes encoding centromere proteins are highly conserved across species. However, it is well established that both centromere DNA and proteins are rapidly evolving in animals and plants [7–15]. This "centromere paradox" was puzzling until Henikoff et al. proposed the centromere drive theory, which makes three predictions: (i) centromere DNA evolve to bias their transmission through female meiosis, (ii) this selfish cheating is associated with fitness costs, and (iii) the fitness costs drive the adaptive evolution of centromere proteins to suppress the biased transmission and restore fitness [16, 17].

Female meiosis provides a unique opportunity for genetic elements to bias their transmission, because only the chromosomes segregated to the egg will transmit to the next generation while the others extruded to the polar body will be degraded [18–20]. There are multiple selfish genetic elements, termed meiotic drivers, that cheat in female meiosis to bias their transmission [21–25]. The original centromere drive theory proposed that the expansion of centromere satellite DNA creates a "stronger" centromere that can bias its segregation to the egg similar to other meiotic drivers [16] (Fig. 1a). Consistent with this model, expanded centromere DNA are associated with biased segregation in house mouse *Mus musculus* and yellow monkeyflower *Mimulus* [9, 26–29]. Time-lapse live imaging of mouse oocytes show that selfish larger centromeres randomly orient on the meiotic spindle in early metaphase I (i.e., half facing the cortical side, which goes to the polar body, and the other half facing the interior side, which remains in the egg), but selfish centromeres tend to detach more often from the cortical side of the spindle, enriched in tyrosinated microtubules [30–33], and re-orient toward the interior side to preferentially segregate to the egg [34] (Fig. 1b). Molecular studies have shown that a selfish centromere recruits more centromere proteins compared to its homologous centromere to build a larger kinetochore, the protein complex bridging centromere DNA and spindle microtubules [28, 35]. Assembling a larger kinetochore leads to the enrichment of higher microtubule-destabilizing activity such as chromosomal passenger complex (CPC) and kinesin-13 MCAK at the peri-centromere region [34, 36, 37]. Microtubule-destabilizing activity is essential for mitosis and meiosis to cancel erroneous spindle microtubule attachments to achieve faithful chromosome segregation [38, 39]. Selfish centromeres are taking advantage of the same destabilizing activity to "correct" their orientation on the meiotic spindle to preferentially segregate to the egg (Fig. 1b).

So far, mechanisms of centromere drive have mainly been studied in mice, but it is interesting to speculate how centromeres can cheat in other species. For example, how does centromeres cheat in plants where female meiosis undergoes symmetric divisions and the most bottom cell becomes the egg cell (i.e., megaspore) [40]. One possibility is that there is a gradient of signaling molecules within the megaspore

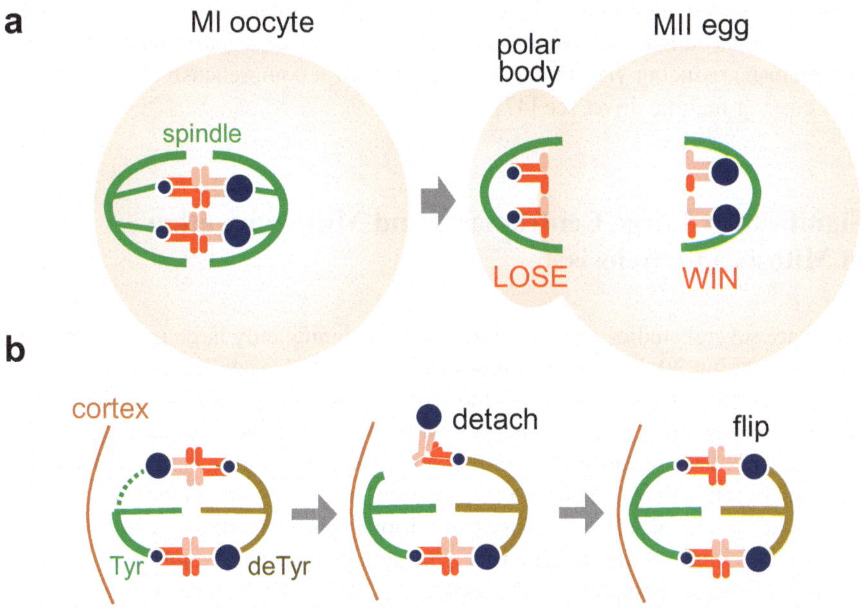

Fig. 1 Schematics of centromere drive in mouse oocytes. (**a**) Asymmetry in female meiosis provides a unique opportunity for selfish centromeres to cheat. The spindle is positioned close to the cortex and orients perpendicular to the cortex. Homologous chromosomes from each parent (pink and red chromosomes) are paired to form meiotic bivalents. Centromeres (blue circles) interact with the spindle and direct chromosome segregation. Selfish centromeres (larger blue circles) have more centromere proteins and preferentially orient to the interior side of the spindle, which remains in the egg. (**b**) The orientation of selfish centromeres (larger blue circles) is initially random. However, microtubule attachments of selfish centromeres with the cortical side (green lines) are unstable and tend to detach. The preferential detachment of selfish centromeres from the cortical side will lead to biased flipping to the interior side (yellow lines), establishing their biased orientation

mother cell (equivalent to meiosis I oocytes), which induces spindle asymmetry to guide selfish centromeres to the bottom side of the cell that eventually becomes the megaspore [40]. Even if there is no obvious structural asymmetry in spindle or cell division itself, signal gradients can create a special cue on the meiotic spindle to guide selfish centromeres. Centromeres are rapidly evolving in different lineages. Such rapid evolution could facilitate the invention of unique cheating strategies that are specifically tailored for female meiosis in each lineage.

Although centromere drive mechanisms and centromere evolution have been investigated and discussed (see [41–46] for comprehensive reviews on centromere drive), less focus was placed on fitness costs associated with centromere drive (i.e., the second prediction of the original centromere drive theory). Especially with the recent advances in understanding the molecular basis underlying centromere drive, it is timely to discuss how selfish centromeres cause fitness costs to the host organism. While there are various possibilities for how centromere drive can cause fitness costs (e.g., selfish centromeres linked to a deleterious allele), here we focus on the possibility that selfish centromeres cause chromosome segregation errors in mitosis and meiosis, reducing viability and/or fertility (for a comprehensive review on fitness costs of meiotic drive, see [47]).

Significantly Large Centromeres and Mis-Segregation in Mitosis and Meiosis

There are several studies reporting species with significantly large regional centromeres (spanning 20–50% of the chromosome) [48–53]. These species would be a great experimental system to study the costs of carrying large centromeres, which can behave selfishly in female meiosis when heterozygous. A study using Indian muntjac cells has elegantly showed that a chromosome with a significantly larger centromere tends to lag in anaphase compared to others with smaller centromeres within the same cell, often leading to the formation of micronuclei [49] (Fig. 2). Despite the chromosome with a larger centromere congressing more efficiently on the mitotic spindle compared to the ones with smaller centromeres, they are biased to establish erroneous merotelic attachments, leading to their anaphase lagging. This study suggests that the expansion of centromeres can induce chromosome mis-segregation in mitosis as a fitness cost of centromere drive. The extra-large centromere in Indian muntjac has an increased merotelic attachment rate despite enriching CPC (the major error-correction machinery) at the centromere, scaling proportionally with the centromere size [49]. The authors discuss an interesting possibility that larger centromeres are prone to merotelic attachments because their chromosomes congress too quickly in a CENP-E motor independent manner.

It would be interesting to extend this analysis to other species with extra-large centromeres such as the red imported fire ant, *Solenopsis invicta* [50], red-necked wallaby, *Macropus rufogriseus* [51], and garden pea, *Pisum sativum* [52]. Centromeres are large for all chromosomes in these species, and therefore, it is

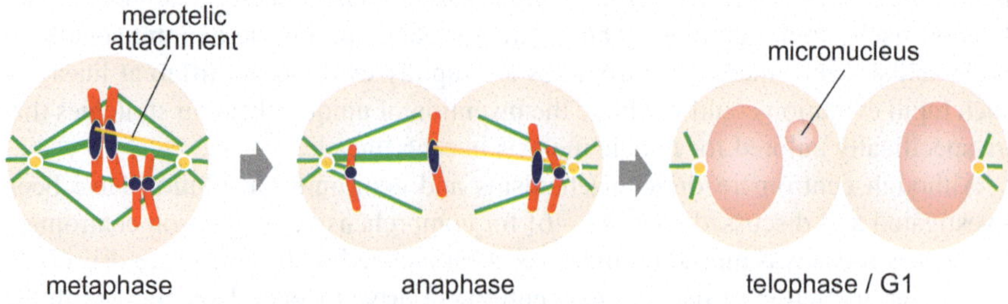

Fig. 2 Schematics of mitosis of Indian Muntjac cells. The chromosome with the larger centromere more efficiently aligns on the metaphase plate compared to the ones with smaller centromeres. However, the larger centromere is prone to mis-attachments, leading to anaphase lagging and the micronucleus formation

possible that these species have evolved a strategy to mitigate any fitness costs associated with carrying large centromeres. An extreme case is holo-centromeric organisms, which has the highest possible centromere/chromosome ratio [54]. Interestingly, an established holocentric model organism, *Caenorhabditis elegans*, has literally no merotelic attachments and only ~1% of anaphase lagging [55], demonstrating its amazing robustness to properly segregate chromosomes with large centromeres. Investigating how centromeres expand and how the host species adaptively evolves to suppress the associated fitness costs in different organisms would be an exciting future direction.

Compared to Indian muntjac chromosomes, human chromosomes have relatively smaller centromeres (6.2% of the genome, 189.9 Mb total genome-wide), mainly composed of alpha satellite arrays (2.8% of the genome, 85.2 Mb total genome-wide) [5, 56, 57]. Because the human centromere size is already smaller in general, ranging from 0.3 Mb in the Y chromosome and chromosome 21 to 4.8 Mb in chromosome 18 [56, 57], human cells seem to have opposite issues than Indian muntjac cells where the smaller centromeres are prone to mis-segregation especially when the chromosome size is relatively large (e.g., chromosome 3, 6, and X) [58]. Therefore, the centromere size needs to correlate with the chromosome size to achieve faithful segregation. Altogether, the centromere size needs to be within a range to properly segregate chromosomes, and if selfish centromeres expand too much, increasing their transmission rate, this could be deleterious to the host by increasing the mis-segregation rate, affecting the viability and promoting tumorigenesis by producing micronuclei and aneuploid cells. Since it is less likely that human chromosomes suffer from carrying too large centromeres, if centromere drive occurs in humans, it might have different kinds of associated fitness costs such as deleterious alleles linked to selfish centromeres.

In Indian muntjac cells, centromere expansion led to the proportional increase of all the centromere components tested, including kinetochore proteins and CPC [49]. In contrast, the mouse centromere DNA, minor satellites, can expand and recruit more microtubule-destabilizing activity (e.g., CPC and MCAK) to increase their selfishness without increasing its kinetochore size [34]. This disproportional increase of centromere components upon centromere expansion can impose different challenges to properly segregate chromosomes. One possible cost of enriching high levels of destabilizing activity relative to their kinetochore size is that centromeres cannot establish stable interactions with the spindle, which is necessary for faithful chromosome segregation. Indeed, constitutively targeting high levels of CPC to centromeres is detrimental during mitosis by preventing centromeres from establishing stable microtubule attachments [59]. Most studies have focused on the kinetochore size when investigating expanded centromeres, but it would be interesting to also analyze these microtubule-destabilizing factors and examine how they respond to centromere expansion.

So far, we discussed how selfish expanded centromere DNA can cause mis-segregation in mitosis, but we think similar segregation errors can occur in meiosis as well, reducing fertility. The only established fitness costs associated with centromere drive comes from the driving "*D*" locus in yellow monkeyflower, *Mimulus*, which carries expanded centromere satellites and causes reduced fertility in both male and female when homozygous [26]. It is likely that the reduced fertility is due to dose-dependent deleterious alleles linked to the *D* locus [9], but it would be interesting to examine chromosome dynamics in meiosis using *Mimulus* homozygous for *D* to test if the expanded centromere satellites build larger kinetochores and/or enrich higher destabilizing activity to affect meiotic divisions, reducing their fertility. Since expanded centromere satellites can be heterochromatic [60, 61], another possibility is the *D* locus forming a large heterochromatic region and affecting homologous chromosome pairing during meiotic prophase as seen for the X chromosome pairing in female meiosis in several rodent species, including *Microtus agrestis*, *Mesocricetus auratus,* and *Nesokia indica* [62, 63].

Centromere Asymmetry Across the Meiotic Bivalent, Causing Segregation Errors

Up to this point, we have discussed how centromere expansion can lead to mis-segregation when segregating centromeres are equal. However, when homologous chromosomes pair in meiosis I, centromeres can be asymmetric across the meiotic bivalent chromosome, imposing further challenges to properly segregate the chromosome. For example, asymmetry in centromere size can lead to asymmetric pulling forces by spindle microtubules, preventing chromosomes to congress on the metaphase plate. We will discuss how centromere asymmetry can cause mis-segregation in female and male meiosis.

Centromere Asymmetry in Female Meiosis

While studying the mechanisms underlying centromere drive in mouse oocytes, researchers have noticed that chromosome dynamics are different when centromeres are asymmetric across the meiotic bivalent. The most notable difference is the off-centerness of meiotic bivalents on the oocyte spindle (Fig. 3a) [34, 35]. This bivalent off-centerness has a significant correlation with the strength of centromere drive in mouse oocytes: the closer selfish centromeres are to the spindle pole, the stronger centromere drive is [8, 34]. In general, mis-aligned chromosomes are prone to segregation errors in mitosis and meiosis, affecting viability and fertility, respectively [64, 65]. Therefore, this off-centered bivalent chromosome could be one of the fitness costs associated with centromere drive. Then, how does selfish mouse centromeres induce the off-centerness of the chromosome? Bivalent chromosomes are off-centered with selfish larger centromeres closer to the spindle pole, suggesting that selfish centromeres are pulled more toward the pole compared to the homologous centromere (Fig. 3a) [34, 35]. Because selfish centromeres build larger

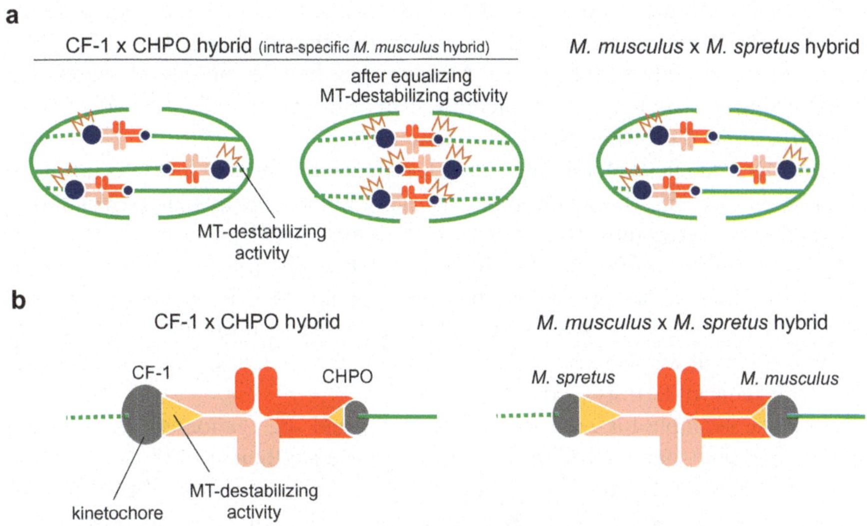

Fig. 3 Centromere asymmetry drives off-centerness of meiotic bivalents in mouse oocytes. (**a**) Both centromere drive systems, the CF-1 × CHPO hybrid and the *M. musculus* × *M. spretus* hybrid, show off-centered bivalents with selfish centromeres closer to the pole. Experimentally equalizing MT-destabilizing activity across the bivalent cancels off-centerness in CF-1 × CHPO hybrids. (**b**) Bivalents in CF-1 × CHPO hybrid oocytes have asymmetry in both the kinetochore size (gray circles) and the destabilizing activity (orange triangles), while only the destabilizing activity is asymmetric in *M. musculus* × *M. spretus* hybrid oocytes

kinetochores and enrich higher levels of microtubule-destabilizing activity, it is likely that one of these features drives the off-centerness. Using two different centromere drive systems in mice, it was shown that the high microtubule-destabilizing activity on the selfish centromere seems to be the major driving force of this asymmetric pulling [34]. The first system, the CF-1 x CHPO hybrid, crossing two *Mus musculus domesticus* strains, shows asymmetry in both kinetochore size and destabilizing activity (Fig. 3b). When the destabilizing activity was experimentally equalized across the bivalent, the off-centerness was canceled, demonstrating that asymmetry in destabilizing activity is required for off-centerness (Fig. 3a). In the second system, crossing *Mus musculus domesticus* with *Mus spretus*, the destabilizing activity is asymmetric while the kinetochore size is symmetric, suggesting that asymmetry in destabilizing activity is sufficient to drive off-centerness (Fig. 3b).

It remains unknown how the high microtubule-destabilizing activity triggers off-centerness, but one possibility is that the high microtubule-destabilizing activity drives microtubule depolymerization at the selfish centromere, pulling the entire bivalent chromosome toward the pole [66]. Consistent with this idea, inducing sister centromere asymmetry in destabilizing activity with an optogenetic approach triggers microtubule depolymerization in mitosis, leading to off-centered chromosomes [67]. On a side note, when the optogenetically recruited destabilizing activity was very high at the centromere, it caused microtubule release [67], which is exactly what selfish centromeres do to re-orient on the spindle to face the interior side of the spindle. What eventually triggers the detachment of the off-centered bivalent is also an open question, but Aurora kinase A at the spindle pole can provide additional destabilizing activity to the off-centered selfish centromeres to facilitate the microtubule release and chromosome re-orientation [68, 69].

High microtubule-destabilizing activity confers selfishness to mouse centromere DNA [8, 34, 41, 45]. This same activity causes chromosome mis-alignment, possibly leading to segregation errors and affecting female fertility. Indeed, in *Mus musculus domesticus* x *Mus spretus* hybrid oocytes, chromosome mis-alignment eventually leads to anaphase chromosome lagging, producing aneuploid eggs [70–72]. This observation supports the idea that selfish centromeres reduce female fertility, although we also need to consider the contribution of hybrid incompatibility on this reduced fertility [73].

A classic example for centromere asymmetry and segregation errors in female meiosis is Robertsonian (Rb) translocations in mice and humans [35, 74–79]. It is now well accepted that Rb translocations impact centromere size when they fuse two telocentric chromosomes at their centromeres, altering "centromere strength" in female meiosis [35]. Studies using female mice heterozygous for an Rb chromosome showed that the centromere of the Rb chromosome has lower levels of centromere and kinetochore proteins compared to its unfused homologous chromosomes, leading to the biased segregation to extrude the Rb chromosome to the polar body [35]. Trivalent chromosomes, composed of an Rb chromosome and two unfused homologous chromosomes, also show off-centerness on the spindle similar to bivalents with centromere asymmetry in destabilizing activity [68]. Destabilizing activity has not been examined for Rb centromeres, but the off-centerness of trivalents implies the asymmetry in destabilizing activity across the trivalent, contributing to their biased segregation and increased segregation errors. Important caveat to note is that segregation errors and reduced female fertility associated with trivalents could be partly due to the trivalent structure itself, which can be mis-folded and causing mis-segregation independently of centromere asymmetry.

Centromere Asymmetry in Male Meiosis

Male heterozygous Rb carriers have reduced fertility as well, although the fertility defects are milder compared to female carriers [78, 79]. Consistent with the fertility data, Rb trivalents segregate significantly better in male meiosis compared to female meiosis at least in humans and pigs [80–85]. It is unknown why male meiosis can deal better with Rb chromosomes, but one possibility is that the large cytoplasm in the egg increases the segregation error rates in female meiosis in general compared to male meiosis [86, 87]. Another possibility is that centromere proteins on Rb trivalents are less asymmetric in male meiosis compared to female meiosis. For meiotic bivalents, it has been shown that larger asymmetry in centromeres leads to more off-centered chromosomes [8], increasing the risk of segregation errors. Because the levels of the core centromere protein, CENP-A, are differentially regulated between male and female germlines [88], it is reasonable to speculate that the degree of centromere asymmetry is different between male and female meiosis. Comparing centromere asymmetry between male and female meiosis would be an exciting future direction not just for Rb systems but also for centromere drive systems discussed in the section "Centromere Asymmetry in Female Meiosis." These analyses would reveal how the same selfish centromere can differentially impact male and female fertility.

The original centromere drive theory proposed that the associated fitness costs could arise in male meiosis, based on a study showing that the deletion of pericentromeric heterochromatin from the X chromosome causes mis-segregation of X-Y meiotic bivalents in *Drosophila* spermatocyte meiosis I, leading to sterility [89]. Since microtubule-destabilizing factors such as CPC localize at pericentromeres [37, 90, 91], it is tempting to speculate that the deletion of pericentromeric heterochromatin created asymmetry in the destabilizing activity across the bivalent, causing chromosome mis-alignment and mis-segregation. Besides this system, there are many examples of X-Y meiotic bivalents showing different dynamics compared to autosomes on the meiotic spindle [92]. Centromere DNA are most dissimilar between X and Y chromosomes, possibly because Y chromosomes do not see female meiosis and are not selected to be selfish. This centromere asymmetry across X-Y bivalents provides a great opportunity to understand how centromere evolution impacts chromosome dynamics and segregation fidelity in meiosis. For example, X-Y bivalents are initially mis-aligned, taking longer time to congress in multiple systems, including the marsupial *Thylamys elegans* [93] and mealworm beetle *Tenebrio molitor* [92]. It would be interesting to analyze if these X-Y bivalents are off-centered with the X centromere, which sees female meiosis, closer to the spindle pole analogous to selfish centromeres being closer to the pole in female meiosis (Fig. 3a).

Multiple lineages, including butterflies, birds, and some reptiles, employ the ZW/ZZ sex determination system where females carry ZW sex chromosomes [94]. Based on the observations for X-Y meiotic bivalents in males, it is tempting to speculate that Z-W bivalents also show off-centerness in females due to centromere asymmetry. Such off-centerness could cause not only mis-segregation but also biased segregation of sex chromosomes in oocytes, distorting the sex ratio of the offspring [95].

As in the case for Rb trivalents, some of these atypical chromosome dynamics of X-Y bivalents could be due to their chromosome structural asymmetry and not due to centromere asymmetry. Nevertheless, these systems provide a great opportunity to investigate how centromere asymmetry can modulate chromosome dynamics, impacting segregation fidelity in meiosis. Analyzing X-Y dynamics in male meiosis could also provide insights into how selfish centromeres can cheat in female meiosis. These insights would especially be valuable for species where female meiosis is less accessible.

Segregation Errors in Embryonic Mitosis

Any cell division that already has a higher error rate makes it an ideal arena to exhibit fitness costs associated with centromere drive, because even a slight defect in chromosome dynamics caused by selfish centromeres can lead to significant segregation errors. Mitosis in early embryos is one such example [96–98]. For example, the mouse Y chromosome mis-segregates more often in early embryogenesis compared to male meiosis, supporting the idea that embryonic mitosis imposes an extra challenge to properly segregate chromosomes [99]. One of the possible reasons why early embryonic mitosis has higher error rates is their error-prone spindle formation process, often leading to dual spindles in a zygote [98, 100] (Fig. 4a). Another possible reason is the significant asymmetry in centromere/kinetochore size between parental chromosomes. During spermiogenesis, centromere proteins will be removed except for the histone H3 variant, CENP-A [101, 102]. Upon fertilization, sperm chromatin uses this remaining CENP-A to rebuild centromeres, resulting in a smaller kinetochore size compared to ones on maternal chromatin even when the paternal and maternal chromatin share the same centromere DNA organization [88, 103] (Fig. 4a). When centromere DNA are asymmetric with the

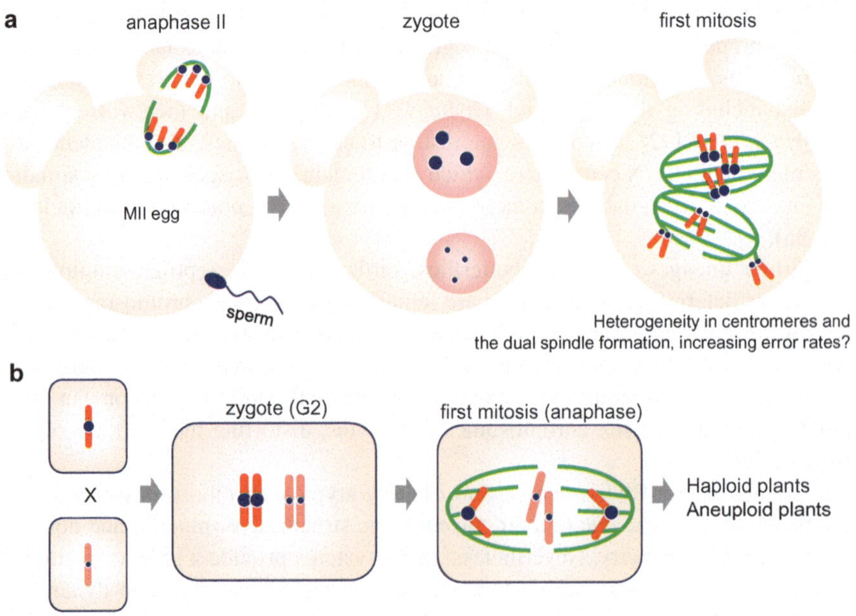

Fig. 4 Centromere heterogeneity in embryo mitosis. (**a**) In mouse embryos, the maternal chromatin builds larger centromeres/kinetochores compared to the paternal chromatin. The zygote spindle can initially assemble privately for each parental chromatin, which together with the centromere heterogeneity, could lead to asynchronous chromosome alignment during the first mitotic division. (**b**) In plants, when parents have significantly different centromere size, smaller centromeres can be inactivated and mis-segregate in the zygote, producing haploid and aneuploid embryos

maternal chromatin having selfish larger centromeres, it enhances the difference in the kinetochore size [88], creating significant functional heterogeneity in centromeres during embryonic mitosis. In human and bovine embryos, such centromere asymmetry between parental chromosomes is a potential driver of chromosome instability termed "heterogoneic division" where fertilized eggs can spontaneously segregate entire parental genomes into different cell lineages during the first embryonic mitosis [104, 105]. In multiple plant species, centromere asymmetry in the zygote leads to inactivating the smaller centromeres, leading to their chromosome mis-segregation and producing aneuploid and haploid embryos [106–109] (Fig. 4b). Zygotes only with smaller centromeres do not exhibit similar mis-segregation phenotype, showing that the heterogeneity in centromeres within the zygote causes mis-segregation. These examples demonstrate embryonic mitosis as a great model system to explore possible fitness costs associated with centromere drive.

Conclusions

We have reviewed how chromosome segregation can go wrong in mitosis and meiosis and discussed how selfish centromeres can contribute to those segregation errors. Centromeres can develop distinct strategies to cheat in female meiosis [29, 30], and therefore, we believe that there would be a variety of fitness costs associated with centromere drive. It is also interesting to consider *when* these fitness costs would arise and drive the adaptive evolution of the genome to restore fitness (the third prediction of the centromere drive theory). For example, if fitness costs appear only when the host carries multiple chromosomes with large centromeres (i.e., becoming homozygous for the selfish centromere or selfish centromeres spreading to other chromosomes in the genome), it would allow the initial spread of selfish centromeres in the population without obvious fitness costs. Once the fitness cost starts affecting the carriers in the population, selection for suppressor mutations will occur. In contrast, if fitness costs are caused by centromere asymmetry (i.e., heterozygous state, section "Centromere Asymmetry Across the Meiotic Bivalent, Causing Segregation Errors"), the fitness costs would appear before selfish centromeres had a chance to spread. In this case, selfish centromeres must be great cheaters while keeping the fitness costs low. Otherwise, severe fitness costs would instantly eliminate such selfish centromeres from the population before any suppressors can be selected.

It is important to note that many of the systems discussed here have some caveats (e.g., chromosome structural asymmetry outside the centromere, hybrid incompatibility, possible adaptive evolution to accommodate large centromeres) that obscure what exactly is the fitness cost caused by selfish centromeres. To directly test what the fitness costs are, we need an experimental system where we can manipulate centromere DNA repeat size without affecting the rest of the genome and investigate its consequence in mitosis and meiosis before the genome adaptively evolves to accommodate large centromeres. Recent advances in the CRISPR/Cas technology shows a potential to make this dream experiment possible [110, 111]. Several studies have demonstrated the ability to CRISPR centromere satellite DNA to shrink/remove centromere DNA repeats, causing chromosome loss and cell death [112, 113]. Experimentally expanding centromere satellites would be the next giant leap to directly test the consequences of carrying large centromeres in the genome [61].

Acknowledgments We thank Zaak Walton for valuable feedback on the manuscript. We apologize to the investigators whose research could not be cited due to space constraints.

Funding Takashi Akera is supported by the Division of Intramural Research at the National Institutes of Health/National Heart, Lung, and Blood Institute (grant number: 1ZIAHL006249).

References

1. Black BE, Cleveland DW (2011) Epigenetic centromere propagation and the nature of CENP—A nucleosomes. Cell 144(4):471–479
2. Musacchio A, Desai A (2017) A molecular view of kinetochore assembly and function. Biology 6(4):5
3. Malik HS, Henikoff S (2009) Major evolutionary transitions in centromere complexity. Cell 138(6):1067–1082
4. Cheeseman IM, Desai A (2008) Molecular architecture of the kinetochore-microtubule interface. Nat Rev Mol Cell Biol 9(1):33–46
5. Logsdon GA, Eichler EE (2022) The dynamic structure and rapid evolution of human centromeric satellite DNA. Genes 14(1):92
6. Sridhar S, Fukagawa T (2022) Kinetochore architecture employs diverse linker strategies across evolution. Front Cell Dev Biol 10:862637
7. Pontremoli C, Forni D, Pozzoli U, Clerici M, Cagliani R, Sironi M (2021) Kinetochore proteins and microtubule-destabilizing factors are fast evolving in eutherian mammals. Mol Ecol 30(6):1505–1515
8. Kumon T, Ma J, Akins RB, Stefanik D, Nordgren CE, Kim J et al (2021) Parallel pathways for recruiting effector proteins determine centromere drive and suppression. Cell 184:4904–4918.e11
9. Finseth FR, Nelson TC, Fishman L (2021) Selfish chromosomal drive shapes recent centromeric histone evolution in monkeyflowers. PLoS Genet 17(4):e1009418
10. Malik HS, Henikoff S (2001) Adaptive evolution of Cid, a centromere-specific histone in drosophila. Genetics 157(3):1293–1298
11. Talbert PB, Bryson TD, Henikoff S (2004) Adaptive evolution of centromere proteins in plants and animals. J Biol 3(4):18
12. van der Lee R, Wiel L, van Dam TJP, Huynen MA (2017) Genome-scale detection of positive selection in nine primates predicts human-virus evolutionary conflicts. Nucleic Acids Res 45(18):10634–10648
13. Cooper JL, Henikoff S (2004) Adaptive evolution of the histone fold domain in centromeric histones. Mol Biol Evol 21(9):1712–1718
14. Malik HS, Vermaak D, Henikoff S (2002) Recurrent evolution of DNA-binding motifs in the drosophila centromeric histone. Proc Natl Acad Sci U S A 99(3):1449–1454
15. Talbert PB, Masuelli R, Tyagi AP, Comai L, Henikoff S (2002) Centromeric localization and adaptive evolution of an Arabidopsis histone H3 variant. Plant Cell 14(5):1053–1066

16. Henikoff S, Ahmad K, Malik HS (2001) The centromere paradox: stable inheritance with rapidly evolving DNA. Science 293(5532):1098–1102
17. Malik HS, Henikoff S (2002) Conflict begets complexity: the evolution of centromeres. Curr Opin Genet Dev 12(6):711–718
18. Gorelick R, Carpinone J, Derraugh LJ (2017) No universal differences between female and male eukaryotes: anisogamy and asymmetrical female meiosis. Biol J Linn Soc 120:1–21. https://doi.org/10.1111/bij.12874
19. Li R, Albertini DF (2013) The road to maturation: somatic cell interaction and self-organization of the mammalian oocyte. Nat Rev Mol Cell Biol 14(3):141–152
20. Clift D, Schuh M (2013) Restarting life: fertilization and the transition from meiosis to mitosis. Nat Rev Mol Cell Biol 14(9):549–562
21. Pardo-Manuel de Villena F, Sapienza C (2001) Nonrandom segregation during meiosis: the unfairness of females. Mamm Genome 12(5):331–339
22. Kruger AN, Mueller JL (2021) Mechanisms of meiotic drive in symmetric and asymmetric meiosis. Cell Mol Life Sci 78(7):3205–3218
23. Chmátal L, Schultz RM, Black BE, Lampson MA (2017) Cell biology of cheating—Transmission of centromeres and other selfish elements through asymmetric meiosis. In: Centromeres and kinetochores. Springer, Cham, pp 377–396. https://doi.org/10.1007/978-3-319-58592-5_16
24. Clark FE, Akera T (2021) Unravelling the mystery of female meiotic drive: where we are. Open Biol 11(9):210074
25. Arora UP, Dumont BL (2022) Meiotic drive in house mice: mechanisms, consequences, and insights for human biology. Chromosom Res 30(2–3):165–186
26. Fishman L, Saunders A (2008) Centromere-associated female meiotic drive entails male fitness costs in monkeyflowers. Science 322(5907):1559–1562
27. Finseth FR, Dong Y, Saunders A, Fishman L (2015) Duplication and adaptive evolution of a key centromeric protein in mimulus, a genus with female meiotic drive. Mol Biol Evol 32(10):2694–2706
28. Iwata-Otsubo A, Dawicki-McKenna JM, Akera T, Falk SJ, Chmátal L, Yang K et al (2017) Expanded satellite repeats amplify a discrete CENP—A nucleosome assembly site on chromosomes that drive in female meiosis. Curr Biol 27(15):2365–2373.e8
29. Wu T, Lane SIR, Morgan SL, Jones KT (2018) Spindle tubulin and MTOC asymmetries may explain meiotic drive in oocytes. Nat Commun 9(1):2952
30. Akera T, Chmátal L, Trimm E, Yang K, Aonbangkhen C, Chenoweth DM et al (2017) Spindle asymmetry drives non-Mendelian chromosome segregation. Science 358(6363):668–672
31. Akera T (2023) Tubulin post-translational modifications in meiosis. Semin Cell Dev Biol 137:38–45. https://doi.org/10.1016/j.semcdb.2021.11.014
32. Magiera MM, Singh P, Gadadhar S, Janke C (2018) Tubulin posttranslational modifications and emerging links to human disease. Cell 173(6):1323–1327
33. Roll-Mecak A (2020) The Tubulin code in microtubule dynamics and information encoding. Dev Cell 54(1):7–20
34. Akera T, Trimm E, Lampson MA (2019) Molecular strategies of meiotic cheating by selfish centromeres. Cell 178(5):1132–1144.e10
35. Chmátal L, Gabriel SI, Mitsainas GP, Martínez-Vargas J, Ventura J, Searle JB et al (2014) Centromere strength provides the cell biological basis for meiotic drive and karyotype evolution in mice. Curr Biol 24(19):2295–2300
36. Kawashima SA, Yamagishi Y, Honda T, Ishiguro K, Watanabe Y (2010) Phosphorylation of H2A by Bub1 prevents chromosomal instability through localizing shugoshin. Science 327(5962):172–177
37. Watanabe Y (2012) Geometry and force behind kinetochore orientation: lessons from meiosis. Nat Rev Mol Cell Biol 13(6):370–382
38. Lampson MA, Grishchuk EL (2017) Mechanisms to avoid and correct erroneous kinetochore-microtubule attachments. Biology 6(1):1

39. Tanaka TU, Zhang T (2022) SWAP, SWITCH, and STABILIZE: mechanisms of kinetochore-microtubule error correction. Cells 11(9):1462
40. Finseth F (2023) Female meiotic drive in plants: mechanisms and dynamics. Curr Opin Genet Dev 82:102101
41. Dudka D, Lampson MA (2022) Centromere drive: model systems and experimental progress. Chromosom Res 30(2–3):187–203. https://doi.org/10.1007/s10577-022-09696-3
42. Kursel LE, Malik HS (2018) The cellular mechanisms and consequences of centromere drive. Curr Opin Cell Biol 52:58–65
43. Lampson MA, Black BE (2017) Cellular and molecular mechanisms of centromere drive. Cold Spring Harb Symp Quant Biol 82:249–257
44. Talbert P, Henikoff S (2022) Centromere drive: chromatin conflict in meiosis. Curr Opin Genet Dev 77:102005
45. Kumon T, Lampson MA (2022) Evolution of eukaryotic centromeres by drive and suppression of selfish genetic elements. Semin Cell Dev Biol 128:51–60
46. Rosin LF, Mellone BG (2017) Centromeres drive a hard bargain. Trends Genet 33(2):101–117
47. Zanders SE, Unckless RL (2019) Fertility costs of meiotic drivers. Curr Biol 29(11):R512–R520
48. Brinkley BR, Valdivia MM, Tousson A, Brenner SL (1984) Compound kinetochores of the Indian muntjac. Evolution by linear fusion of unit kinetochores. Chromosoma 91(1):1–11
49. Drpic D, Almeida AC, Aguiar P, Renda F, Damas J, Lewin HA et al (2018) Chromosome segregation is biased by kinetochore size. Curr Biol 28(9):1344–1356.e5
50. Huang YC, Lee CC, Kao CY, Chang NC, Lin CC, Shoemaker D et al (2016) Evolution of long centromeres in fire ants. BMC Evol Biol 16(1):189
51. Metcalfe CJ, Bulazel KV, Ferreri GC, Schroeder-Reiter E, Wanner G, Rens W et al (2007) Genomic instability within centromeres of interspecific marsupial hybrids. Genetics 177(4):2507–2517
52. Neumann P, Pavlíková Z, Koblížková A, Fuková I, Jedličková V, Novák P et al (2015) Centromeres off the hook: massive changes in centromere size and structure following duplication of cenh3 gene in fabeae species. Mol Biol Evol 32(7):1862–1879
53. Dutrillaux B, Dutrillaux AM (2023) Why are X autosome rearrangements so frequent in beetles? A study of 50 cases. Genes 14(1):150
54. Senaratne AP, Cortes-Silva N, Drinnenberg IA (2022) Evolution of holocentric chromosomes: drivers, diversity, and deterrents. Semin Cell Dev Biol 127:90–99
55. Stear JH, Roth MB (2002) Characterization of HCP-6, a C. elegans protein required to prevent chromosome twisting and merotelic attachment. Genes Dev 16(12):1498–1508
56. Rudd MK, Willard HF (2004) Analysis of the centromeric regions of the human genome assembly. Trends Genet 20(11):529–533
57. Altemose N, Logsdon GA, Bzikadze AV, Sidhwani P, Langley SA, Caldas GV et al (2022) Complete genomic and epigenetic maps of human centromeres. Science 376(6588):eabl4178
58. Dumont M, Gamba R, Gestraud P, Klaasen S, Worrall JT, De Vries SG et al (2020) Human chromosome-specific aneuploidy is influenced by DNA-dependent centromeric features. EMBO J 39(2):e102924
59. Liu D, Vader G, Vromans MJM, Lampson MA, Lens SMA (2009) Sensing chromosome bi-orientation by spatial separation of aurora B kinase from kinetochore substrates. Science 323(5919):1350–1353
60. Louzada S, Vieira-da-Silva A, Mendes-da-Silva A, Kubickova S, Rubes J, Adega F et al (2015) A novel satellite DNA sequence in the Peromyscus genome (PMSat): evolution via copy number fluctuation. Mol Phylogenet Evol 92:193–203
61. Liu ZW, Liu J, Liu F, Zhong X (2023) Depositing centromere repeats induces heritable intragenic heterochromatin establishment and spreading in Arabidopsis. Nucleic Acids Res 51:6039–6054

62. Singh AP, Raman R (1993) Mammalian sex chromosomes. VI. Synapsis in the heterochromatin-rich X chromosomes of four rodent species, Mus dunni, Bandicota bengalensis, Mesocricetus auratus, and Nesokia indica. Genome 36(1):195–198
63. Singh A, Henschel S, Sperling K, Kalscheuer V, Neitzel H (2000) Differences in the meiotic pairing behavior of gonosomal heterochromatin between female and male Microtus agrestis: implications for the mechanism of heterochromatin amplification on the X and Y. Cytogenet Cell Genet 91(1–4):253–260
64. Vukušić K, Tolić IM (2022) Polar chromosomes-challenges of a risky path. Cells 11(9):1531
65. Ferreira LT, Maiato H (2021) Prometaphase. Semin Cell Dev Biol 117:52–61
66. Lampson M, Renduchitala K, Khodjakov A, Kapoor TM (2004) Correcting improper chromosome-spindle attachments during cell division. Nat Cell Biol 6(3):232–237
67. Chen GY, Renda F, Zhang H, Gokden A, Wu DZ, Chenoweth DM et al (2021) Tension promotes kinetochore-microtubule release by Aurora B kinase. J Cell Biol 220(6):e202007030
68. Chmátal L, Yang K, Schultz RM, Lampson MA (2015) Spatial regulation of kinetochore microtubule attachments by destabilization at spindle poles in meiosis I. Curr Biol 25(14):1835–1841
69. Ye AA, Deretic J, Hoel CM, Hinman AW, Cimini D, Welburn JP et al (2015) Aurora A kinase contributes to a pole-based error correction pathway. Curr Biol 25(14):1842–1851
70. Sodek M, Kovacovicova K, Anger M (2017) True nondisjunction of whole bivalents in oocytes with attachment and congression defects. Cytogenet Genome Res 151(1):10–17
71. Sebestova J, Danylevska A, Novakova L, Kubelka M, Anger M (2012) Lack of response to unaligned chromosomes in mammalian female gametes. Cell Cycle 11(16):3011–3018
72. Koehler KE, Schrump SE, Cherry JP, Hassold TJ, Hunt PA (2006) Near-human aneuploidy levels in female mice with homeologous chromosomes. Curr Biol 16(15):R579–R580
73. El Yakoubi W, Akera T (2023) Condensin dysfunction is a reproductive isolating barrier in mice. Nature 623(7986):347–355
74. Pardo-Manuel de Villena F, Sapienza C (2001) Female meiosis drives Karyotypic evolution in mammals. Genetics 159(3):1179–1189
75. Mack H, Swisshelm K (2013) Robertsonian translocations. In: Brenner's encyclopedia of genetics. Elsevier, pp 301–305. https://doi.org/10.1016/B978-0-12-374984-0.01357-7
76. Piálek J, Hauffe HC, Searle JB (2005) Chromosomal variation in the house mouse. Biol J Linn Soc 84(3):535–563
77. Britton-Davidian J, Catalan J, da Graça Ramalhinho M, Ganem G, Auffray JC, Capela R et al (2000) Rapid chromosomal evolution in Island mice. Nature 403(6766):158
78. Gardner RJM, Amor DJ (2018) Gardner and Sutherland's chromosome abnormalities and genetic counseling, 5th edn. Oxford University Press, Oxford
79. Boué A, Gallano P (1984) A collaborative study of the segregation of inherited chromosome structural rearrangements in 1356 prenatal diagnoses. Prenat Diagn 4 Spec No:45–67
80. Morel F, Laudier B, Guérif F, Couet ML, Royère D, Roux C et al (2007) Meiotic segregation analysis in spermatozoa of pericentric inversion carriers using fluorescence in-situ hybridization. Hum Reprod 22(1):136–141
81. Pinton A, Calgaro A, Bonnet N, Ferchaud S, Billoux S, Dudez AM et al (2009) Influence of sex on the meiotic segregation of a t(13;17) Robertsonian translocation: a case study in the pig. Hum Reprod 24(8):2034–2043
82. Anton E, Vidal F, Blanco J (2007) Role of sperm FISH studies in the genetic reproductive advice of structural reorganization carriers. Hum Reprod 22(8):2088–2092
83. Munné S, Escudero T, Sandalinas M, Sable D, Cohen J (2000) Gamete segregation in female carriers of Robertsonian translocations. Cytogenet Cell Genet 90(3–4):303–308
84. Durban M, Benet J, Boada M, Fernández E, Calafell JM, Lailla JM et al (2001) PGD in female carriers of balanced Robertsonian and reciprocal translocations by first polar body analysis. Hum Reprod Update 7(6):591–602

85. Pujol A, Boiso I, Benet J, Veiga A, Durban M, Campillo M et al (2003) Analysis of nine chromosome probes in first polar bodies and metaphase II oocytes for the detection of aneuploidies. Eur J Hum Genet 11(4):325–336
86. Kyogoku H, Kitajima TS (2017) Large cytoplasm is linked to the error-prone nature of oocytes. Dev Cell 41(3):287–298.e4
87. Lane SIR, Jones KT (2017) Chromosome biorientation and APC activity remain uncoupled in oocytes with reduced volume. J Cell Biol 216(12):3949–3957
88. Das A, Iwata-Otsubo A, Destouni A, Dawicki-McKenna JM, Boese KG, Black BE et al (2022) Epigenetic, genetic and maternal effects enable stable centromere inheritance. Nat Cell Biol 24(5):748–756
89. McKee BD, Wilhelm K, Merrill C, Ren X (1998) Male sterility and meiotic drive associated with sex chromosome rearrangements in Drosophila. Role of X-Y pairing. Genetics 149(1):143–155
90. Carmena M, Wheelock M, Funabiki H, Earnshaw WC (2012) The chromosomal passenger complex (CPC): from easy rider to the godfather of mitosis. Nat Rev Mol Cell Biol 13(12):789–803
91. Godek KM, Kabeche L, Compton DA (2015) Regulation of kinetochore–microtubule attachments through homeostatic control during mitosis. Nat Rev Mol Cell Biol 16(1):57–64
92. John B, Lewis KR (1965) Meiotic behaviour of special chromosomes. In: The meiotic system, protoplasmatologia, vol VI. Springer-Verlag, Vienna, pp 167–201
93. Page J, Viera A, Parra MT, de la Fuente R, Suja JA, Prieto I et al (2006) Involvement of synaptonemal complex proteins in sex chromosome segregation during marsupial male meiosis. PLoS Genet 2(8):e136
94. Bachtrog D, Mank JE, Peichel CL, Kirkpatrick M, Otto SP, Ashman TL et al (2014) Sex determination: why so many ways of doing it? PLoS Biol 12(7):e1001899
95. Rutkowska J, Badyaev AV (2008) Review. Meiotic drive and sex determination: molecular and cytological mechanisms of sex ratio adjustment in birds. Philos Trans R Soc Lond Ser B Biol Sci 363(1497):1675–1686
96. Currie CE, Ford E, Benham Whyte L, Taylor DM, Mihalas BP, Erent M et al (2022) The first mitotic division of human embryos is highly error prone. Nat Commun 13(1):6755
97. Cavazza T, Takeda Y, Politi AZ, Aushev M, Aldag P, Baker C et al (2021) Parental genome unification is highly error-prone in mammalian embryos. Cell 184(11):2860–2877.e22
98. Reichmann J, Nijmeijer B, Hossain MJ, Eguren M, Schneider I, Politi AZ et al (2018) Dual-spindle formation in zygotes keeps parental genomes apart in early mammalian embryos. Science 361(6398):189–193
99. Bean CJ, Hunt P, Millie E, Hassold TJ (2001) Analysis of a malsegregating mouse Y chromosome: evidence that the earliest cleavage divisions of the mammalian embryo are non-disjunction-prone. Hum Mol Genet 10(9):963–972
100. Schneider I, de Ruijter-Villani M, Hossain MJ, Stout TAE, Ellenberg J (2021) Dual spindles assemble in bovine zygotes despite the presence of paternal centrosomes. J Cell Biol 220(11):e202010106
101. Das A, Smoak EM, Linares-Saldana R, Lampson MA, Black BE (2017) Centromere inheritance through the germline. Chromosoma 126(5):595–604
102. Palmer DK, O'Day K, Margolis RL (1990) The centromere specific histone CENP-A is selectively retained in discrete foci in mammalian sperm nuclei. Chromosoma 100(1):32–36
103. De La Fuente R, Baumann C, Viveiros MM (2015) ATRX contributes to epigenetic asymmetry and silencing of major satellite transcripts in the maternal genome of the mouse embryo. Development 142(10):1806–1817
104. Destouni A, Vermeesch JR (2017) How can zygotes segregate entire parental genomes into distinct blastomeres? The zygote metaphase revisited. Bioessays 39(4):1–7
105. Destouni A, Zamani Esteki M, Catteeuw M, Tšuiko O, Dimitriadou E, Smits K et al (2016) Zygotes segregate entire parental genomes in distinct blastomere lineages causing cleavage-stage chimerism and mixoploidy. Genome Res 26(5):567–578

106. Comai L, Tan EH (2019) Haploid induction and genome instability. Trends Genet 35(11):791–803
107. Maheshwari S, Tan EH, West A, Franklin FCH, Comai L, Chan SWL (2015) Naturally occurring differences in CENH3 affect chromosome segregation in zygotic mitosis of hybrids. PLoS Genet 11(1):e1004970
108. Marimuthu MPA, Maruthachalam R, Bondada R, Kuppu S, Tan EH, Britt A et al (2021) Epigenetically mismatched parental centromeres trigger genome elimination in hybrids. Sci Adv 7(47):eabk1151
109. Ravi M, Chan SWL (2010) Haploid plants produced by centromere-mediated genome elimination. Nature 464(7288):615–618
110. Hsu PD, Lander ES, Zhang F (2014) Development and applications of CRISPR-Cas9 for genome engineering. Cell 157(6):1262–1278
111. Wang JY, Doudna JA (2023) CRISPR technology: a decade of genome editing is only the beginning. Science 379(6629):eadd8643
112. Schindele A, Gehrke F, Schmidt C, Röhrig S, Dorn A, Puchta H (2022) Using CRISPR-kill for organ specific cell elimination by cleavage of tandem repeats. Nat Commun 13(1):1502
113. Adikusuma F, Williams N, Grutzner F, Hughes J, Thomas P (2017) Targeted deletion of an entire chromosome using CRISPR/Cas9. Mol Ther 25(8):1736–1738

Cell Cycle Regulation of Vertebrate Female Meiotic Divisions

Aude Dupré and Katja Wassmann

Abstract Cell division is the basis of development and gametogenesis, its misregulation can lead to growth defects and, if affecting germ cells, infertility. During female meiosis, germ cells undergo two successive divisions, meiosis I and meiosis II, without intervening S-phase. Moreover, they segregate the homologous chromosomes in meiosis I and then sister chromatids in meiosis II to reduce by half the genome content. Besides controlling ploidy, female meiosis must halt at the proper cell-cycle stage to avoid parthenogenesis. Moreover, oocytes stockpile maternal transcripts and proteins required for early embryogenesis after fertilization. These processes are orchestrated by oocyte-specific mechanisms operating between the two divisions, under the control of protein turnover and protein phosphorylation. Here, we will discuss how key molecular meiotic events of female meiosis are intertwined with the regulation of maternal proteins to form a functional oocyte in vertebrates. Although some clear differences exist between vertebrates, the core machinery controlling meiosis is conserved, and therefore provides an encompassing view of the orchestration of female meiosis.

Keywords Meiosis · Oocyte · Cdk1 · Cyclins · Phosphatases

Abbreviations

APC	Anaphase-Promoting Complex
CaMKII	Calcium/Calmodulin Kinase II
cAMP	Cyclic AMP
CaN	Calcineurin
cArpp19	cAMP-regulated phosphoprotein of 19 kDa
CDK	Cyclin-dependent Kinase

A. Dupré (✉) · K. Wassmann
Université Paris Cité, CNRS, Institut Jacques Monod, Paris, France
e-mail: aude-isabelle.dupre@ijm.fr; katja.wassmann@ijm.fr

B. Mogessie (ed.), *Cellular Architecture and Dynamics in Female Meiosis*,
https://doi.org/10.1007/978-3-031-97173-0_4

Cdk1	Cyclin-dependent kinase 1
CSF	Cytostatic
CycBs	Cyclin B
DDK	Dbf4-dependent kinase
GV	Germinal Vesicle
GVBD	Germinal Vesicle Break Down
MCMs	Mini Maintenance Chromosomes
MI-MII transition	Meiosis I-to-Meiosis II transition
MPs	Maternal Proteins
ORI	origin of replication
PKA	Protein kinase dependent of cyclic AMP
PP1	Protein phosphatase 1
PP2A	Protein phosphatase 2A
PPases	Phosphatases
Pre-RC	Pre-replicative complex
SAC	Spindle Assembly Checkpoint
T14	Threonine 14
Y15	Tyrosine 15

Cell division is the basis of development and gametogenesis, its misregulation can lead to growth defects and, if affecting germ cells, infertility. To maintain genome stability, each step of the cell cycle is precisely orchestrated. DNA is first duplicated in S-phase and then equally distributed among two daughter cells during M-phase (mitosis).

In meiosis, this coordinated series of events is changed to generate haploid gametes for sexual reproduction. Meiosis is initiated from mitotically dividing germ stem cells. After one last round of pre-meiotic DNA replication, germ cells undergo meiotic recombination and two successive divisions, meiosis I and meiosis II, without intervening S-phase. Crucially, in meiosis I, it is the homologous chromosomes that segregate and, in meiosis II, sister chromatids, similar to a mitotic division. The sequence of a reductional division (meiosis I) followed by an equational division (meiosis II) results in the formation of haploid gametes harboring only half the genome content.

The formation of a functional oocyte that can be fertilized is more complex than just reducing cell ploidy. While spermatozoids morphologically differentiate after meiosis to support fertilization, oocytes divide asymmetrically, accumulate and orderly arrange organelles, preparing to transmit their cytoplasm and mitochondria to embryos and, in some species, lay the groundwork for embryonic patterning. Moreover, oocytes must resume cell division upon fertilization while avoiding parthenogenesis. They have to stockpile maternal transcripts and proteins required for early embryogenesis after fertilization. During female meiosis, this coordination is primarily regulated by arrest and release at two cell cycle stages: the first one at prophase of meiosis I and the second at metaphase of meiosis II (metaphase II) in vertebrates. Furthermore, some of the maternal proteins required for embryonic development are synthesized only after metaphase of meiosis I. As these proteins regulate cell division, oocyte-specific mechanisms operate between the two divisions (the MI-MII transition) to target their turnover and/or activities, thus avoiding deleterious effects on meiosis progression if they are present. Hence, oocytes implement regulatory mechanisms, which are either shared with mitosis or specific, to orchestrate the proper execution of two functionally distinct divisions, without intervening DNA replication.

Detailed reviews on prophase arrest, chromosome segregation, and regulation of protein translation in vertebrates have been published recently. Here, we will discuss how key molecular meiotic events of the MI-MII transition are intertwined with the regulation of maternal proteins in order to form a functional oocyte. We focus on two vertebrate species, mouse and frog (*Xenopus laevis*). Although some clear differences exist between these two species, the core machinery controlling meiosis is conserved, and therefore provides an encompassing view of the orchestration of female meiosis.

Basic Control of Female Meiosis

Overview of Oocyte Cell Division

In all species, oocytes remain arrested in prophase of meiosis I in the ovaries (Fig. 1). This arrest, which can last up to 40 years in humans, corresponds to a growth phase called vitellogenesis; it allows oocytes to accumulate nutrients, organelles, maternal RNAs, and proteins needed not only for early development but also for the resumption of meiosis [1, 2]. At the time of ovulation, fully-grown oocytes resume meiosis upon external stimulation; *e.g.* hormones in vertebrates (Fig. 1). The nucleus, called germinal vesicle (GV) in oocytes, breaks down (GVBD for germinal vesicle break down) and the metaphase I spindle is formed. In vertebrates, once chromosomes are properly aligned in metaphase I, oocytes segregate homologs and extrude half of them in an extremely asymmetric division, forming the first polar body, marking exit from meiosis I. Almost immediately, oocytes enter meiosis II without reformation of the nucleus or genomic DNA replication. The spindle of meiosis II is organized and vertebrate oocytes arrest again at metaphase II, this time to await fertilization. Meiosis II completion is triggered by fertilization, which induces sister-chromatid segregation in anaphase II and extrusion of the second polar body (Fig. 1). Fertilized eggs then enter interphase with decondensation of their DNA and formation of a female pronucleus. Male and female pronuclei fuse, thus re-establishing the diploid genome content of the zygote, the first cell of the embryo.

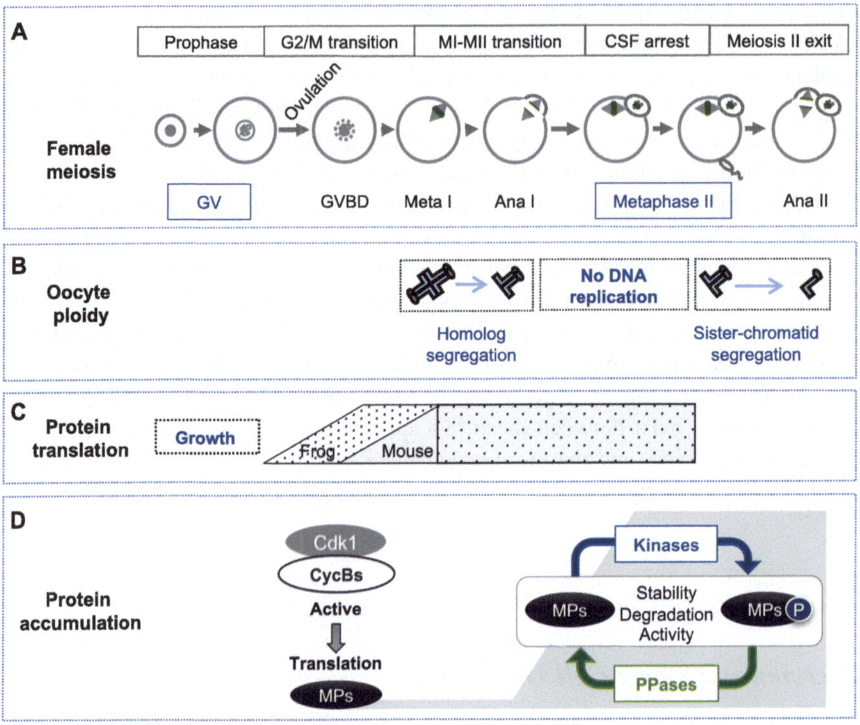

Fig. 1 Control of vertebrate female meiosis. (**a**) Oocytes are arrested in prophase of the first meiotic division (GV for germinal vesicle). At the time of ovulation, they resume meiosis and progress through two meiotic divisions, meiosis I and meiosis II. In vertebrates, oocytes arrest again at metaphase II, this time to await fertilization. Meiosis II completion is triggered by fertilization, which induces sister-chromatid segregation in anaphase II and extrusion of the second polar body. (**b**) Control of oocyte ploidy. Female meiosis occurs with no DNA replication in between the two meiotic divisions (MI-MII transition). Moreover, homologous chromosomes are segregated during the first meiotic division and sister chromatids during the second meiotic division, hence reducing by half the genome content. (**c**) Protein translation during female meiosis. The arrest in prophase corresponds to a growth phase during which maternal RNAs and proteins are stockpiled. Upon entry into meiosis, translation is initiated from maternal RNAs before GVBD in *Xenopus*. In the mouse, translation starts after GVBD. After metaphase I, protein translation is necessary to drive female meiosis in all species. (**d**) Interplay between protein turnover and phosphorylation during female meiosis. During the MI-MII transition, some of the maternal proteins (MPs) required for embryonic development are synthesized. These proteins start to be translated at metaphase I following the activation of Cdk1-Cyclin B complexes. However, they only accumulate when oocytes enter meiosis II. As these proteins regulate cell division, specific mechanisms have evolved in oocytes to target their turnover, stability and/or activities to ensure proper meiosis progression. Most of these mechanisms are controlled by protein phosphorylation/dephosphorylation mediated by various kinases and their antagonistic phosphatases (PPases). *GV* germinal vesicle, *GVBD* germinal vesicle break down, *Meta I* metaphase of meiosis I, *CSF* cytostatic, *CycBs* cyclin B, *Ana I* anaphase of meiosis I, *Ana II* anaphase of meiosis II, *MPs* maternal proteins

Protein Translation and Degradation

As is the case in mitosis, female meiosis is driven by dynamic changes in protein abundance and post-translational modifications, in particular phosphorylations (Fig. 1). These two intertwined mechanisms precisely control in time and space the activity, the stability and/or localization of a broad range of substrates orchestrating the structural changes of meiosis as well as its progression.

Changes in protein composition result from regulated protein translation and degradation. Protein translation is the only way to control gene expression in meiosis as it occurs in nearly total absence of transcription and relies on stockpiled maternal mRNAs (Fig. 1). In all vertebrates, translation drives female meiosis, although it is not necessary to resume meiosis in small rodents. Protein degradation is mainly controlled by two E3 ubiquitine ligases during female meiosis. The first one is the Anaphase-Promoting Complex associated with either Cdh1 (APC^{Cdh1}) or Cdc20 (APC^{Cdc20}). As in mitosis, APC^{Cdc20} is the main driver of the metaphase-to-anaphase transition during meiosis I and meiosis II. The second one, the Skp1-Cul1-Fbox protein complex (SCF) is less characterized but, associated with its activator βTrCP (β-transducin repeat containing protein) ($SCF^{\beta TrCP}$), contributes to lifting the oocyte arrest in metaphase II upon fertilization in vertebrates [3, 4].

Cdk1 and Serine/Threonine Phosphatases

Protein phosphorylation is regulated by various kinases and phosphatases; the latter's ensuring that phosphorylations are dynamic and reversible [5]. At the heart of this network stand the universal engines of cell division, Cdk1-Cyclin complexes. During cell division, Cdk1-Cyclin complexes phosphorylate hundreds of proteins through a dual-mechanism that modulates both Cdk1 enzymatic levels and substrate specificity. Levels of Cdk1 activity, or thresholds, are determined by transient association of Cdk1 with two mitotic Cyclin isoforms, Cyclin A and B, whose expression levels are governed by their periodic synthesis and degradation [6]. In addition, Cdk1 complexes are inhibited by Cdk1 phosphorylation at T14/Y15, which preferentially targets Cdk1-Cyclin B complexes to avoid premature entry in mitosis [7]. These inhibitory phosphorylations are catalyzed by two kinases, Wee1/Myt1, and removed by their opposing phosphatase, Cdc25 [8–10]. Wee1 is a nuclear protein that phosphorylates Y15, whereas Myt1 is a membrane-associated kinase that phosphorylates both residues Y15/T14 and further sequesters Cdk1 in the cytoplasm [11, 12]. During G2/M transition, Wee1/Myt1 are inhibited whereas Cdc25 is activated; hence allowing the dephosphorylation and activation of Cdk1 complexes [8, 13–15]. The specificity of Cdk1's activity toward different substrates is governed by the identity of the associated Cyclin, which recognizes CDK substrates using highly selective docking sites, in combination with Cyclin-distinct specific partners whose identity has not been clearly established yet in many cases [8, 13–15]. Facing the kinases, the main phosphatases regulating cell division belong to the Serine/Threonine phosphatase family, including PP1, PP2A and more recently described

PP4 and PP6 [3, 16–21]. These enzymes function through combinatorial interactions between conserved catalytic "C" subunits and various scaffolding and/or regulatory subunits. For example, catalytic PP2A subunit (PP2A-C) is associated with one scaffold A subunit (PP2A-A) and four distinct families of regulating B subunits (PP2A-B), including such as B55 or B56 for PP2A [22]. These subunits govern phosphatase localization, activity and specificity by recognizing substrates through short linear interacting motifs and/or polybasic-acidic regions in the case of B55 [22, 23]. By adjusting the level of protein phosphorylation, PP1, PP2A-B55, and PP2A-B56 control numerous structural events of meiosis, such as chromosome shape, dynamic and segregation, as well as meiotic cell cycle arrest [18, 19, 24]. However, in big contrast to kinases, their specific function during the MI-MII transition remains poorly defined.

Maintaining Vertebrate Oocytes Arrested in Prophase and Releasing the Arrest

The Prophase Arrest

Cdk1 and Its Regulators in Prophase-Arrested Oocytes

Vertebrate prophase-arrested oocytes (also called GV oocytes) express Cyclin A2 as well as Cyclin B1 (Cyclin B1/B4 in *Xenopus*), Cyclin B2 (Cyclin B2/B5 in *Xenopus*), and Cyclin B3 [25] (Fig. 2). These cyclins exhibit various expression

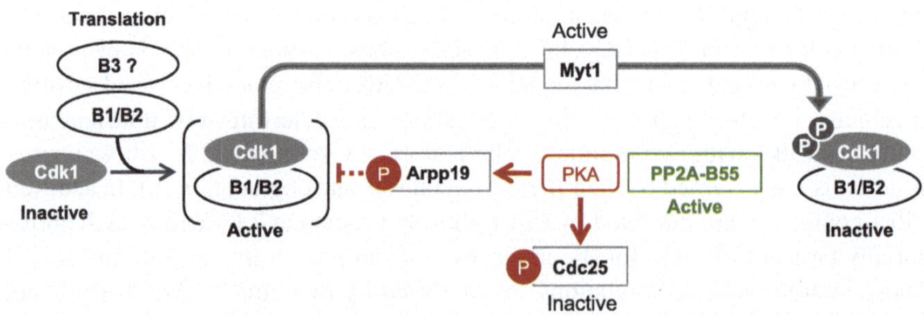

Fig. 2 Control of the prophase arrest in frog oocytes. In *Xenopus*, prophase-arrested oocytes express Cyclin B1 (Cyclin B1/B4), Cyclin B2 (Cyclin B2/B5) and Cyclin B3. Cdk1 is inactive but present in different pools. Oocytes contain free monomeric Cdk1 as well as Cdk1-Cyclin B2 and little amounts of Cdk1-Cyclin B1 complexes. Whether Cyclin B3 is associated with Cdk1 in prophase-arrested oocytes is currently unknown. Cdk1 complexes are inhibited by phosphorylation at T14/Y15, under the control of the kinase Myt1. The phosphatase Cdc25, which antagonizes Myt1, is inactive. In *Xenopus*, Wee1 is not present [31]. As for all vertebrates, PKA (protein kinase dependent on cAMP) is the key inhibitor of Cdk1. In *Xenopus*, PKA inhibits various signaling pathways required for Cdk1 activation by targeting at least two substrates, Cdc25 and Arpp19 (cAMP-regulated phosphoprotein). Moreover, in this species, the PKA-counteracting phosphatase PP2A-B55 is active, indicating that the prophase arrest is controlled by reversible phosphorylation events

profiles in mouse and *Xenopus;* the most striking differences are the very low levels of Cyclin B1 and the absence of Cyclin A in *Xenopus* oocytes (based on studies using western blots). In both species, Cdk1 is inactive but present in different pools (Fig. 2). *Xenopus* oocytes contain free monomeric Cdk1 as well as Cdk1-Cyclin B2 and little amounts of Cdk1-Cyclin B1 complexes, which are formed during the prophase arrest by the basal translation and accumulation of Cyclin B [1, 2, 26]. In contrast, Cyclins are in large excess compared to Cdk1 in mouse prophase-arrested oocytes, suggesting that Cdk1 is fully associated with Cyclin A2, B1, and B2 [27]. Given the recent discovery of Cyclin B3, its biochemical characterization is still needed to ascertain whether it is associated with Cdk1 in prophase-arrested oocytes.

In prophase-arrested oocytes, Cdk1-Cyclin B1 and Cdk1-Cyclin B2 complexes are inhibited by T14/Y15 phosphorylation, catalyzed by Wee1 and/or Myt1 activities [28–32] (Fig. 2). These kinases exhibit various expression profiles in *Xenopus* and mouse oocytes. Myt1 is present in both species while Wee1 is present in mouse but not in *Xenopus* prophase-arrested oocytes [28–32]. In *Xenopus*, Myt1 activity contributes to the prophase arrest by preventing premature activation of Cdk1-Cyclin B complexes as its pharmacological inhibition stimulates or even promotes meiosis resumption [1, 2]. In mouse oocytes, inhibition of Cdk1 complexes is equally essential to preserve germ cells in ovaries. Indeed, the expression of a mutant of Cdk1 that cannot be phosphorylated at T14/Y15 (Cdk1-AF) in a knock-in mouse model causes premature loss of growing oocytes by apoptosis and further triggers meiosis resumption in fully-grown oocytes [33]. Whether Wee1, Myt1, or both, are responsible for inhibiting Cdk1 complexes in the mouse has not been clarified yet as both kinases are present in prophase-arrested oocytes. They likely display redundant functions as both kinases must be inhibited to fully abolish meiosis resumption in fully-grown mouse oocytes [29, 32].

PKA, the Universal Inhibitor of Cdk1 Activation in Vertebrate Oocytes

In vertebrate prophase-arrested oocytes, the key mechanism keeping Cdk1 inactive relies on the intracellular concentration of cyclic AMP (cAMP) that is produced by follicular cells and enters oocytes through gap junctions [34, 35] (Fig. 2). High levels of cAMP activate PKA (cAMP-dependent protein kinase) that prevents Cdk1 activation. In the mouse, PKA directly phosphorylates both Wee1 and Cdc25 [29, 36]. It stimulates Wee1 activity and keeps Cdc25 inactive as a result of its sequestration in the cytoplasm [29, 36]. In *Xenopus*, PKA inhibits multiple signaling pathways required for Cdk1 activation by targeting at least two substrates, the small protein Arpp19 (cAMP-regulated regulated phosphoprotein) and Cdc25 [37–39]. However, the molecular details on how PKA indirectly prevents Cdk1 activation in this species remain partly unknown. The Arpp19-dependent mechanism inhibiting Cdk1 has not been elucidated yet [18, 19]. Moreover, PKA phosphorylates a still unidentified substrate to inhibit protein translation, a process necessary for Cdk1 activation in this species [37–39]. Finally, PP2A-B55 counteracts PKA and is active in prophase; implying that the prophase arrest is maintained by transient phosphorylation resulting from both PKA and PP2A-B55 activities [18, 19]. As for PKA, the substrates that are dephosphorylated by PP2A-B55 to avoid precocious resumption of meiosis have not been identified yet.

Release from Prophase Arrest

In all vertebrates, PKA downregulation is both necessary and sufficient for meiosis resumption [18, 39] (Fig. 3). This process allows PP2A-B55 to dephosphorylate PKA substrates, at least in *Xenopus* oocytes and further initiates protein synthesis in this species [18, 19]. In contrast, in the mouse, PKA inhibition directly switches the balance between Myt1/Wee1 and Cdc25 activities, thus promoting activation of Cdk1 complexes by dephosphorylating T14/Y15 [36] (Fig. 3). Likewise, as Cdk1 activation is thought to be initiated in the nucleus in this species, the shuttling of these actors into and out of the nucleus further contributes to meiosis resumption, with Cyclin B1 and Cdc25 being imported into the nucleus whereas Wee1 is exported to the cytoplasm.

In *Xenopus* and mouse oocytes, one of the main mechanisms triggering Cdk1 activation is the synthesis, the accumulation and the association of Cdk1 with Cyclin B1. In *Xenopus*, two other cell cycle regulators are *de novo* synthesized for Cdk1 activation: Ringo, an allosteric activator of monomeric Cdk1 and the kinase Mos, an oocyte-specific and indirect activator of MAPK (the Mos/MAPK pathway) [40, 41]. Although any of these proteins can activate Cdk1 independently of protein synthesis, Cyclin B1 is considered as the key activator of Cdk1. This protein is the only one that accumulates before and independently of Cdk1 activation [28, 42–46]. Furthermore, upon its association with monomeric Cdk1, Cyclin B1 generates low levels of Cdk1 activity [28] (Fig. 3). Ringo and Mos further contribute to timely activation of Cdk1 by inhibiting Myt1 and activating Cdc25 [40, 47–51]. Subsequently, Cdk1 activity reaches the levels necessary for GVBD induction through positive and negative feedbacks, also called the "auto-amplification loop."

Fig. 3 (continued) complexes without the requirement for protein synthesis by inducing Cdk1 dephosphorylation at T14/Y15. In this species, Cdk1 is therefore thought to be activated through an "auto-amplification loop," in which Cdk1 phosphorylates its direct regulators. As a result, Myt1/Wee1 are fully inhibited, whereas Cdc25 is activated. This process generates low levels of Cdk1 activity for GVBD induction. Thereafter, full activation of Cdk1 is achieved in a protein synthesis-dependent manner for proper progression through prometaphase I. Furthermore, Cyclin B1 is stabilized, most likely by the inhibition of anaphase-promoting complex (APC). (**b**) Cdk1 activation in *Xenopus* oocytes. Upon PKA downregulation, the substrates of PKA are dephosphorylated, most likely by the phosphatase PP2A-B55, which initiates multiple signaling pathways to activate Cdk1. Among them, protein translation is activated, thereby allowing the accumulation of Cyclin B1. Upon its association with Cdk1, Cyclin B1 generates low levels of Cdk1 activity. Subsequently, Cdk1 activity reaches the levels necessary for GVBD induction by the "auto-amplification loop" as in the mouse oocyte. Besides Cdk1, the "auto-amplification loop" further requires inactivation of the Cdk1-counteracting phosphatase PP2A-B55 for GVBD induction in this species

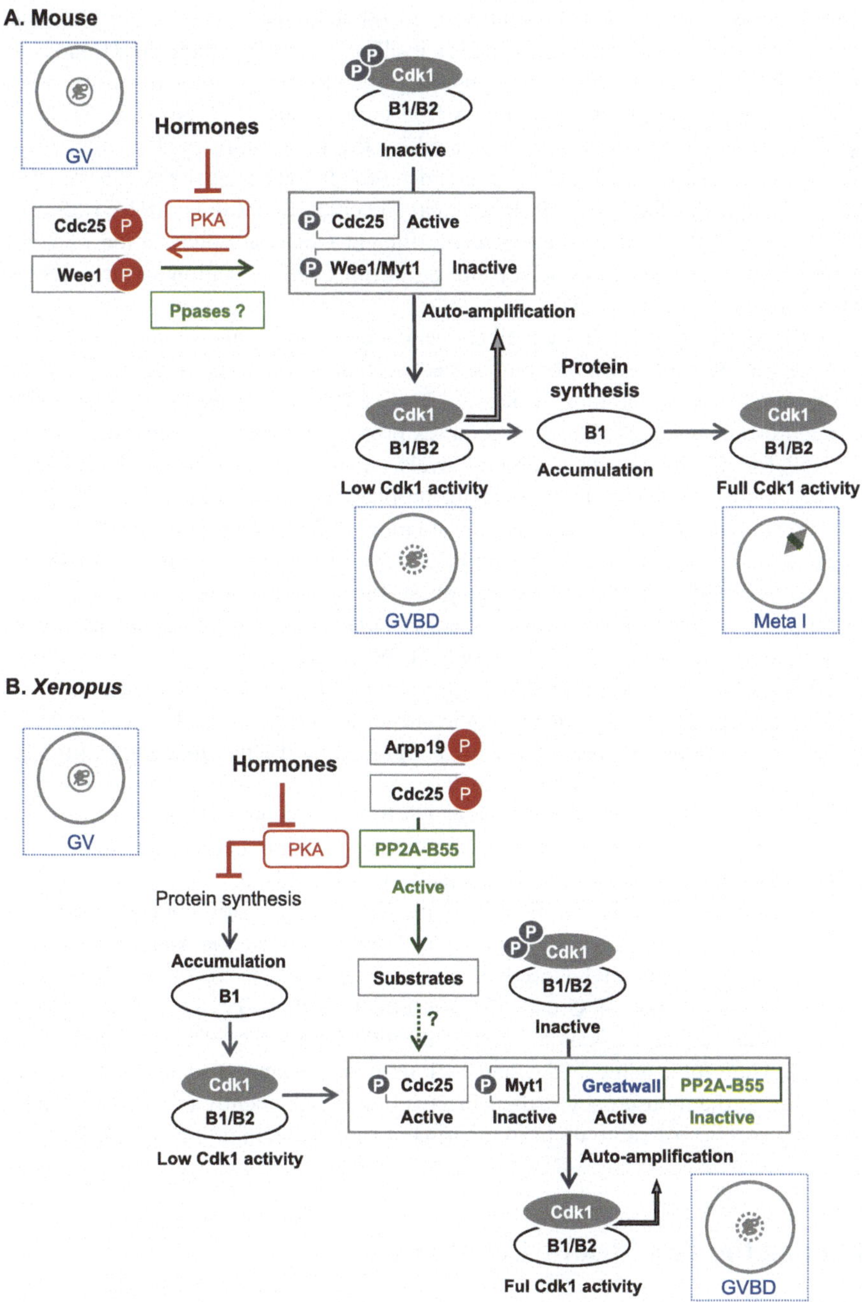

Fig. 3 Proposed mechanism of Cdk1 activation in *Xenopus* and mouse oocytes. In vertebrates, the resumption of meiosis depends on the downregulation of PKA, which promotes the dephosphorylation of PKA substrates in order to activate Cdk1. (**a**) Mouse oocytes. PKA inhibition directly switches the balance between Myt1/Wee1 and Cdc25 activities, owing to the activity of a still unknown phosphatase (PPase). This process promotes the activation of stockpiled Cdk1-Cyclin B

This process relies on Cdk1's ability, with other kinases, to phosphorylate and thereby control direct regulators of Cdk1 itself, *e.g.* to fully inhibit Myt1 and activate Cdc25 [18]. In addition, the "auto-amplification loop" only functions if the Cdk1-counteracting phosphatase PP2A-B55 is inactivated [52–56] (Fig. 3). In all species, PP2A-B55 inhibition is triggered by the kinase Greatwall, which phosphorylates Arpp19 on a highly conserved motif distinct from PKA [39, 55–58]. Once phosphorylated, Arpp19 binds and specifically inhibits PP2A-B55 [39, 55–58]. PP2A-B55 inactivation ensures efficient Cdk1-phosphorylation of Cdk1 regulators and meiotic substrates, resulting in full Cdk1 activation and irreversible entry into meiosis [18, 39, 57].

In the mouse, Cdk1 activity is likely generated by the "auto-amplification loop" as stockpiled Cdk1-Cyclin B complexes are sufficient for GVBD induction without the requirement for protein synthesis [59–61] (Fig. 3). Thereafter, progression through prometaphase I, however, requires protein synthesis for full Cdk1 activation [59, 61]. In this species, Cyclin B1 levels, and to a lesser extent Cyclin B2, must be continuously repressed to maintain the prophase arrest [44, 62]. Indeed, the pharmacological inhibition or the genetic ablation ("Knock-out") of APC^{Cdh1} either induces or stimulates GVBD by stabilizing Cyclin B1 [63–68]. After GVBD, the exact role of Cdh1 for timing prometaphase I progression still needs to be determined [65]. However, Cyclin B1 accumulates thanks to its translational activation, leading to an increase of Cdk1 activity [62]. Whether, besides Cyclins, PP1 and/or PP2A must be regulated for meiosis resumption in mouse oocytes remains unclear. On the one hand, there is evidence that inactivation of PP1 and/or PP2A contributes to meiosis resumption, because GVBD is induced by the pharmacological inhibition of PP1 and PP2A with okadaic acid or Calyculin A [69–71]. The specific inhibition of PP1 using antibodies promotes GVBD [72]. Upregulating PP2A activity abolishes meiosis resumption whereas preventing the degradation of the scaffolding PP2A subunit, PP2A-A, results in delayed meiosis entry and failure to complete meiosis I [69–71, 73]. On the other hand, the genetic invalidation of catalytic PP2A subunits does not affect GVBD but oocytes fail to complete meiosis I due to chromosome misalignment and abnormal spindle assembly [24, 70, 74, 75]. Moreover, conditional knock-out of Greatwall does affect neither GVBD nor progression through prometaphase I, which occurs with wild type levels of Cdk1 activity [52, 76, 77]. Therefore, while PP2A and PP1 inhibition seems dispensable for Cdk1 activation in the mouse oocyte, their inactivation independently of Greatwall function may still be critical to orchestrate proper progression in prometaphase I.

Linking the Two Meiotic Cell Divisions

After metaphase I, protein synthesis from stockpiled mRNAs is stimulated upon Cdk1 activation and is required to proceed through the MI-MII transition in *Xenopus* and mouse oocytes [40, 47, 48, 62, 78]. This process produces cell cycle regulators that drive (1) the MI-MII transition by allowing Cyclin B1 and B2 accumulation after exit from meiosis I, (2) arrest of oocytes at metaphase II through Emi2 in vertebrates, and (3) even support embryonic cell cycles after fertilization by allowing the accumulation Cdc6 and Wee1, at least in *Xenopus*. Notably, not only must the translation of these specific proteins be tightly controlled, their accumulation and activity must also be properly regulated at all stages of the MI-MII transition to avoid deleterious effects on female meiotic progression.

Exit from Meiosis I

The MI-MII transition is mainly driven by Cdk1 activity that transiently drops after metaphase I to exit meiosis I and rises again for meiosis II entry (Fig. 4). The overall level of kinase activity results exclusively from translation and degradation of Cyclin B1 and Cyclin B2, as Cdk1 is thought not to be inhibited by T14/Y15 phosphorylation, based on studies in *Xenopus* oocytes. The metaphase I-to-anaphase I transition is induced by activation of APC^{Cdc20}, which ubiquitinates and promotes proteolysis of Cyclin B1 and B2 to inactivate Cdk1 [6, 8, 25, 79–81] (Fig. 4). Simultaneously, the protein securin, an inhibitor of the protease separase that cleaves cohesin complexes on chromosome arms in meiosis I, is degraded; thereby allowing the segregation of homologous chromosomes [80–87]. Centromeric cohesion between sister chromatids, however, is maintained to avoid their precocious separation in meiosis I. As in mitosis, the spindle assembly checkpoint (SAC) is present and active during mouse female meiosis to control APC^{Cdc20} activation [88, 89]. In mouse oocytes, stable kinetochore-microtubule attachments are established during prometaphase I under control of the gradually increasing Cdk1-Cyclin B1 and B2 activity [90]. The SAC detects missing spindle-kinetochore attachments, and together with a mechanism called error correction (ER) recognizes whether kinetochores are attached to correct pole [89, 91]. As long as chromosomes are not correctly attached and bi-orientated on the meiosis I spindle, SAC proteins accumulate at kinetochores and inhibit APC^{Cdc20} to prevent anaphase I onset [88, 89, 92]. In contrast to the mouse and even though the molecular SAC machinery is present in *Xenopus* oocytes, the SAC seems to be inactive as disrupting spindle microtubules does not delay homolog segregation during anaphase I [93].

Besides Cdk1, exit from meiosis I most probably also relies on the reactivation of PP1, PP2A-B55 or both in order to temporally dephosphorylate Cdk1 substrates and/or regulate protein degradation (Fig. 4). In *Xenopus* and mouse oocytes, the catalytic subunit of PP1 is partly dephosphorylated during the MI-MII transition, suggesting its transient activation [75, 94, 95]. In the mouse, active PP1 would maintain chromosome shape and allows exit from meiosis I [75, 95, 96]. PP2A-B55 activity is quite likely contributing to proper exit of meiosis I, because loss of Greatwall, which targets PP2A-B55, leads to a delay in anaphase I completion due to low levels of APC^{Cdc20} activity [52, 77]. Finally, phosphoproteomic approaches in sea star oocytes have recently revealed that PP2A-B55 reactivation after metaphase I leads to preferential dephosphorylation of threonine residues [97]. Interestingly,

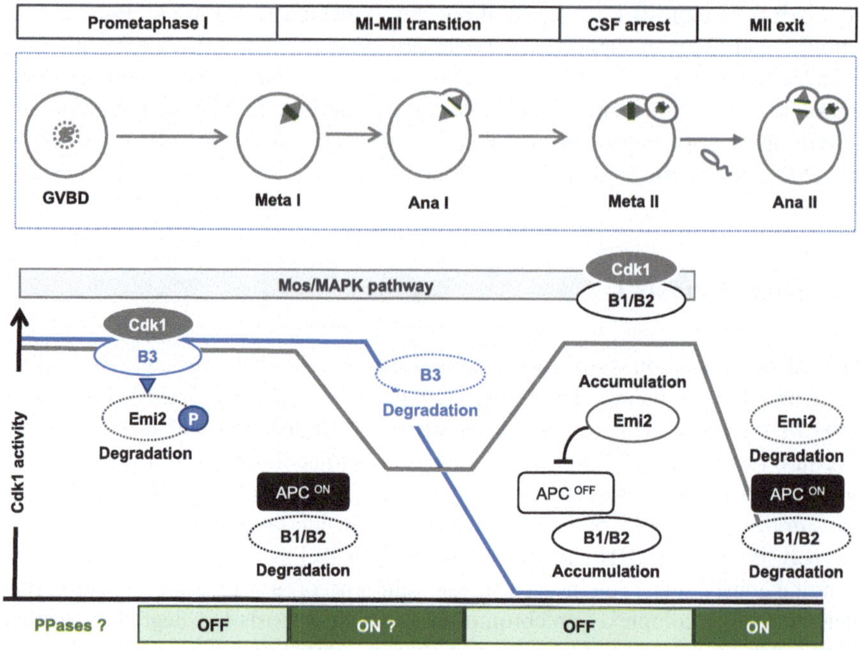

Fig. 4 Regulatory circuits of the MI-MII transition and arrest in metaphase II. The MI-MII transition is mainly driven by Cdk1 activity that transiently drops after metaphase I to exit meiosis I and rises again for meiosis II entry. The overall levels of kinase activity result exclusively from translation and degradation of Cyclin B1 and Cyclin B2. The metaphase I-to-anaphase I transition is induced by activation of APC^{Cdc20}, which promotes proteolysis of Cyclin B1 and B2 to inactivate Cdk1 and of securin, an inhibitor of separase and thus, chromosome segregation. Thereafter, entry into meiosis II depends on the accumulation of Cyclin B1 and B2 thanks to their translation and the gradual inhibition of APC^{Cdc20}. This process depends on the accumulation of its oocyte-specific inhibitor, Emi2, in vertebrates. Emi2 starts to be translated in metaphase I but only accumulates during meiosis II. This pattern depends on the activity of Cdk1-Cyclin B3. In metaphase I, Cdk1-Cyclin B3 activity phosphorylates Emi2 and promotes its degradation, thus ensuring activation of APC^{Cdc20} for meiosis I exit. Upon degradation of Cyclin B3 in anaphase of meiosis I (Ana I), Cdk1-Cyclin B3 is inactivated. Emi2 can thus accumulate, allowing the arrest in metaphase II. After fertilization, the mechanism inhibiting APC^{Cdc20} is relieved, allowing once again the degradation of Cyclin B1, Cyclin B2 and securin for sister-chromatid segregation in anaphase II and completion of meiosis II

these dephosphorylation events are also observed in *Xenopus* oocytes [97], suggesting that PP2A-B55 reactivation during the MI-MII transition is an evolutionary conserved mechanism in vertebrates. By antagonizing threonine phosphorylation specifically, PP2A-B55 would thus create distinct patterns of phosphorylation in order to control specific structural events of the MI-MII transition such as the inhibition of DNA replication [97].

Entry into Meiosis II

Upon exit from meiosis I, Cyclin B1 and B2 levels rise again, thanks to their translation initiated at GVBD and loss of APC^{Cdc20} activity (Fig. 4). This leads to reactivation of Cdk1 for meiosis II entry, without intervening interphase. As the oocyte reaches metaphase II, Cyclin B1 and B2 are stabilized, thus keeping Cdk1 active at metaphase II until fertilization. At this stage, PP1 and PP2A-B55 are both inhibited, respectively by Cdk1 phosphorylation and Greatwall, to avoid precocious dephosphorylation of meiotic substrates [54, 56]. After fertilization, the mechanism inhibiting APC^{Cdc20} is relieved, allowing once again the degradation of Cyclin B1, Cyclin B2 and securin for sister-chromatid segregation in anaphase II and completion of meiosis II. Concomitantly, PP1 is activated to turn off Greatwall; thereby inducing PP2A-B55 reactivation [54–56, 98–100]. As a result, Cdk1 substrates are dephosphorylated to drive exit from meiosis II and interphase entry [54, 56]. Importantly, as the ordered dephosphorylation of threonines and serines orchestrates mitosis exit [22, 23], an important question will be to elucidate whether differential phosphorylation patterns, implemented by PP1, PP2A or both, drive meiosis II completion.

Redundant and Non-redundant Functions of Cyclin B

Any treatment interfering with Cdk1 reactivation after meiosis I leads to entry into an interphase-like state instead of meiosis II. In *Xenopus*, inhibiting protein synthesis with the drug cycloheximide (CHX), expressing a kinase dead mutant of Cdk1 (K33R) or reducing all Cyclin B1 and B2 isoforms with a knock-down approach after GVBD triggers interphase [43, 94, 101]. Importantly, expressing *Xenopus* Cyclin B1 in either Cyclin B-null or CHX-treated oocytes rescues meiosis II entry [43], showing that Cyclin B1 is the key Cyclin for the transition. More recently, the study of individual knock-outs of Cyclin B1 and Cyclin B2 in mouse oocytes has shown that oocytes devoid of either Cyclin B1 or Cyclin B2 undergo GVBD but are unable to resume meiosis when both Cyclins are depleted, arguing that Cyclin B1 and Cyclin B2 carry out redundant functions to control Cdk1 activation and GVBD [25, 102, 103]. Cyclin B1 knock-out oocytes progress through meiosis I correctly but enter an interphase-like state after meiosis I, whereas those lacking Cyclin B2 undergo prometaphase I with a reduced level of Cdk1 activity, resulting in meiotic spindle defects and delayed entry into anaphase I [25, 102, 103]. Of note, since Cdk1-Cyclin B2 activity has been shown to enhance Cyclin B1 translation in mouse oocytes [25, 103], it is likely that the reduced level of Cdk1 activity is at least partially due to inefficient Cyclin B1 translation in Cyclin B2-depleted mouse oocytes. As in *Xenopus* oocytes, proper meiosis progression is restored by the ectopic expression of either Cyclin B1 or Cyclin B2 in both Cyclin B1 and Cyclin B2 knock-out mouse oocytes [25, 103, 104]. Therefore, Cyclin B1 or Cyclin B2 can compensate for each other, suggesting that Cdk1-Cyclin B1 and Cdk1-Cyclin B2 phosphorylate common substrates to drive proper progression through meiosis after metaphase I in vertebrate oocytes.

The redundant functions of Cyclin B1 and Cyclin B2 are not shared by Cyclin B3. Cyclin B3's role has been elusive in vertebrate female meiosis as it was for many years wrongly considered to remain untranslated, at least in *Xenopus* oocytes [43]. Conserved in all metazoans, the expression of Cyclin B3 varies among cell types and species, but is restricted to female gametes in vertebrates [105]. Consistent with this expression pattern, Cyclin B3 knock-out mice are viable and males are fertile [105]. In contrast, females are infertile, showing that Cyclin B3 is dispensable for mitosis or spermatogenesis but plays a critical function during oogenesis or early embryogenesis [105]. Without Cyclin B3, mouse and *Xenopus* oocytes prematurely arrest at metaphase I [105–107]. This phenotype is associated with the inhibition of APC^{Cdc20} as the APC substrates Cyclin B1, Cyclin B2 and securin are stabilized, independently of SAC activation [105–108]. The arrest at metaphase I in oocytes devoid of Cyclin B3 is relieved upon expression of wild type Cyclin B3 but not a mutant that does not result in Cyclin B3-Cdk1 kinase activity [105, 106]. Crucially, exit from the metaphase I arrest is not rescued by the expression of wild type Cyclin B1 [105]. Hence, Cdk1-Cyclin B3 phosphorylates substrates distinct from those phosphorylated by Cdk1-Cyclin B1 or Cdk1-Cyclin B2 to promote anaphase I and meiosis I exit.

Cyclin B3 also exhibits a particular expression profile during female meiosis (Fig. 4). Present throughout meiosis I, Cyclin B3 is degraded in anaphase I but with a delay compared to Cyclin B1 and B2 [105, 108]. Thereafter, Cyclin B3 does not reappear during meiosis II, unlike Cyclin B1 and Cyclin B2 that both re-accumulate [105, 108] (Fig. 4). In both mouse and *Xenopus*, Cyclin B3 degradation is essential to implement the arrest at metaphase II. When ectopically expressed in mouse metaphase II-arrested oocytes, a stage when Cyclin B3 is normally absent, Cyclin B3 promotes meiosis II exit independently of fertilization [108]. Thus, interfering with the expression profiles of Cyclin B1 and Cyclin B3 has opposite effects in oocytes: Cdk1-Cyclin B3 activity promotes anaphase onset, whereas Cyclin B1-Cdk1 activity prevents it [82]. Cyclin B1 has to be degraded for anaphase I and II onset, whereas the presence of Cyclin B3 in meiosis I and its absence in meiosis II results in anaphase I onset but metaphase II arrest, respectively. Cdk1-Cyclin B3 activity would thus oppose that of Cdk1-Cyclin B1 and Cdk1-Cyclin B2 for entry into anaphase I, and also for anaphase II onset in case of artificial expression at this stage. Why Cyclin B3 does not accumulate in meiosis II is currently unknown. In mouse oocytes, Cyclin B3 is degraded by a mechanism involving a "D-box" (Degradation-box), a conserved motif ubiquitinated by APC^{Cdc20}, and found in all B-type Cyclins [105]. Moreover, Cyclin B3 mRNA likely decreases during the MI-MII transition in mouse oocytes [62], thus no new Cyclin B3 protein can be synthezised in meiosis II. Additionally, ectopically expressed Cyclin B3 is highly unstable in meiosis II and degraded in a D-box dependent manner [105]. Given the essential role of Cyclin B3 in controlling female meiosis, a major challenge will be to elucidate how Cyclin B3 degradation is regulated during the MI-MII transition, where Cyclin B3 is degraded in a delayed manner compared to Cyclin B1 and Cyclin B2; and in meiosis II, when Cyclin B1 and B2 are stable and Cyclin B3 is not, suggesting distinct degradation mechanisms.

Implementing Fertilization Arrest at the Right Moment, at Metaphase II, in Vertebrates

In vertebrates, metaphase II arrest depends on an activity called "CSF" (cytostatic factor) that appears during the MI-MII transition and keeps Cdk1-Cyclin B1 and B2 complexes active at finely-tuned activity levels. Since its discovery, numerous studies have been conducted to identify and understand how "CSF" activity prevents meiosis progression in metaphase II. In short, the key protein responsible for arresting oocytes at metaphase II is Emi2, a direct inhibitor of APC^{Cdc20} [109, 110]. Emi2 is only expressed in vertebrate oocytes and starts to be translated in metaphase I [110–116]. As soon as Emi2 is present, its stability and activity have to be precisely controlled during the MI-MII transition to ensure proper timing of APC^{Cdc20} activation and thus, arrest of oocyte at the correct metaphase, in meiosis II and not meiosis I (Fig. 4). Indeed, overexpressing Emi2 in metaphase I stops meiosis progression at this stage by inhibiting APC^{Cdc20} [112–116]. Conversely, depleting Emi2 using knock-down approaches abolishes progression into meiosis II by impairing Cyclin B1 and Cyclin B2 accumulation in both mouse and *Xenopus* oocytes [117, 118]. As Cdk1 is not reactivated without Emi2, oocytes enter an interphase replicative state [110–116]. Finally, depleting Emi2 in metaphase II lifts the arrest without fertilization in both species [4, 119, 120].

Hence, vertebrate oocytes have evolved precise mechanisms to ensure that Emi2 only accumulates during meiosis II. As Emi2 is a direct inhibitor of APC^{Cdc20}, its accumulation gradually inhibits protein degradation, hence allowing Cyclin B1 and Cyclin B2 accumulation in meiosis II and their stabilization in metaphase II [108, 109, 117, 121]. The regulation of Emi2 is conserved in mouse and *Xenopus* oocytes. It involves multiple phosphorylation events resulting from various kinases and phosphatases in order to target both its accumulation and activity. Among the best characterized regulators of Emi2 is the Mos/MAPK pathway. As mentioned earlier, this pathway is activated at GVBD following Mos accumulation and operates during the MI-MII transition until fertilization. In the mouse, Mos-deficient oocytes reach metaphase II but fail to maintain the arrest, as they spontaneously exit meiosis II [122–124]. In *Xenopus* oocytes, knocking down Mos with morpholino oligos causes oscillatory activity of Cdk1-Cyclin B1 and Cdk1-Cyclin B2 after metaphase I [125], indicating that Cyclin B1 and Cyclin B2 are not properly stabilized for meiosis II entry. In agreement with these results, the Mos/MAPK pathway indirectly controls APC^{Cdc20} by positively regulating both the stability and the activity of Emi2, a process that operates only during meiosis II [112–115]. The Mos/MAPK pathway does so by the downstream substrate $p90^{RSK}$, a kinase which directly phosphorylates Emi2 [112–115]. In *Xenopus* oocytes, phosphorylation of Emi2 by p90^{RSK} leads to the recruitment of PP2A-B56, which protects Emi2 from degradation by removing inhibitory phosphorylation catalyzed by Cdk1 complexes and other kinases [109, 112, 115, 126–128]. Hence, the activity of the Mos/MAPK pathway and Emi2 work together to inhibit APC^{Cdc20}, keeping Cdk1-Cyclin B active in metaphase II. At fertilization, a transient rise in calcium levels leads to the release from the metaphase II arrest, by activating the Calcium/Calmodulin Kinase II (CaMKII) and the phosphatase calcineurin (CaN), which cause Emi2 and Mos destruction [113, 114, 121, 129, 130]. In *Xenopus*, CaMKII phosphorylates Emi2 on a docking site for Polo-like kinase 1 (Plx1), thereby creating a phosphodegron motif for SCF$^{\beta TRCP}$ [119, 121, 131]. Simultaneously, CaN dephosphorylates Emi2 within the PP2A-B56 binding site, thus allowing Cdk1 inhibitory phosphorylation [132]. In addition to Emi2, CaN also dephosphorylates Cdc20 on inhibitory phosphorylation sites mediated by Cdk1-Cyclin B1 [6, 132, 133], thus promoting both Emi2's destruction and APC^{Cdc20} activity for meiosis II exit.

Despite extensive work regarding Emi2's regulation in metaphase II, a long unresolved question has been to understand how Emi2 expression is restricted at metaphase I and during exit of meiosis I to allow APC^{Cdc20} activation. Indeed, Cdk1-Cyclin B1 and Cdk1-Cyclin B2 were the only proteins known to destabilize Emi2 and they are present in both meiosis I and II. A breakthrough was made recently with the discovery that Cdk1-Cyclin B3 preferentially phosphorylates in vitro and in vivo Emi2 at a highly conserved site, resulting in its degradation in metaphase I [108]. In vivo, phosphorylation of this single site has major functional consequences on meiotic progression in both *Xenopus* and mouse oocytes. Cyclin B3 knock-out oocytes accumulate Emi2 early in metaphase I, resulting in APC^{Cdc20} inhibition and thus precocious arrest at metaphase I [108]. Remarkably, mouse Cyclin B3-deficient oocytes can complete meiosis I when the Mos/MAPK pathway is inactivated; thereby suppressing Emi2 activity [108]. Moreover, mimicking fertilization overcomes the metaphase I arrest in Cyclin B3-null oocytes, indicating that the metaphase I arrest corresponds to the fertilization arrest usually implemented in meiosis II. Thus, vertebrate oocytes can arrest at metaphase I in absence of Cyclin B3 by setting up the CSF pathway that arrests oocyte in metaphase II. However, in wild type oocytes, this network is not functional in metaphase I because Cdk1-Cyclin B3 activity promotes the degradation of Emi2, hence ensuring meiosis I exit (Fig. 4). Upon degradation of Cyclin B3 at exit from meiosis I, Cdk1-Cyclin B3 is inactivated and metaphase II arrest can be implemented (Fig. 4). Thus, vertebrate oocytes arrest for fertilization only once cellular ploidy has been reduced, ensuring correct ploidy of the offspring. Cyclin B3 therefore belongs to an intricate network that distinguishes the two meiotic metaphases in order to halt meiosis at the proper stage in vertebrates, after the genome content has been halved. Besides Emi2, no other potential substrates of Cdk1-Cyclin B3 have been identified yet, and it remains attractive to speculate that Cyclin B3 additionally drives meiosis I through other, still unknown substrates in vertebrate oocytes.

Supporting Embryonic Development Without Being Functional During Oocyte Meiosis

After fertilization, embryonic cell cycles differ in many ways from the canonical mitotic cell cycle. They are devoid of Gap phases and occur without zygotic gene transcription; they are thus entirely controlled by maternal mRNAs and proteins provided by the oocyte until activation of the zygotic genome; *i.e.* 13 and 4 cycles in *Xenopus* and in the mouse, respectively. A key challenge of female meiosis is the fact that mRNAs and proteins required for the rapid mitosis in the embryo are already present in the oocyte, but must not interfere with progression through meiosis I and II. Among these embryonic cell cycle regulators, only few examples have been documented so far, most of which concern non-mammalian species including *Xenopus*. In this species, two maternal proteins have been studied: the DNA replication licensing factor, Cdc6, and the Cdk1 inhibitory kinase, Wee1.

Being Competent for DNA Replication Without Replicating

In eukaryotes, DNA replication must be tightly regulated to ensure that the whole genome is duplicated exactly once per cell cycle. This process depends on a two-step mechanism that regulates the initiation of DNA replication. Replication origins (ORI) are licensed in G1, followed by the firing of replicative forks at the G1/S transition [134]. In short, the licensing step depends on the loading of Cdc6 and Cdt1 to ORI, which recruits the replicative helicases mini maintenance chromosomes (MCMs) to form the pre-replicative complex (Pre-RC). At the G1/S transition, activation of CDKs and DDKs (Dbf4-dependent kinase) fires the replicative forks by phosphorylating Cdt1 and Cdc6, thereby causing their dissociation from ORIs and subsequent loading of the replisome [134]. By preventing pre-RC formation, high CDK activity guarantees that ORIs are not licensed again until the next mitosis. The absence of DNA replication is a hallmark of meiosis. However, and in order to control their correct ploidy, oocytes face a major threat during female meiosis. They must be competent for DNA replication in metaphase II in order to support S-phase after fertilization, but inhibit replication after meiosis I (Fig. 5). The competence for replication is acquired during the MI-MII transition, when Cdk1 activity drops. In both mouse and *Xenopus* oocytes, it depends on the accumulation of Cdc6, the only factor of pre-RCs that is missing in these oocytes, after metaphase I [94, 101, 135–137]. Once Cdc6 is expressed, pre-RCs become immediately functional and must be inhibited for correct oocyte ploidy [94, 135, 137]. The mechanism that suppresses DNA replication is implemented after metaphase I and depends on protein synthesis [94, 101]. In *Xenopus*, DNA replication is inhibited by two kinases, Cdk1 and the Mos/MAPK pathway, which target several but still elusive pathways in this process [31, 101, 125, 138–140]. The residual activity of Cdk1 could be sufficient to prevent nuclear membrane reformation, a prerequisite to initiate DNA

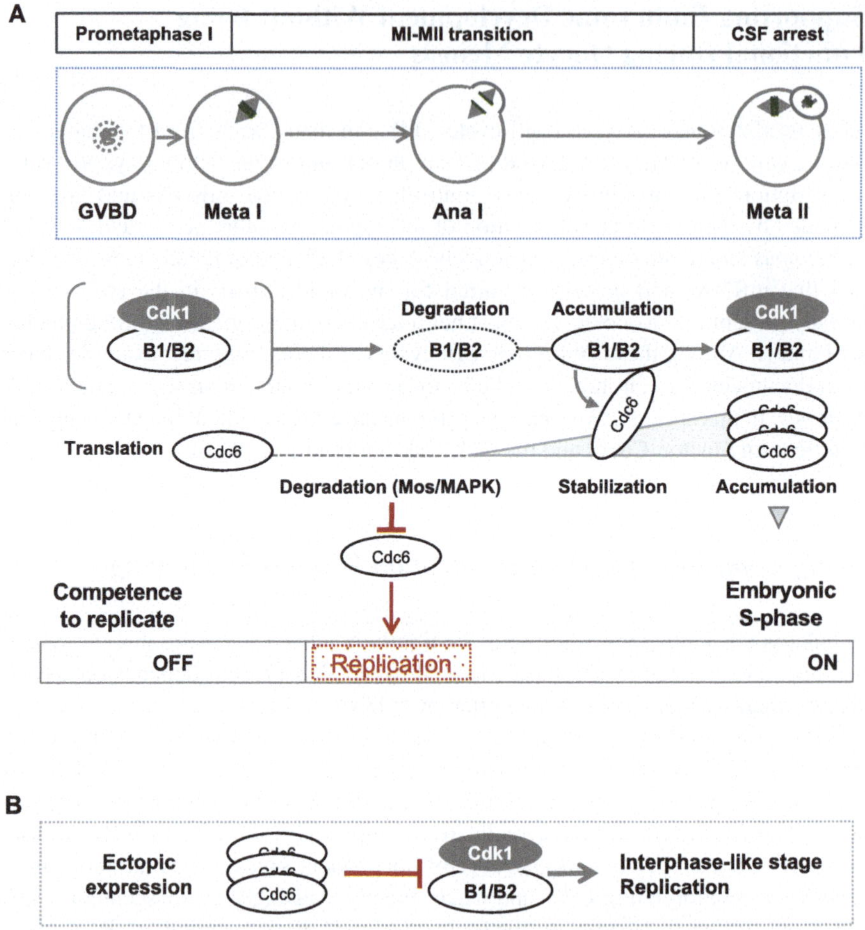

Fig. 5 Control of Cdc6 accumulation to suppress DNA replication in *Xenopus* oocytes. (**a**) Competence and inhibition of DNA replication. In both *Xenopus* and the mouse, prophase-arrested oocytes are unable to replicate DNA. The competence for replication is acquired during the MI-MII transition, when Cdk1 activity drops, in order to support S-phase after fertilization. In both species, the competence to replicate DNA depends on the accumulation of Cdc6 after metaphase I. As Cdc6 is the only factor of the replicative machinery that is missing in these oocytes, the replicative machinery becomes immediately functional once Cdc6 is expressed and must be inhibited for correct oocyte ploidy. in *Xenopus,* one way to inhibit DNA replication is to control the timing of Cdc6 accumulation after meiosis I. Cdc6 starts to be translated at GVBD but only accumulates in meiosis II. During the MI-MII transition, the stability of Cdc6 results from two antagonistic mechanisms: its indirect degradation mediated by the Mos/MAPK pathway and its stabilization upon its interaction with Cyclin B. After metaphase I, Cyclin B1 and B2 are degraded, which leads to the Mos/MAPK pathway-dependent degradation of Cdc6, through a still unknown mechanism, thus preventing replication. In meiosis II, Cdc6 is stabilized by interacting with newly-synthesized Cyclin B. (**b**) Overexpressing Cdc6 in metaphase I inhibits meiosis II entry. The regulation of Cdc6 turnover is not only essential to inhibit DNA replication, it also ensures proper progression through meiosis. Indeed, not only does Cdc6 bind Cyclin B, its ectopic expression in metaphase I also efficiently inhibits Cdk1 reactivation and thus meiosis II entry, hence leading to a replicative interphase. Thus, by limiting the expression of Cdc6 during the MI-MII transition, oocytes synchronize meiotic progression with the inhibition of DNA replication

replication in somatic cells. Furthermore, some of the pre-RC components including ORC1 and MCM4 are phosphorylated after metaphase I while others, such as Cdt1, exhibit cytoplasmic localization at least in prophase-arrested oocytes, suggesting impairment of pre-RC assembly [135, 137]. More recently, it has been proposed that the timing of Cdc6 accumulation is key to suppress DNA replication after meiosis I in *Xenopus* [94]. In this species but also in mouse oocytes, Cdc6 starts to be translated at GVBD but only accumulates in meiosis II [94, 135, 137]. Crucially, Cdc6 stability is regulated by the Mos/MAPK pathway and the expression levels of Cyclin B during exit from meiosis I [94, 135, 137]. After metaphase I, the activity of Cdk1 decreases and the Mos/MAPK pathway promotes the degradation of Cdc6 through a still unknown mechanism (Fig. 5). In meiosis II, Cdc6 is then stabilized by interacting with newly-synthesized Cyclin B, a mechanism that bypasses the negative effect of the Mos/MAPK pathway on Cdc6 stability [94]. Consequently, Cdc6 accumulates only in meiosis II in order to support embryonic S-phase. Thus, the regulation of Cdc6 turnover is essential to inhibit DNA replication and ensure proper progression through meiosis. Indeed, not only does Cdc6 bind Cyclin B, it also efficiently inhibits Cdk1 activity in several organisms, including yeast, human cells and *Xenopus* early embryonic divisions [134, 141–143]. In *Xenopus* oocytes, overexpressing Cdc6 in meiosis I, at GVBD, inhibits the reactivation of Cdk1 and thus meiosis II entry, hence leading to a replicative interphase [94] (Fig. 5). These results reveal that both Cyclin B degradation and the Mos/MAPK pathway restrict Cdc6 expression during the MI-MII transition in order to synchronize meiotic progression with the inhibition of DNA replication.

Cdk1 Inhibitory Phosphorylation During Female Meiosis

Wee1 is a key regulator of the mitotic cell cycle. Its activity coordinates S-phase with the mitotic G2/M transition by slowing down S-phase upon DNA replication errors or DNA damages [144]. Furthermore, in G2, Wee1 prevents premature entry into mitosis by phosphorylating Cdk1-Cyclin complexes [144]. In vertebrate oocytes, Wee1 is inhibited in metaphase II-arrested oocytes. In the mouse oocyte, Wee1 reactivation upon fertilization enhances the inhibition of Cdk1 promoted by the proteolysis of Cyclins [145]. In *Xenopus*, Wee1 reactivation further slows down the first mitotic cell cycle triggered by fertilization [30, 146] to allow completion of meiosis II, decondensation of sperm chromatin and fusion of male and female pronuclei to form the zygote [30, 145–148].

In *Xenopus* oocytes, Wee1 accumulates during early vitellogenesis and then disappears in fully-grown oocytes, most likely due to translation repression [30, 31]. The absence of Wee1 is required for efficient Cdk1 activation because Wee1 is a much better inhibitor of Cdk1 complexes than the related Myt1 kinase, at least in *Xenopus* prophase-arrested oocytes. Accordingly, overexpressing Myt1 slightly delays GVBD while ectopic expression of Wee1 almost completely abolishes meiotic resumption [31]. As described earlier for Emi2 and Cdc6, Wee1 starts to be translated in metaphase I but only accumulates during meiosis II onset [31]. Not surprisingly, the tight control of its accumulation is critical for proper progression through the MI-MII transition. When overexpressed in metaphase I, Wee1 strongly delays or even abolishes entry in meiosis II by promoting the inhibitory phosphorylation of Cdk1 complexes at Y15 [31, 140]. As a result, oocytes enter interphase and replicate their DNA [31, 140]. Of note, while the untimely presence of Wee1 impairs both meiosis resumption and progression into meiosis II, its depletion by knockdown does not interfere with meiosis progression [31], indicating that translation of this protein is not required for the MI-MII transition during female meiosis.

How Wee1 is kept under control during the MI-MII transition in mouse oocytes is currently unknown. Most studies to understand Wee1 regulation have been focused on its activity during the metaphase II arrest and upon fertilization in vertebrate oocytes. In *Xenopus*, Wee1 is thought to be inhibited by multiple phosphorylations, mostly resulting from Cdk1 activity [15]. At fertilization, it has been proposed that the Mos/MAPK positively regulates Wee1 by direct phosphorylation to lengthen the first mitotic cell cycle [15, 30, 146, 149]. Besides the Mos/MAPK pathway, Wee1 is reactivated in both *Xenopus* and mouse by a mechanism involving CaMKII activation as well as Cdk1 inactivation [145]. In contrast to Wee1's activity, the mechanism that restricts its expression to occur only after metaphase I in meiosis has not been determined yet. In mitosis, Wee1 is downregulated by both decreased synthesis and proteolytic degradation. This regulation depends on its phosphorylation by Plk1 and Cdk1, which create a phosphodegron motif for two forms of SCF, $SCF^{\beta\text{-TrCP}}$ and $SCF^{\text{Tome-1}}$ [144]. Given the impact of Wee1 overexpression on meiosis II entry, it will be interesting to determine whether SCF ubiquitin ligases orchestrate the MI-MII transition in vertebrate oocytes by preventing precocious Wee1 accumulation.

Conclusion

Detailed mechanistic studies in *Xenopus* oocytes combined with genetic invalidation approaches in mice have provided valuable insights into the control of the MI-MII transition, highlighting the functional conservation and specificity of meiotic cell cycle regulators. Despite extensive studies, critical questions regarding female meiosis remain open to fully understand orchestration of this specialized form of cell division. The choreography of the molecular events that synchronize meiotic progression, in particular those controlled by Cdk1-Cyclin B3 activity, remains largely unknown. Moreover, protein phosphorylation is one of the most significant post-translational modifications controlling female meiosis. The interplay of kinases and phosphatases regulates protein stability and activity at the transition from meiosis I into meiosis II. Their functional specificity and the mechanisms they regulate have not been clearly established. Global proteomic and phosphoproteomic approaches will lead to new insights into how this very special cell cycle transition is regulated.

Acknowledgements We thank lab members for discussion, and Roger Karess for critical reading of an earlier version of the manuscript. K.W. is supported by an ANR grant (ANR-23-CE13-0015-01) and a grant by the Fondation de la Recherche Médicale (EquipeDEQ 202103012574). A.D. is supported by an ANR grant (ANR-24-CE13-2359) and the Association pour la recherche sur le cancer (ARC, PJA2022060005161) and CNRS.

References

1. Christou-Kent M et al (2020) Diversity of RNA-binding proteins modulating post-transcriptional regulation of protein expression in the maturing mammalian oocyte. Cells 9(3):662
2. Jessus C, Ozon R (2004) How does Xenopus oocyte acquire its competence to undergo meiotic maturation? Biol Cell (Paris) 96(3):187–192
3. Hansen DV et al (2006) CaMKII and polo-like kinase 1 sequentially phosphorylate the cytostatic factor Emi2/XErp1 to trigger its destruction and meiotic exit. Proc Natl Acad Sci U S A 103(3):608–613
4. Tung JJ et al (2005) A role for the anaphase-promoting complex inhibitor Emi2/XErp1, a homolog of early mitotic inhibitor 1, in cytostatic factor arrest of *Xenopus* eggs. Proc Natl Acad Sci U S A 102(12):4318–4323
5. Nilsson J (2019) Protein phosphatases in the regulation of mitosis. J Cell Biol 218(2):395–409
6. Yamano H (2019) APC/C: current understanding and future perspectives. F1000Research 23(8):F1000
7. Vigneron S et al (2018) Cyclin A-cdk1-dependent phosphorylation of bora is the triggering factor promoting mitotic entry. Dev Cell 45(5):637–650.e637
8. Crncec A, Hochegger H (2019) Triggering mitosis. FEBS Lett 593(20):2868–2888
9. Morgan DO (1997) Cyclin-dependent kinases: engines, clocks, and microprocessors. Annu Rev Cell Dev Biol 13:261–291
10. Parker LL, Piwnica-Worms H (1992) Inactivation of the $p34^{cdc2}$-cyclin B complex by the human WEE1 tyrosine kinase. Science 257(5078):1955–1957
11. Fattaey A, Booher RN (1997) Myt1: a Wee1-type kinase that phosphorylates Cdc2 on residue Thr14. Prog Cell Cycle Res 3:233–240
12. Mueller PR et al (1995b) Myt1: a membrane-associated inhibitory kinase tht phosphorylates cdc2 on both threonine-14 and tyrosine-15. Science 270:86–90
13. Hoffmann I et al (1993) Phosphorylation and activation of human Cdc25-C by Cdc2-cyclin B and its involvement in the self-amplification of MPF at mitosis. EMBO J 12(1):53–63
14. Kumagai A, Dunphy WG (1992) Regulation of the Cdc25 protein during the cell cycle in *Xenopus* extracts. Cell 70:139–151
15. Mueller PR et al (1995a) Cell cycle regulation of a Xenopus wee1-like kinase. Mol Biol Cell 6:119–134
16. Hu MW et al (2016) Protein phosphatase 6 protects prophase I-arrested oocytes by safeguarding genomic integrity. PLoS Genet 12(12):e1006513
17. Hu MW et al (2015) Loss of protein phosphatase 6 in oocytes causes failure of meiosis II exit and impaired female fertility. J Cell Sci 128(20):3769–3780
18. Lemonnier T et al (2020) The G2-to-M transition from a phosphatase perspective: a new vision of the meiotic division. Cell Div 15:9
19. Lemonnier T et al (2021) The M-phase regulatory phosphatase PP2A-B55delta opposes protein kinase A on Arpp19 to initiate meiotic division. Nat Commun 12(1):1837
20. Wurzenberger C et al (2012) Sds22 and Repo-Man stabilize chromosome segregation by counteracting Aurora B on anaphase kinetochores. J Cell Biol 198(2):173–183
21. Zeng K et al (2010) Protein phosphatase 6 regulates mitotic spindle formation by controlling the T-loop phosphorylation state of Aurora A bound to its activator TPX2. J Cell Biol 191(7):1315–1332

22. Brautigan DL, Shenolikar S (2018) Protein serine/threonine phosphatases: keys to unlocking regulators and substrates. Annu Rev Biochem 87:921–964
23. Cundell MJ et al (2016) A PP2A-B55 recognition signal controls substrate dephosphorylation kinetics during mitotic exit. J Cell Biol 214(5):539–554
24. Lei WL et al (2021) Critical functions of PP2A-like protein phosphotases in regulating meiotic progression. Front Cell Dev Biol 9:638559
25. Bouftas N, Wassmann K (2019) Cycling through mammalian meiosis: B-type cyclins in oocytes. Cell Cycle 18(14):1537–1548
26. De Smedt V et al (2002) Thr-161 phosphorylation of monomeric Cdc2. Regulation by protein phosphatase 2C in Xenopus oocyte. J Biol Chem 277(32):28592–28600
27. Kanatsu-Shinohara M et al (2000) Acquisition of meiotic competence in mouse oocytes: absolute amounts of p34(cdc2), cyclin B1, cdc25C, and wee1 in meiotically incompetent and competent oocytes. Biol Reprod 63(6):1610–1616
28. Gaffré M et al (2011) A critical balance between cyclin B synthesis and Myt1 activity controls meiosis entry in Xenopus oocytes. Development 138(17):3735–3744
29. Han SJ et al (2005) Wee1B is an oocyte-specific kinase involved in the control of meiotic arrest in the mouse. Curr Biol 15(18):1670–1676
30. Murakami MS, VandeWoude GF (1998) Analysis of the early embryonic cell cycles of Xenopus; regulation of cell cycle length by Xe-wee1 and Mos. Development 125(2):237–248
31. Nakajo N et al (2000) Absence of Wee1 ensures the meiotic cell cycle in Xenopus oocytes. Genes Dev 14(3):328–338
32. Oh JS et al (2010) Wee1B, Myt1, and Cdc25 function in distinct compartments of the mouse oocyte to control meiotic resumption. J Cell Biol 188(2):199–207
33. Adhikari D et al (2016) Inhibitory phosphorylation of Cdk1 mediates prolonged prophase I arrest in female germ cells and is essential for female reproductive lifespan. Cell Res 26(11):1212–1225
34. Filatov M et al (2019) Molecular mechanisms of prophase I meiotic arrest maintenance and meiotic resumption in mammalian oocytes. Reprod Sci 26(11):1519–1537
35. Pan B, Li J (2019) The art of oocyte meiotic arrest regulation. Reprod Biol Endocrinol 17(1):8
36. Han SJ, Conti M (2006) New pathways from PKA to the Cdc2/cyclin B complex in oocytes: Wee1B as a potential PKA substrate. Cell Cycle 5(3):227–231
37. Duckworth BC et al (2002) G2 arrest in *Xenopus* oocytes depends on phosphorylation of cdc25 by protein kinase A. Proc Natl Acad Sci U S A 99(26):16794–16799
38. Dupré A et al (2014) Phosphorylation of ARPP19 by protein kinase A prevents meiosis resumption in Xenopus oocytes. Nat Commun 5:3318
39. Dupré AI et al (2017) The greatwall kinase is dominant over PKA in controlling the antagonistic function of ARPP19 in Xenopus oocytes. Cell Cycle 16(15):1440–1452
40. Dupré A et al (2011) Mos in the oocyte: how to use MAPK independently of growth factors and transcription to control meiotic divisions. J Signal Transduct 2011:350412
41. Haccard O, Jessus C (2006a) Oocyte maturation, Mos and cyclins—A matter of synthesis: two functionally redundant ways to induce meiotic maturation. Cell Cycle 5(11):1152–1159
42. Frank-Vaillant M et al (1999) Two distinct mechanisms control the accumulation of cyclin B1 and mos in *Xenopus* oocytes in response to progesterone. Mol Biol Cell 10(10):3279–3288
43. Hochegger H et al (2001) New B-type cyclin synthesis is required between meiosis I and II during *Xenopus* oocyte maturation. Development 128(19):3795–3807
44. Ledan E et al (2001) Meiotic maturation of the mouse oocyte requires an equilibrium between cyclin B synthesis and degradation. Dev Biol 232(2):400–413
45. McGuinness BE et al (2009) Regulation of APC/C activity in oocytes by a Bub1-dependent spindle assembly checkpoint. Curr Biol 19(5):369–380
46. Reis A et al (2007) Prometaphase APCcdh1 activity prevents non-disjunction in mammalian oocytes. Nat Cell Biol 9(10):1192–1198
47. Dinarina A et al (2009) Cell cycle regulation of the mammalian CDK activator RINGO/speedy A. FEBS Lett 583(17):2772–2778

48. Haccard O, Jessus C (2006b) Redundant pathways for Cdc2 activation in *Xenopus* oocyte: either cyclin B or Mos synthesis. EMBO Rep 7(3):321–325
49. Lenormand JL et al (1999) Speedy: a novel cell cycle regulator of the G2/M transition. EMBO J 18(7):1869–1877
50. Peter M et al (2002) A new role for Mos in *Xenopus* oocyte maturation: targeting Myt1 independently of MAPK. Development 129(9):2129–2139
51. Ruiz EJ et al (2008) Meiotic inactivation of Xenopus Myt1 by CDK/XRINGO, but not CDK/cyclin, via site-specific phosphorylation. Mol Cell 32(2):210–220
52. Adhikari D et al (2014) Mastl is required for timely activation of APC/C in meiosis I and Cdk1 reactivation in meiosis II. J Cell Biol 206(7):843–853
53. Alvarez-Fernandez M et al (2013) Greatwall is essential to prevent mitotic collapse after nuclear envelope breakdown in mammals. Proc Natl Acad Sci U S A 110(43):17374–17379
54. Castilho PV et al (2009) The M phase kinase Greatwall (Gwl) promotes inactivation of PP2A/B55delta, a phosphatase directed against CDK phosphosites. Mol Biol Cell 20(22):4777–4789
55. Gharbi-Ayachi A et al (2010) The substrate of Greatwall kinase, Arpp19, controls mitosis by inhibiting protein phosphatase 2A. Science 330(6011):1673–1677
56. Mochida S et al (2009) Regulated activity of PP2A-B55delta is crucial for controlling entry into and exit from mitosis in Xenopus egg extracts. EMBO J 28(18):2777–2785
57. Dupré A et al (2013) The phosphorylation of ARPP19 by Greatwall renders the auto-amplification of MPF independently of PKA in Xenopus oocytes. J Cell Sci 126(Pt 17):3916–3926
58. Yamamoto TM et al (2011) Regulation of Greatwall kinase during Xenopus oocyte maturation. Mol Biol Cell 22(13):2157–2164
59. Choi T et al (1991) Activation of $p34^{cdc2}$ protein kinase activity in meiotic and mitotic cell cycles in mouse oocytes and embryos. Development 113:789–795
60. Fulka J Jr et al (1986) Effect of cycloheximide on nuclear maturation of pig and mouse oocytes. J Reprod Fertil 77(1):281–285
61. Hampl A, Eppig JJ (1995) Translational regulation of the gradual increase in histone H1 kinase activity in maturing mouse oocytes. Mol Reprod Dev 40:9–15
62. Han SJ et al (2017) The translation of cyclin B1 and B2 is differentially regulated during mouse oocyte reentry into the meiotic cell cycle. Sci Rep 7(1):14077
63. Bai GY et al (2020) Mis12 controls cyclin B1 stabilization via Cdc14B-mediated APC/C(Cdh1) regulation during meiotic G2/M transition in mouse oocytes. Development 147(8):dev185322
64. Holt JE et al (2010) Spatial regulation of APCCdh1-induced cyclin B1 degradation maintains G2 arrest in mouse oocytes. Development 137(8):1297–1304
65. Holt JE et al (2012) APC(FZR1) prevents nondisjunction in mouse oocytes by controlling meiotic spindle assembly timing. Mol Biol Cell 23(20):3970–3981
66. Holt JE et al (2011) The APC/C activator FZR1 coordinates the timing of meiotic resumption during prophase I arrest in mammalian oocytes. Development 138(5):905–913
67. Marangos P et al (2007) Prophase I arrest and progression to metaphase I in mouse oocytes are controlled by Emi1-dependent regulation of APCCdh1. J Cell Biol 176(1):65–75
68. Reis A et al (2006) APCcdh1 activity in mouse oocytes prevents entry into the first meiotic division. Nat Cell Biol 8(5):539–540
69. Alexandre H et al (1991) Pleiotropic effect of okadaic acid on maturing mouse oocytes. Development 112(4):971–980
70. Gavin A-C et al (1991) Induction of M-phase entry of prophase-blocked mouse oocytes through microinjection of okadaic acid, a specific phosphatase inhibitor. Exp Cell Res 192:75–81
71. Su YQ et al (2012) Meiosis arrest female 1 (MARF1) has nuage-like function in mammalian oocytes. Proc Natl Acad Sci U S A 109(46):18653–18660
72. Swain JE et al (2003) Specific inhibition of mouse oocyte nuclear protein phosphatase-1 stimulates germinal vesicle breakdown. Mol Reprod Dev 65(1):96–103

73. Yu C et al (2015) CRL4-DCAF1 ubiquitin E3 ligase directs protein phosphatase 2A degradation to control oocyte meiotic maturation. Nat Commun 6:8017
74. Tang A et al (2016) PP2A regulates kinetochore-microtubule attachment during meiosis I in oocyte. Cell Cycle 15(11):1450–1461
75. Wang X et al (2004) Endogenous regulators of protein phosphatase-1 during mouse oocyte development and meiosis. Reproduction 128(5):493–502
76. Kang M et al (2021) Discovery and characterization of a novel MASTL inhibitor MKI-2 targeting MASTL-PP2A in breast cancer cells and oocytes. Pharmaceuticals (Basel) 14(7):647
77. Zhao X et al (2014) Role of Greatwall kinase in release of mouse oocytes from diplotene arrest. Dev Growth Differ 56(9):669–678
78. Meneau F et al (2020) Translational control of Xenopus oocyte meiosis: toward the genomic era. Cells 9(6):1502
79. Levasseur MD et al (2019) Aneuploidy in oocytes is prevented by sustained CDK1 activity through Degron masking in cyclin B1. Dev Cell 48(5):672–684.e675
80. Peter M et al (2001) The APC is dispensable for first meiotic anaphase in *Xenopus* oocytes. Nat Cell Biol 3(1):83–87
81. Taieb FE et al (2001) Activation of the anaphase-promoting complex and degradation of cyclin B is not required for progression from Meiosis I to II in Xenopus oocytes. Curr Biol 11(7):508–513
82. Gorr IH et al (2006) Essential CDK1-inhibitory role for separase during meiosis I in vertebrate oocytes. Nat Cell Biol 8(9):1035–1037
83. Kudo NR et al (2009) Role of cleavage by separase of the Rec8 kleisin subunit of cohesin during mammalian meiosis I. J Cell Sci 122(Pt 15):2686–2698
84. Kudo NR et al (2006) Resolution of chiasmata in oocytes requires separase-mediated proteolysis. Cell 126(1):135–146
85. Nikalayevich E et al (2018) Detection of separase activity using a cleavage sensor in live mouse oocytes. Methods Mol Biol 1818:99–112
86. Terret ME et al (2003) The meiosis I-to-meiosis II transition in mouse oocytes requires separase activity. Curr Biol 13(20):1797–1802
87. Wassmann K (2022) Separase control and Cohesin cleavage in oocytes: should I stay or should I go? Cells 11(21):3399
88. Mihajlovic AI, FitzHarris G (2018) Segregating chromosomes in the mammalian oocyte. Curr Biol 28(16):R895–R907
89. Touati SA, Wassmann K (2016) How oocytes try to get it right: spindle checkpoint control in meiosis. Chromosoma 125(2):321–335
90. Nakagawa S, FitzHarris G (2017) Intrinsically defective microtubule dynamics contribute to age-related chromosome segregation errors in mouse oocyte meiosis-I. Curr Biol 27(7):1040–1047
91. Vallot A et al (2018) Tension-induced error correction and not kinetochore attachment status activates the SAC in an Aurora-B/C-dependent manner in oocytes. Curr Biol 28(1):130–139.e133
92. Marston AL, Wassmann K (2017) Multiple duties for spindle assembly checkpoint kinases in meiosis. Front Cell Dev Biol 5:109
93. Shao H et al (2013) Xenopus oocyte meiosis lacks spindle assembly checkpoint control. J Cell Biol 201(2):191–200
94. Daldello EM et al (2015) Fine-tuning of Cdc6 accumulation by Cdk1 and MAP kinase is essential for completion of oocyte meiotic divisions. J Cell Sci 128(14):2482–2496
95. Swain JE et al (2007) Proper chromatin condensation and maintenance of histone H3 phosphorylation during mouse oocyte meiosis requires protein phosphatase activity. Biol Reprod 76(4):628–638
96. Camlin NJ et al (2022) M-phase oscillations in PP1 activity are essential for accurate progression through mammalian oocyte meiosis. bioRxiv. https://doi.org/10.1101/2021.09.28.461976
97. Swartz SZ et al (2021) Selective dephosphorylation by PP2A-B55 directs the meiosis I-meiosis II transition in oocytes. eLife 10:e70588

98. Heim A et al (2015) Protein phosphatase 1 is essential for Greatwall inactivation at mitotic exit. EMBO Rep 16(11):1501–1510
99. Ma S et al (2016) Greatwall dephosphorylation and inactivation upon mitotic exit is triggered by PP1. J Cell Sci 129(7):1329–1339
100. Mochida S et al (2010) Greatwall phosphorylates an inhibitor of protein phosphatase 2A that is essential for mitosis. Science 330(6011):1670–1673
101. Furuno N et al (1994) Suppression of DNA replication via Mos function during meiotic divisions in *Xenopus* oocytes. EMBO J 13(10):2399–2410
102. Daldello EM et al (2019) Cyclin B2 is required for progression through meiosis in mouse oocytes. Development 146(8):dev172734
103. Li J et al (2018) Cyclin B2 can compensate for cyclin B1 in oocyte meiosis I. J Cell Biol 217(11):3901–3911
104. Brandeis M et al (1998) Cyclin B2-null mice develop normally and are fertile whereas cyclin B1- null mice die in utero. Proc Natl Acad Sci U S A 95(8):4344–4349
105. Karasu ME et al (2019) Cyclin B3 promotes anaphase I onset in oocyte meiosis. J Cell Biol 218(4):1265–1281
106. Li Y et al (2019) Cyclin B3 is required for metaphase to anaphase transition in oocyte meiosis I. J Cell Biol 218(5):1553–1563
107. Zhang T et al (2015) Cyclin B3 controls anaphase onset independent of spindle assembly checkpoint in meiotic oocytes. Cell Cycle 14(16):2648–2654
108. Bouftas N et al (2022) Cyclin B3 implements timely vertebrate oocyte arrest for fertilization. Dev Cell 57(19):2305–2320.e2306
109. Hormanseder E et al (2013) Modulation of cell cycle control during oocyte-to-embryo transitions. EMBO J 32(16):2191–2203
110. Schmidt A et al (2006) Cytostatic factor: an activity that puts the cell cycle on hold. J Cell Sci 119(Pt 7):1213–1218
111. Belloc E, Mendez R (2008) A deadenylation negative feedback mechanism governs meiotic metaphase arrest. Nature 452(7190):1017–1021
112. Inoue D et al (2007) A direct link of the Mos-MAPK pathway to Erp1/Emi2 in meiotic arrest of Xenopus laevis eggs. Nature 446(7139):1100–1104
113. Nishiyama T et al (2007a) Phosphorylation of Erp1 by p90rsk is required for cytostatic factor arrest in Xenopus laevis eggs. Nature 446(7139):1096–1099
114. Nishiyama T et al (2007b) Transient activation of calcineurin is essential to initiate embryonic development in Xenopus laevis. Nature 449(7160):341–345
115. Ohe M et al (2007) Erp1/Emi2 is essential for the meiosis I to meiosis II transition in Xenopus oocytes. Dev Biol 303(1):157–164
116. Tung JJ et al (2007) Translational unmasking of Emi2 directs cytostatic factor arrest in meiosis II. Cell Cycle 6(6):725–731
117. Madgwick S et al (2006) Mouse Emi2 is required to enter meiosis II by reestablishing cyclin B1 during interkinesis. J Cell Biol 174(6):791–801
118. Tang W et al (2008) Cdc2 and Mos regulate Emi2 stability to promote the meiosis I-meiosis II transition. Mol Biol Cell 19(8):3536–3543
119. Rauh NR et al (2005) Calcium triggers exit from meiosis II by targeting the APC/C inhibitor XErp1 for degradation. Nature 437(7061):1048–1052
120. Shoji S et al (2006) Mammalian Emi2 mediates cytostatic arrest and transduces the signal for meiotic exit via Cdc20. EMBO J 25(4):834–845
121. Suzuki T et al (2010) Mouse Emi2 as a distinctive regulatory hub in second meiotic metaphase. Development 137(19):3281–3291
122. Colledge WH et al (1994) Disruption of c-mos causes parthenogenetic development of unfertilized mouse eggs. Nature 370(6484):65–68
123. Hashimoto N et al (1994) Parthenogenetic activation of oocytes in c-mos-deficient mice. Nature 370(6484):68–71
124. Verlhac M et al (1996) Mos is required for MAP kinase activation and is involved in microtubule organization during meiotic maturation in the mouse. Development 122(3):815–822

125. Dupré A et al (2002) Mos is not required for the initiation of meiotic maturation in *Xenopus* oocytes. EMBO J 21(15):4026–4036
126. Isoda M et al (2011) Dynamic regulation of Emi2 by Emi2-bound Cdk1/Plk1/CK1 and PP2A-B56 in meiotic arrest of Xenopus eggs. Dev Cell 21(3):506–519
127. Wu JQ et al (2007a) Control of Emi2 activity and stability through Mos-mediated recruitment of PP2A. Proc Natl Acad Sci U S A 104(42):16564–16569
128. Wu Q et al (2007b) A role for Cdc2- and PP2A-mediated regulation of Emi2 in the maintenance of CSF arrest. Curr Biol 17(3):213–224
129. Lorca T et al (1993) Calmodulin-dependent protein kinase II mediates inactivation of MPF and CSF upon fertilization of Xenopus eggs. Nature 366(6452):270–273
130. Mochida S, Hunt T (2007) Calcineurin is required to release Xenopus egg extracts from meiotic M phase. Nature 449(7160):336–340
131. Schmidt A et al (2005) *Xenopus* polo-like kinase Plx1 regulates XErp1, a novel inhibitor of APC/C activity. Genes Dev 19(4):502–513
132. Heim A et al (2018) Calcineurin promotes APC/C activation at meiotic exit by acting on both XErp1 and Cdc20. EMBO Rep 19(12):e46433
133. Kataria M, Yamano H (2019) Interplay between phosphatases and the anaphase-promoting complex/cyclosome in mitosis. Cells 8(8):814
134. Petropoulos M et al (2019) Replication licensing aberrations, replication stress, and genomic instability. Trends Biochem Sci 44(9):752–764
135. Lemaitre JM et al (2004) The regulation of competence to replicate in meiosis by Cdc6 is conserved during evolution. Mol Reprod Dev 69(1):94–100
136. Swiech L et al (2007) Accumulation and dynamics of proteins of the MCM family during mouse oogenesis and the first embryonic cell cycle. Int J Dev Biol 51(4):283–295
137. Whitmire E et al (2002) Cdc6 synthesis regulates replication competence in Xenopus oocytes. Nature 419(6908):722–725
138. Castro A et al (2001a) c-Mos and cyclin B/cdc2 connections during Xenopus oocyte maturation. Biol Cell (Paris) 93(1–2):15–25
139. Castro A et al (2001b) Cyclin B/cdc2 induces c-Mos stability by direct phosphorylation in xenopus oocytes. Mol Biol Cell 12(9):2660–2671
140. Iwabuchi M et al (2000) Residual Cdc2 activity remaining at meiosis I exit is essential for meiotic M-M transition in Xenopus oocyte extracts. EMBO J 19(17):4513–4523
141. Archambault V et al (2003) Genetic and biochemical evaluation of the importance of Cdc6 in regulating mitotic exit. Mol Biol Cell 14(11):4592–4604
142. Calzada A et al (2001) Cdc6 cooperates with Sic1 and Hct1 to inactivate mitotic cyclin-dependent kinases. Nature 412(6844):355–358
143. El Dika M et al (2014) CDC6 controls dynamics of the first embryonic M-phase entry and progression via CDK1 inhibition. Dev Biol 396(1):67–80
144. Esposito F et al (2021) Wee1 kinase: a potential target to overcome tumor resistance to therapy. Int J Mol Sci 22(19):583
145. Oh JS et al (2011) Protein tyrosine kinase Wee1B is essential for metaphase II exit in mouse oocytes. Science 332(6028):462–465
146. Murakami MS et al (1999) Mos positively regulates Xe-wee1 to lengthen the first mitotic cell cycle of Xenopus. Genes Dev 13(5):620–631
147. Gerhart J et al (1984) Cell cycle dynamics of an M-phase-specific cytoplasmic factor in *Xenopus lævis* oocytes and eggs. J Cell Biol 98:1247–1255
148. Sikora-Polaczek M et al (2006) The first mitosis of the mouse embryo is prolonged by transitional metaphase arrest. Biol Reprod 74(4):734–743
149. Walter SA et al (2000) Activation of Wee1 by p42 MAPK in vitro and in cycling Xenopus egg extracts. Mol Biol Cell 11(3):887–896

Preserving Genomic Integrity During Female Meiosis: Detecting, Repairing, and Responding to Maternal DNA Damage

Jiyeon Leem and Jeong Su Oh

Abstract Damage to maternal DNA during meiosis can lead to genetic abnormalities that cause complications such as infertility, birth defects, and miscarriage. To maintain oocyte quality, it is crucial to detect and repair damaged DNA. Despite the negative impact of DNA damage on the reproductive capacity and genomic integrity of embryos, the ability of oocytes to respond to DNA damage and the influence of DNA repair on oocyte quality have received relatively little research attention until recently. In this chapter, we discuss the current state of knowledge regarding the preservation of genomic integrity in the face of DNA damage and the mechanisms of DNA repair during female meiosis.

Keywords Meiosis · DNA-damage response · DNA repair · Genome integrity · Fertility

Introduction

The DNA of all living organisms faces constant threats from various sources, including environmental factors, cellular processes, and oxidative stress generated by cellular metabolism. When left unrepaired, damaged DNA can lead to mutations, genetic instability, and disease. The DNA-damage response (DDR) is complex collection of mechanisms that cells have evolved to counter these assaults [1–4]. The DDR is an intricately orchestrated set of cellular processes that serves as the first line of defense against DNA damage. It involves a sequence of steps, beginning with the detection of damage, followed by a signaling cascade and repair

J. Leem
Department of Molecular, Cellular and Developmental Biology, Yale University, New Haven, CT, USA

J. S. Oh (✉)
Department of Integrative Biotechnology, Sungkyunkwan University, Suwon, South Korea
e-mail: ohjs@skku.edu

B. Mogessie (ed.), *Cellular Architecture and Dynamics in Female Meiosis*,
https://doi.org/10.1007/978-3-031-97173-0_5

mechanisms. The biological significance of the DDR is evident in the adverse consequences of inherited defects in the process, which can result in various diseases, such as immune deficiency, neurological degeneration, premature aging, and heightened susceptibility to cancer [5, 6].

The DDR is orchestrated through the activation of two pivotal kinases, ataxia-telangiectasia mutated (ATM) kinase and ATM- and Rad3-related (ATR) kinase [2, 7–12]. The activation of ATM occurs at the sites of double-strand breaks (DSBs), where it interacts with the MRE11-RAD50-NBS1 (MRN) complex, which serves as a DSB sensor [11]. Once activated by DSBs, ATM phosphorylates the histone variant H2AX (referred to as γ-H2AX) and the mediator of DNA damage checkpoint 1 (MDC1), which triggers a cascade of DDR events in the vicinity of DSBs. Moreover, ATM phosphorylates several substrates, including p53 binding protein 1 (53BP1), breast cancer susceptibility gene 1 (BRCA1), checkpoint kinase 2 (CHK2), and p53. These substrates mediate the effects of ATM on DNA repair, cell-cycle arrest, apoptosis, and other downstream processes [7, 9, 12–17]. In contrast, ATR responds to DNA damage and stress induced by nucleotide depletion, DNA polymerase inhibition, base-alkylating agents, or ultraviolet radiation [8, 12]. In these cases, single-strand breaks (SSBs) in DNA can become coated with replication protein A (RPA) [18]. ATR is recruited to the sites of SSBs through the association of its partner, ATR-interacting protein (ATRIP), with RPA and is activated by interaction with TOPBP1. Activated ATR recognizes and phosphorylates checkpoint kinase 1 (CHK1), leading to activation of the ATR-CHK1 pathway [19, 20]. A growing list of DNA replication, DNA repair, and cell-cycle proteins has been shown to be substrates and effectors of ATR and CHK1. For example, in the presence of unrepaired DNA damage, CHK1 phosphorylates and inhibits CDC25 phosphatases, maintaining a high level of inhibitory phosphorylation in cyclin-dependent kinase 1 (CDK1) and subsequently preventing cell-cycle progression [12, 16, 17]. ATR also regulates several DNA repair pathways, such as homologous recombination, DNA interstrand crosslink repair, and nucleotide excision repair [21–24]. Although the ATM and ATR pathways differ in DDR signaling, they typically work cooperatively and share common components. In addition to DNA repair and cell-cycle arrest, the DDR regulates many physiological processes involved in apoptosis, terminal differentiation through senescence, heightened immune surveillance, and DNA-damage prophylaxis [9].

Mammalian oocytes are particularly vulnerable to accumulation of DNA damage due to their prolonged arrest at the G2/prophase of meiosis within the ovary [25, 26]. This prolonged arrest exposes them to potential chromosomal abnormalities, contributing to embryonic defects that can lead to miscarriage or infertility [27, 28]. A deep understanding of the DDR in oocytes is crucial to ensure successful reproduction. While extensive research has been carried out on DDR mechanisms in somatic cells, particularly in the context of cancer, the complexity of the DDR machinery in oocytes has been suggested only recently. This chapter summarizes the intricate machinery that protects the female genome during meiosis.

DNA-Damage Checkpoint and Apoptosis in Mouse Oocytes

In eukaryotic organisms, the maintenance of genetic integrity relies on the activation of cell-cycle checkpoints. These regulatory mechanisms ensure the proper sequence of events in the mitotic cell cycle, including DNA replication, repair of DNA damage, preservation of chromosomal integrity, maintenance of cell size and function, and appropriate chromosome segregation during mitosis [29]. These checkpoints are not exclusive to mitotic cells but also play a crucial role in meiosis, a specialized chromosome segregation process that gives rise to haploid gametes from diploid parental cells [30]. However, the DNA damage checkpoint process during female meiosis has several unique features.

Meiotic Recombination Checkpoint in Mouse Oocyte

The early stages of meiosis include homologous chromosome synapsis and genetic recombination [31, 32]. Meiotic recombination supports the faithful segregation of homologous chromosomes and reshuffles maternal and paternal chromosomes, enhancing the genetic diversity of progeny [33]. SPO11, a meiosis-specific endonuclease belonging to the topoisomerase family, is central in this process. After premeiotic DNA replication, programmed DSBs are catalyzed by SPO11. The central role of SPO11 becomes evident in SPO11-deficient oocytes, where homologous pairing is abolished [34, 35]. The subsequent removal of SPO11 and the 5′ end resection of DSBs result in 3′ single-stranded DNA (ssDNA) ends, which are then rapidly coated by the heterotrimeric protein complex RPA stabilizing the ssDNA and preventing the formation of secondary structures. Subsequently, the specialized recombinases RAD51 and DMC1 are recruited to the ssDNA and catalyze strand-exchange events [33, 36, 37]. While some DSBs in meiosis are repaired via sister chromatids, recombination occurs primarily between homologous chromosomes, promoting homology pairing and segregation [33]. Additionally, meiotic recombination establishes chiasmata, holding homologs together after recombination has occurred and chromosomes have desynapsed [38]. Chiasmata help co-orientation of sister chromatids of each homolog to opposite spindle, thereby ensuring that they segregate to opposite poles during meiosis I [30, 39]. Because meiosis I with unsynapsed chromosomes would be detrimental, this process must be carefully monitored, and oocytes have evolved to employ a specialized surveillance program known as the "meiotic recombination checkpoint." Oocytes that fail to complete recombination are eliminated in the primordial follicles. The absence of SPO11-dependent DSB formation and homolog pairing leads to apoptosis during early follicular development [35, 40]. This elimination process, which is crucial for preventing the transmission of genetic defects, is regulated tightly by the CHK2-dependent DNA damage checkpoint [41].

Oocytes displaying significant deficiencies in chromosome synapsis accumulate spontaneous DSBs, triggering the CHK2-dependent DNA-damage checkpoint and leading to their elimination [41]. Given the well-established role of the ATM-CHK2 pathway in the DDR, it is possible that ATM plays a role in SPO11-mediated recombination. The absence of ATM results in programmed cell death in prophase oocytes, rendering ATM-null mice infertile, as in SPO11-null mice [26, 42]. However, oocytes lacking ATM do not become enclosed in follicles and degenerate through apoptosis earlier than SPO11-null oocytes do. This phenotypic difference in ATM mutants is possibly the result of persistent, unrepaired DNA damage. Recent findings suggest that ATM regulates multiple steps in the initial processing of SPO11-bound DSBs, and this activity could be hindered by the binding of PRDM9, meiosis-specific histone H3 lysine 4 (H3K4) methyltransferase to hotspots, which are known sites for histone modification following SPO11 cleavage (Fig. 1) [40, 42–44]. This suggests that ATM and PRDM9 may act as local regulators of SPO11 processing during meiotic recombination. Although ATM has been demonstrated to regulate DSB numbers and locations, further studies are required to clarify the roles of ATM in SPO11 processing and recombination checkpoints [43, 45].

Evidence reveals that DSBs emerge within chromosomes during synapsis and subsequent homologous pairing initiation. It must be determined how oocytes detect these DSBs and initiate DNA repair at such an early stage. A leading hypothesis suggests that DSB repair may be downregulated deliberately to promote genetic recombination between homologous non-sister chromatids. The rationale behind this speculation lies in the potential harm that could result from activation of the DNA-damage response on oocytes during this process, leading to failure of the proper number of crossover formations during recombination. It is crucial that at least one crossover occurs per chromosome to ensure proper segregation and, consequently, to increase genetic diversity. Notably, this phenomenon is linked with the absence of the G2/M checkpoint in mouse oocytes, as discussed in the next section. In summary, the CHK2-dependent DNA-damage checkpoint during meiotic recombination serves to eliminate oocytes that fail to complete both meiotic recombination and chromosome synapsis. This checkpoint is a critical safeguard against the transmission of genetic anomalies.

Declines in Apoptosis During Follicular Development

Oocytes with numerous unsynapsed chromosomes inevitably accumulate DSBs. If these are not properly repaired, the CHK2-dependent checkpoint is triggered, initiating apoptosis and eliminating oocytes with unsynapsed chromosomes. This apoptotic process is triggered by the Trans-activation p63 (TAp63)-dependent signaling pathway [46]. As a homolog of p53, TAp63 is expressed abundantly and plays a pivotal role in orchestrating apoptosis in the oocytes in primordial follicles. Transcription of BH3-only pro-apoptotic BCL2 family proteins, PUMA and NOXA, is stimulated by TAp63, which is activated through sequential phosphorylation mediated by CHK2 signals. In response, apoptosis is initiated by these proteins through activation of BAX (BCL-2-associated X protein) and BAK (BCL-2 antagonist killer 1), which are pro-apoptotic and commit the cells to apopsosis [47–49]. Subsequent translocation of BAX and BAK to the mitochondria triggers apoptotic reactions, which includes the release of cytochrome c and activation of caspase 9, leading to oocyte apoptosis (Fig. 1B) [49, 50]. TAp63-null mice oocytes are resistant to irradiation, cisplatin, and other thermotherapeutic agents at doses fatal to wild-type oocytes in p53-null ovaries [51, 52]. This provides strong support for the hypothesis that TAp63 is a crucial regulator of apoptosis in oocytes. PUMA- or PUMA/NOXA-null mice also have a high capacity for reducing cell residue in wild-type oocytes, improving their ability to tolerate genotoxins [53]. Both TAp63 and PUMA/NOXA play critical roles in the initiation of apoptosis during the primary phase of oocyte development. However, as follicles develop, a decline in TAp63 expression has been reported (Fig. 1B). This limits the ability of oocytes to respond to signals related to apoptosis signals [51, 54]. As a result, survival rates in oocytes in late-stage follicles are relatively high, and development can proceed even as DNA damage accumulates. This leaves the DNA repair pathway as the primary safeguard against DNA damage in oocytes.

The finite nature of the female germ cell from birth raises an intriguing question: Why is apoptosis favored at the primary stages but gradually gives way to the DNA repair pathway? Preserving high-quality oocytes from the initial stages is essential because female reproductive lifespans are established during fetal development [25]. During the early development of follicles and oocytes, apoptosis is the primary checkpoint for identifying and eliminating, rather than repairing, cells that have been severely damaged, preventing genetic mutations from being passed on to future generations. Future studies that provide a comprehensive picture of the mechanism responsible for reduction in apoptosis during follicle and oocyte development may provide valuable insights into maintaining high-quality oocytes throughout the reproductive lifespan and ensure the transmission of healthy genctic material to progeny.

Lack of a Robust G2/M DNA-Damage Checkpoint

Accurate duplication and segregation of genetic materials is ensured by a strictly controlled cell cycle [55–57]. The G2/M DNA damage checkpoint is a crucial part of this process, monitoring DNA integrity before cells enter mitosis. This checkpoint acts as a protective barrier against mitotic entry for cells with DNA damage, allowing enough time for DNA repair while maintaining genomic stability [58, 59]. Based on reports that oocytes arrested at prophase correspond to G2/M transition, the mechanisms responsible for triggering the G2/M checkpoint following DNA damage in oocytes may be comparable to those in somatic cells [60]. However, DNA damage, which can cause G2 arrest in somatic cells, reportedly does not arrest G2/prophase-stage in fully grown oocytes unless the damage is severe [26, 48,

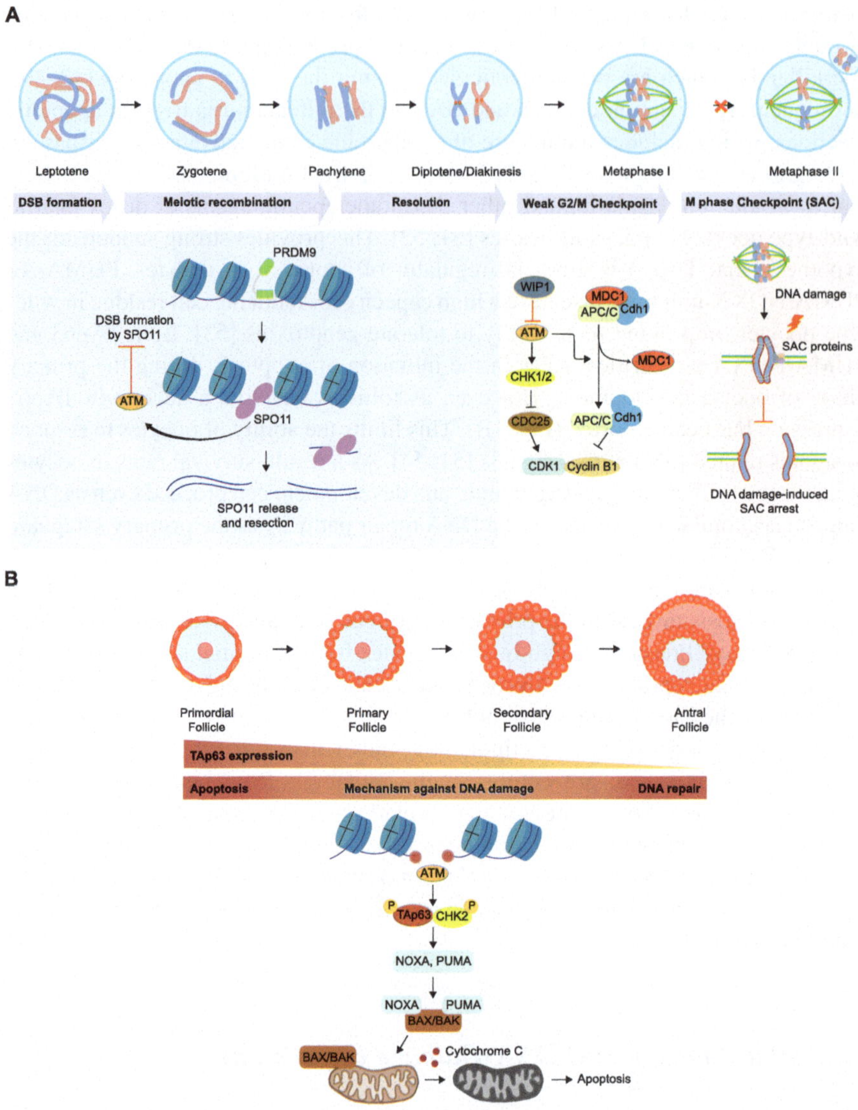

Fig. 1 A model of DNA damage checkpoints and apoptosis in mammalian oocytes. (**A**) The early stages of meiosis involve homologous chromosome synapsis and genetic recombination, essential for proper chromosome segregation and genetic diversity. This process begins with programmed DSBs catalyzed by SPO11 and is tightly regulated by surveillance mechanisms such as the meiotic recombination checkpoint in oocytes. ATM and PRDM9 potentially serve as localized regulators of SPO11 processing during meiotic recombination. Deficiencies in chromosome synapsis trigger spontaneous DSB accumulation, activating the CHK2 pathway for oocyte elimination. The absence of a robust G2/M checkpoint in mouse oocytes suggests deliberate downregulation of DSB repair to promote genetic recombination. The canonical ATM-CHK1/2-CDC25 pathway is downregulated in fully grown oocytes, suppressed by WIP1 activity. Oocytes utilize a noncanonical G2/M DNA damage checkpoint, associated with the regulation of APC/C-Cdh1 activity. DNA damage

61–63], suggesting that the G2/M checkpoint in oocytes is relatively weak. A lack of G2/prophase arrest can be linked to low levels of ATM kinase activation and reduced expression of ATM. As a result, ATM activation requires large amounts of DNA damage, resulting in CHK1-dependent inhibitory phosphorylation of CDC25 and preventing meiotic resumption in mouse oocytes [63].

The enigma behind the compromised G2/M DNA-damage checkpoint in oocytes becomes more captivating when considering the involvement of WIP1, a key regulator in terminating the DNA damage response. Recent studies have revealed that inhibiting WIP1 following DNA damage increases the level of γ-H2AX through ATM activation, establishing the DNA damage G2/M checkpoint and delaying or inhibiting the resumption of meiosis (Fig. 1A) [64]. The intriguing role of WIP1 in this context adds a layer of complexity to our understanding of the DDR in oocytes. Because WIP1 generally is downregulated during mitosis [65], the significance of its presence and the suppression of ATM activity during prophase arrest in oocytes remain ambiguous. One intriguing possibility is that the suppression of ATM by WIP1 facilitates the generation of programmed DSBs for HR during early prophase in oocytes.

This unique interplay between WIP1 and ATM in mouse oocytes adds another layer to the complex orchestration of the G2/M DNA-damage checkpoint. Understanding the intricacies of WIP1 regulation and its connection to ATM activity may generate novel insights into the distinctive response of oocytes to DNA damage during prophase arrest.

The absence of canonical G2/M DNA-damage checkpoints in oocytes does not imply a lack of vulnerability to DNA damage during resumption of meiosis. Recent studies show that oocytes have adopted a noncanonical G2/M DNA-damage checkpoint that is activated by increased anaphase-promoting complex/cyclosome (APC/C) activity, promoting cyclin B1 proteolysis [66]. This effect can be attributed to increased activity of a protein phosphatase, CDC14B, and reduced activity of an APC/C inhibitor, EMI1 [66]. MDC1 also regulates the APC/C-Cdh1–mediated G2/M checkpoint by directly interacting with APC/C-Cdh1 (Fig. 1A) [67]. Damaging DNA can disrupt this interaction, increasing APC/C-Cdh1 activity and delaying resumption of meiosis. However, the APC/C-mediated G2/M checkpoint does not respond rapidly to DNA damage. In addition to these mechanisms, enclosing the oocyte within a layer of cumulus cells enhances the G2/prophase

Fig. 1 (continued) induces MDC1 dissociation from APC/C-Cdh1, leading to increased APC/C-Cdh1 activity. During DNA damage-induced SAC arrest, the SAC is activated independently of unattached kinetochores during meiosis I in oocytes, contrasting with its activation mechanism in mitotic cells. (**B**) During early follicular development, both primordial and primary follicles exhibit high expression levels of TAp63, enabling them to respond to DNA damage by undergoing apoptosis. This apoptotic process mediated by TAp63 occurs in oocytes from primordial and primary follicles. DNA damage triggers the activation of TAp63, which subsequently induces the transcription of PUMA and NOXA, leading to the inhibition of BAX and BAK, thus facilitating apoptosis. However, as follicles progress to the late antral stage, the expression of TAp63 diminishes. Consequently, fully grown oocytes become less prone to undergo apoptosis in response to DNA damage

DNA-damage checkpoint. The promotion of cAMP production in cumulus cells, which can be transferred to oocytes through gap junctions between the oocyte and cumulus cells [68]. This communication may contribute to strengthening the G2/M checkpoint in oocytes. In summary, mammalian oocytes do not use a robust checkpoint to respond immediately to damaged DNA but instead rely on a weak and less efficient checkpoint that involves proteolysis mediated by APC/C-Cdh1.

DNA Damage-Induced Spindle Assembly Checkpoint

Fully grown oocytes with DNA damage typically undergo meiotic maturation when the damage is not substantial due to the absence of a robust G2/M checkpoint. However, when exposed to DNA-damaging agents, they arrest at metaphase I (MI) and fail to progress to metaphase II (MII) [62, 69, 70]. This MI arrest does not rely on canonical ATM/ATR pathways alone but is triggered by activation of the spindle assembly checkpoint (SAC) [62, 70].

The SAC coordinates chromosome segregation by ensuring proper kinetochore-microtubule (K-MT) attachments [71, 72]. Activation of the SAC occurs upon detection of abnormal K-MT attachments and inhibits anaphase onset by triggering a cascade of events. Unattached kinetochores form a platform for the recruitment of MAD2 by binding the SAC proteins MAD1 and MAD2. MAD2 is then released into the cytoplasm, where it sequesters CDC20, creating a reversible 'wait-anaphase' signal that inhibits the APC/C [73–75]. This sequential process stabilizes APC/C substrates, including Aurora kinases, Polo-like kinase-1 (PLK1), cyclin B1, and securin. In turn, stabilized cyclin B1 maintains CDK1 activity, while securing prevents separase activation, thereby preserving sister chromatid cohesion. These events collectively prevent the onset of anaphase until all chromosomes are correctly attached to spindle microtubules [74, 76, 77]. However, SAC activity induced by DNA damage should not be confused with canonical SAC activity triggered by unattached kinetochores. Two critical distinctions can be made between the two activities. First, oocytes treated with etoposide, a topoisomerase II inhibitor that induces DSBs, show no signs of losing K-MT attachment or tension across the bivalents [75]. Second, if the level of DNA damage is low, oocytes can still extrude the polar body, albeit with a delay in the timing of the polar body extrusion, which may lead to aneuploidy [71]. Recent studies suggest that exogenous DNA damage does not trigger SAC activation in human oocytes, allowing the damage to persist through meiotic maturation. This results in damaged chromosomes and abnormal spindles in egg [78]. The onset of anaphase I under these conditions is determined by the loss of CDK1 activity rather than by non-aligned bivalents [74]. A small number of abnormal K-MT attachments does not appear to trigger a strong SAC response in mammalian oocytes, and a certain threshold of anomalies may be required for effective SAC activation [79].

Aging oocytes have been shown to have reduced SAC efficiency [80]. DNA damage in females of advanced reproductive age may lead to severe chromosomal anomalies, adversely affecting embryonic development. Nonetheless, significant recruitment of SAC components to MII kinetochores after DNA damage has not been observed, and DNA damage does not provoke SAC activation at the MII stage, even though egg can typically activate SAC in response to MII spindle disruptions or misalignment of chromosomes [75]. As oocytes are susceptible to accumulating DNA damage, it is imperative that they detect DNA damage and delay meiotic maturation to allow adequate time for DNA repair. In this context, the SAC remains a critical and final checkpoint to ensure oocyte quality and genome integrity prior to fertilization.

DNA Repair During Meiotic Maturation

DSBs pose a significant challenge to the ordered progression of mitosis and meiosis, which are key processes for chromosomal segregation [81, 82]. While cellular responses to DSBs during interphase are well understood, the mechanism through which cells respond to such lesions during mitosis and meiosis is less clear.

Differential Response to DNA Breaks in Mitosis and Meiosis

Cells can halt the cell cycle when they traverse DSBs during interphase and wait for the breaks to be repaired before entering mitosis [83]. However, when DSBs occur during mitosis, cells prioritize completion of cell division over arresting the cell cycle and repairing DNA damage instead of stopping cell cycle [84–86]. In mitotic cells, initial DNA damage responses, such as H2AX phosphorylation and MDC1 recruitment, are active, but recruitment of downstream repair actors such as RNF8, BRCA1, and 53BP1 to DSB sites is deficient [87–89]. This impediment is linked to mitosis-specific phosphorylation that prevents these proteins from effectively regulating the DSB repair pathway. Specifically, CDK1 and PLK1 mediate the phosphorylation of 53BP1, inhibiting its association with γ-H2AX. Additionally, CDK1 phosphorylates RNF8, obstructing its interaction with MDC1 and hampering its recruitment to DSB sites [87, 90–93]. Therefore, the DNA damage response to DSBs in mitosis is hindered only partially, permitting upstream events such as H2AX phosphorylation and MDC1 recruitment. Despite dampened downstream events during mitosis, DSBs are efficiently marked and stabilized. Recent insights reveal a critical role for the TOPBP1 in this process. After marking DSBs with γ-H2AX and MDC1, MDC1 recruits TOPBP1 to damaged chromosomes and forms filamentous structures at mitotic DSB sites that may help to tether and stabilize broken chromosomes until they can be repaired in the following G1 phase [89]. This suggests that the inability of mitotic cells to recruit RNF8, BRCA1, and 53BP1 to DSB sites may be compensated by recruiting TOPBP1, thereby maintaining genomic stability. However, the mechanism by which MDC1 selectively recruits

TOPBP1, instead of downstream factors like 53BP1 and BRCA1, remains unclear. Emerging evidence suggest that CIP2A might play a role in the recruitment of TOPBP1 specifically to sites of DSBs during mitosis [94].

In contrast, oocytes have demonstrated a proficient response to DSBs during meiosis, efficiently recruiting BRCA1 and 53BP1 to DSB sites and utilizing both the HR and non-homologous end-joining (NHEJ) pathways for repair (Fig. 2) [95–97]. The unique challenges facing oocytes, such as the need to maintain genomic integrity for prolonged periods, highlight the need for such a response. The precise mechanisms that facilitate recruitment of BRCA1 and 53BP1 and enable efficient DSB repair in meiotic oocytes are not fully understood. However, the regulatory phosphorylation events in mitosis may be modulated in oocytes to permit repair, which is a testament to the resilience and adaptability of meiosis. Indications

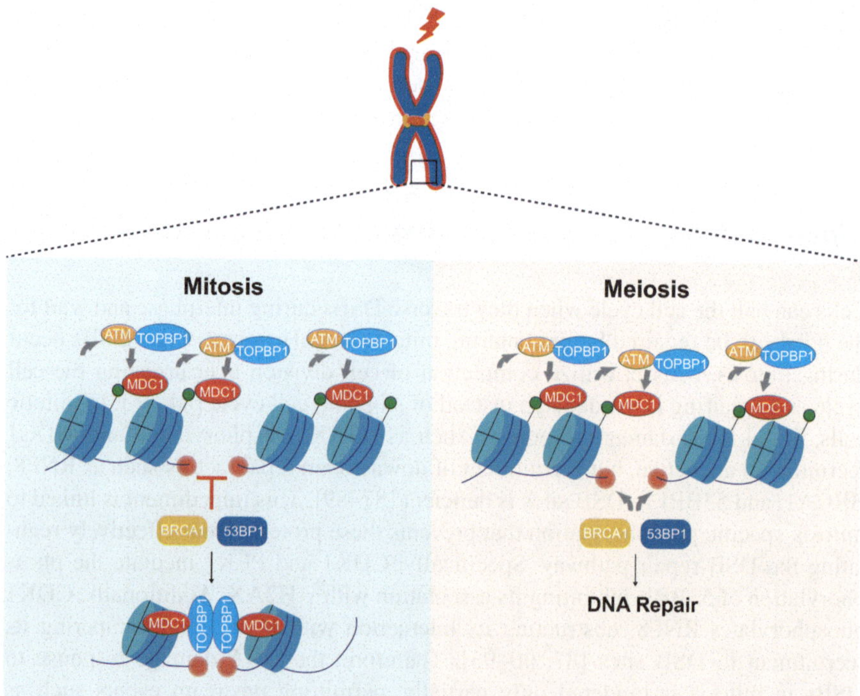

Fig. 2 Distinct response to DNA breaks in mitosis and meiosis. Cells exhibit distinct responses to DNA breaks in mitosis and meiosis. In mitosis, completing cell division takes precedence over repairing DNA damage, resulting in insufficient recruitment of repair factors such as BRCA1 and 53BP1 to DSB sites. Chromatin responses to DSBs in mitosis are partially impeded, allowing for upstream events such as H2AX phosphorylation and MDC1 recruitment. TOPBP1, recruited by MDC1, forms filamentous structures at mitotic DSB sites, marking them for repair during the G1 phase. Conversely, oocytes exhibit a proficient response to DSBs during meiosis, efficiently recruiting BRCA1 and 53BP1 to DSB sites and utilizing both HR and NHEJ pathways for repair. The unique challenges faced by oocytes, such as the need to maintain genomic integrity for prolonged periods, necessitate such a robust response

that PLK1 activity [95], which tends to inhibit 53BP1 in repair, is inactivated at spindle poles in response to DSBs, underscoring the robust reparative responses during oocyte meiosis. This targeted DNA repair capability underlines the importance of oocyte quality and fidelity in successful reproduction and represents a critical facet of oocyte resilience as it ensures DSB repair, chromosomal integrity, and reproduction.

Distinct Roles of HR and NHEJ During Meiotic Maturation

DSBs are critical lesions that can induce genomic instability. They are repaired primarily through two significant pathways in mammalian cells: NHEJ and HR [98–101]. Although each pathway has its own fidelity and template requirements, both are integral to genomic integrity.

NHEJ can rejoin broken ends with limited homology, often resulting in deletions or insertions at the repair site. This process begins when the Ku heterodimer recognizes and binds to DSBs, subsequently recruiting the DNA-dependent protein kinase catalytic subunit (DNA-PKcs) [102, 103]. The DNA-PKcs coordinates the repair process, signaling other proteins to halt DNA end resection and preserve the substrate for HR [104]. Following DNA-PKcs phosphorylation, the XRCC4-DNA ligase IV complex, aided by XLF, completes ligation of the break [102, 103]. Despite a reputation for being error-prone, NHEJ can repair DSBs with complementary overhangs [102, 105]. In contrast, HR meticulously restores the original sequence using an undamaged DNA template, often sourced from a sister chromatid or homologous chromosome [98–101, 106]. HR, which is preferred during the S and G2 cell-cycle phases, begins with DSB end recognition and resection, producing a 3′-OH single-stranded tail [96, 102, 107]. This step, initiated by the CtIP and the MRN complex, allows RPA to bind to DNA and subsequently be replaced by RAD51 to form a presynaptic filament [107]. The filament invades homologous DNA, facilitating template-driven DNA synthesis and ultimately resolving breakthrough ligation, often creating a double Holliday junction [108, 109].

Recent insights into oocyte meiosis have highlighted the use of both HR and NHEJ in a departure from mitotic DNA repair strategies [96]. Inhibiting NHEJ components has been linked to increased DNA damage and reduced developmental competence [96, 97]. Inhibition of NHEJ in oocytes with DNA damage leads to SAC-mediated arrest with partially impaired K-MT attachments, which is contradicting the notion that DNA damage-induced SAC arrest is independent of K-MT attachment [75, 96]. Given the significant increase in DNA damage levels after NHEJ inhibition, it is probably due to the accumulation of DNA damage beyond the repair capacity of oocytes after NHEJ inhibition. In contrast to NHEJ, HR inhibition does not impede meiotic progression. It can result in chromosomal fragmentation and compromised centromere integrity, suggesting a sophisticated interplay between HR and centromeric sequences [96]. The role of HR as a guardian of centromeric stability in somatic cells indicates a similar protective function during oocyte maturation. As oocytes progress from prophase to MII stages, a shift from HR to NHEJ occurs, dictated by the accessibility of repair components [107, 110]. A move toward HR is signaled by the onset of end resection, given reduced Ku affinity for binding long and single-stranded DNA overhangs [111]. HR is more effective in the absence of NHEJ, suggesting a compensatory function for HR [112].

Beyond these established mechanisms, microhomology-mediated end-joining (MMEJ) has attracted attention from researchers of DNA repair. MMEJ uses short homologous sequences to align the ends of a DSB for ligation in a process that may lead to deletions adjacent to the break [104, 110]. Further research is necessary to fully understand the interplay between these DSB repair pathways and their role in maintaining genome integrity in oocytes.

Role of Spindle Microtubules in DNA Repair During Oocytes Meiosis

The movement and reorganization of chromatin within the nucleus are dynamic processes that intensify in response to DNA damage [113, 114]. Enhanced chromatin mobility at DSB sites is thought to facilitate the search for donor sequences necessary for HR and improve access to repair factors, promoting effective DNA repair [113–116]. The cytoskeleton, including cytoplasmic actin and microtubules, plays a pivotal role in driving chromatin movement at these damaged sites [117, 118]. Research in both yeast and mammalian cells has demonstrated that cytoplasmic actin and microtubules bring about significant changes in chromatin structure following DNA damage [119–121]. Inhibiting microtubule polymerization with nocodazole leads to abnormal chromosome structures, underscoring the essential role of spindle microtubules in maintaining chromosome architecture during repair [95]. Additionally, the interaction between γ-tubulin, RAD51, and BRCA1 is evidence of a link between DNA repair and the microtubule network [122]. This interaction is crucial for maintaining genomic stability, particularly during interphase, though its efficacy during mitosis remains uncertain. Overall, these processes contribute to the accurate positioning and repair of DNA damage. Recent discoveries have shed light on the unique dynamics of spindle microtubules in response to DNA damage in mouse oocytes during meiosis [95]. DNA damage prompts spindle microtubules to shrink and stabilize, a process that includes the inactivation of PLK1 at spindle poles and the subsequent dephosphorylation of CIP2A at S904. Dephosphorylated CIP2A, along with phosphorylated MDC1 and TOPBP1, traverses spindle microtubules and congregates at chromosomes via kinetochores and centromeres. This migration is vital for the recruitment of downstream DNA repair factors, such as BRCA1 and 53BP1, during meiosis. Although the biological significance of spindle microtubule shrinkage and stabilization is not fully understood, it is known that proteins engaged in DDR pathways, including the CIP2A-MDC1-TOPBP1 complex, either associate with spindle microtubules or localize at spindle poles (Fig. 3) [67, 95, 123]. Intriguingly, the forces behind these spindle alterations

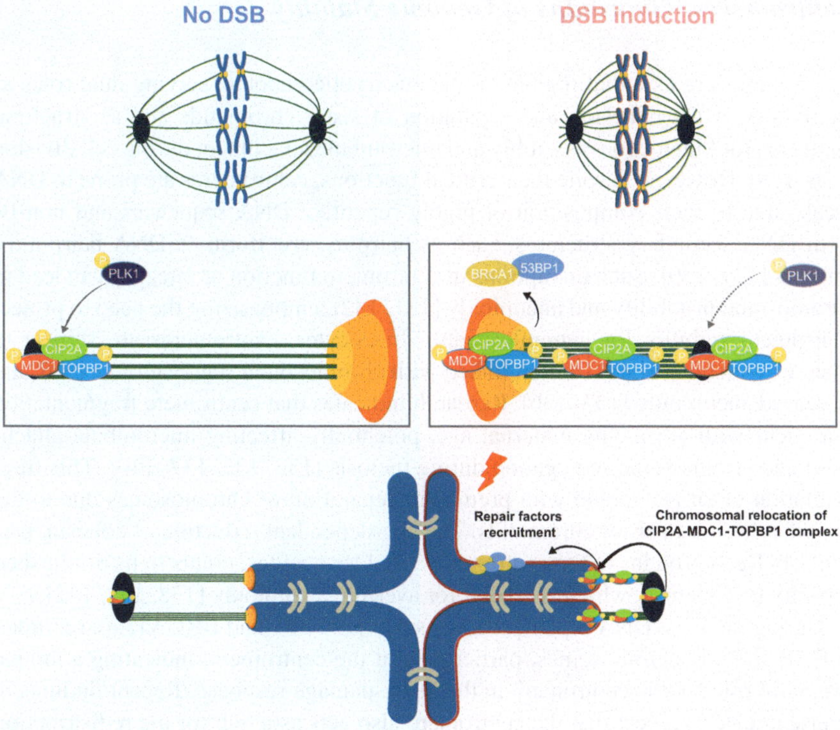

Fig. 3 DNA repair during meiosis. In response to DNA damage, MI oocytes orchestrate a series of coordinated events to repair DSBs. During meiosis, spindle microtubules undergo unique changes in response to DNA damage, including rapid shrinkage and stabilization, along with temporary PLK1 inactivation at spindle poles. This leads to the dephosphorylation of CIP2A, promoting the relocation of CIP2A-MDC1-TOPBP1 complex from spindle poles to chromosomes. These alterations facilitate the recruitment of DNA repair factors such as BRCA1 and 53BP1. The centromere serves as a focal point for redistributing the CIP2A-MDC1-TOPBP1 complex, relying on microtubules. Depletion of centromere and kinetochore components impedes the recruitment of DNA damage-response proteins to chromosomes, highlighting the centromere's role in maintaining genomic integrity. These mechanisms ensure efficient and accurate DNA damage repair, safeguarding genomic integrity during oocyte meiosis

in oocytes cannot be attributed to ATM, PLK1, or K-MT attachments [95]. Given the intricate interaction between actin filaments and microtubules during spindle formation in oocytes [124], actin filaments also might regulate the spindle changes induced by DSBs. This is supported by the reported assembly of actin filaments following DNA damage in somatic cells [125]. In response to DNA damage, actin regulators, such as FMN2 and Spire 1/2, have been shown to create various actin structures in the nucleus, and depletion of these proteins impairs DNA repair. These findings underscore the necessity for more detailed studies to unravel the mechanisms of chromosome-spindle pole communication in oocytes in response to DNA damage.

Centromeres: Guardians of Genome Stability

Centromeres are essential regions of the eukaryotic genome, serving dual roles as facilitators of the cohesion and separation of sister chromatids and as structural platforms for kinetochore assembly and microtubule attachment during cell division [126–128]. However, despite their critical functions, centromeres are prone to DNA breaks due to their composition of highly repetitive DNA sequences and non-B-form DNA secondary structures, such as hairpins, cruciform, Z-DNA hairpin, or i-motifs [129, 130]. Such disruptions in centromere function or integrity can lead to chromosome instability and aneuploidy [131, 132], emphasizing the need to protect centromere stability for genomic health. In oocytes, centromeres are subject to changes with age, which is often linked with chromosomal segregation defects and increased aneuploidy [133–136]. Research indicates that centromere fragmentation can occur with advancing maternal age, potentially affecting microtubule attachment and chromosome segregation during meiosis [132, 135, 137–140]. This fragmentation often is coupled with premature separation of chromosomes due to the loss of cohesion in sister chromatids. The age-dependent reduction of cohesin, particularly Rec8 proteins in the centromeres of older oocytes, points to its diminished capacity to replenish, which is crucial for averting aneuploidy [138, 141, 142].

During meiosis, DNA repair proteins, such as 53BP1 and BRCA1, are recruited to DSB sites on chromosomes, particularly at the centromere, indicating a unique structural role for the centromere in the DNA-damage response. Recent findings in mouse oocytes suggest that the centromere also acts as a hub for the redistribution of the CIP2A-MDC1-TOPBP1 complex during spindle assembly, relying on a microtubule-dependent pathway. Depletion of centromere and kinetochore components, such as CENP-A or HEC1, reportedly impairs recruitment of DNA damage-response proteins to the chromosome [95], emphasizing the role of the centromere in genomic integrity (Fig. 3).

Maintaining centromere integrity is not only critical to preventing chromosomal instability, but also plays a key role in DNA-damage response and repair mechanisms in oocytes. To gain more insights into the aging process and DNA-damage response in centromeres, more in-depth studies of mechanisms associated with centromere assembly and structure are warranted.

Challenges for Preserving Oocyte Quality and Fertility Against DNA Damage

Quality, fertility, and embryo viability

Our understanding of the DNA damage response and repair mechanisms in oocytes underscores the significant impact of such damage on oocyte quality, fertility, and embryo viability. Preserving DNA integrity in oocytes is crucial, as damage can lead to compromised cellular functions, meiotic errors, and an increased risk of aneuploidy. These adverse effects may result in decreased embryo viability, a higher risk of miscarriage, and the potential for genetic disorders in offspring. The extended prophase I arrest to which oocytes are subjected increases their susceptibility to the accumulation of DNA damage [139, 140, 143]. The abundance of mitochondria within oocytes, which provide energy during fertilization, and embryo development, becomes a vulnerability when mitochondrial DNA is damaged. Such damage can disrupt oxidative phosphorylation, impair mitochondrial function, and degrade oocyte quality to reduce embryo viability and increase the risk of infertility [143–147]. DNA damage in oocytes also can induce epigenetic changes that affect the regulation of gene expression in embryos, potentially causing developmental complications and an elevated risk of diseases later in life [148]. The damage is not limited to natural causes; cancer therapies such as chemotherapy and radiation also can corrupt the DNA of healthy cells, including oocytes. The extent of the impact of cancer treatments on oocyte quality and fertility is dependent on the type of therapy, as well as the dosage and duration. Oocytes that suffer DNA damage during such treatments may experience temporary or permanent loss of fertility, premature ovarian insufficiency, and an increased probability of genetic abnormalities in offspring conceived post-treatment [149–151].

To enhance oocyte quality and address the fertility challenges presented by DNA damage, a comprehensive approach is essential. Antioxidant supplementation can provide a defense against oxidative stress and its harmful effects. Glutathione, melatonin, vitamin E, resveratrol, and coenzyme Q10 have been found to bolster the cellular defenses of oocytes against DNA damage [152–158]. Regulation of WIP1 plays a vital role in the DDR and repair in mammalian oocytes. Inhibiting WIP1 may have the potential to reverse age-related decline in the repair of damaged DNA in mammalian oocytes, thereby improving oocyte quality and developmental competence [159]. WIP1 inhibition also might lessen sperm-derived DNA damage in fertilized oocytes [139], suggesting implications for assisted-reproductive technologies. However, as clinical trials of the effectiveness of WIP1 inhibitors continue, careful consideration is paramount. Prolonged WIP1 inhibition could affect early embryo development, and a fine balance must be struck between the therapeutic benefits for and potential risks to patients with fertility issues.

As we continue to uncover the oocyte DNA repair mechanisms, the potential to improve assisted-reproductive technologies and safeguard fertility increases. Our research aims not only to find tools to shield oocytes from damage, but also to boost their resilience and secure healthier futures for the next generations.

Concluding Remarks

The delicate balance between biological intricacies and progressive strides in reproductive medicine has never been more evident. Our investigation into the DDR and repair in oocytes has uncovered vulnerabilities to genetic insults. This understanding opens the door to novel techniques to preserve fertility, particularly as we explore the nuanced differences in the DDR among species, such as mice and humans, and the need for tailored research. The current wave of innovation in high-throughput sequencing, gene editing, and advanced experimental methods offers unprecedented opportunities to improve our knowledge of mammalian oocyte DNA repair mechanisms. These advances will not only broaden our scientific horizons, but also forge new fertility treatment tools.

Declaration of Interests The authors declare no competing interests.

Funding This work was supported by the Basic Science Research Program through the National Research Foundation of Korea (NRF) grant funded by the Ministry of Education (NRF-2017R1A6A1A03015642 to JSO; RS-2022-NR070884 to JL).

References

1. Abuetabh Y et al (2022) DNA damage response revisited: the p53 family and its regulators provide endless cancer therapy opportunities. Exp Mol Med 54:1658–1669. https://doi.org/10.1038/s12276-022-00863-4
2. Jackson SP, Bartek J (2009) The DNA-damage response in human biology and disease. Nature 461:1071–1078. https://doi.org/10.1038/nature08467
3. Ashour ME, Mosammaparast N (2021) Mechanisms of damage tolerance and repair during DNA replication. Nucleic Acids Res 49:3033–3047. https://doi.org/10.1093/nar/gkab101
4. Gartner A, Engebrecht J (2022) DNA repair, recombination, and damage signaling. Genetics 220:iyab178. https://doi.org/10.1093/genetics/iyab178
5. Hoeijmakers JH (2007) Genome maintenance mechanisms are critical for preventing cancer as well as other aging-associated diseases. Mech Ageing Dev 128:460–462. https://doi.org/10.1016/j.mad.2007.05.002
6. Hoeijmakers JH (2009) DNA damage, aging, and cancer. N Engl J Med 361:1475–1485. https://doi.org/10.1056/NEJMra0804615
7. Lavin MF (2008) Ataxia-telangiectasia: from a rare disorder to a paradigm for cell signalling and cancer. Nat Rev Mol Cell Biol 9:759–769. https://doi.org/10.1038/nrm2514
8. Smits VA, Gillespie DA (2015) DNA damage control: regulation and functions of checkpoint kinase 1. FEBS J 282:3681–3692. https://doi.org/10.1111/febs.13387
9. Ciccia A, Elledge SJ (2010) The DNA damage response: making it safe to play with knives. Mol Cell 40:179–204. https://doi.org/10.1016/j.molcel.2010.09.019
10. Marechal A, Zou L (2013) DNA damage sensing by the ATM and ATR kinases. Cold Spring Harb Perspect Biol 5:a012716. https://doi.org/10.1101/cshperspect.a012716
11. Lee JH, Paull TT (2005) ATM activation by DNA double-strand breaks through the Mre11-Rad50-Nbs1 complex. Science 308:551–554. https://doi.org/10.1126/science.1108297
12. Smith J, Tho LM, Xu N, Gillespie DA (2010) The ATM-Chk2 and ATR-Chk1 pathways in DNA damage signaling and cancer. Adv Cancer Res 108:73–112. https://doi.org/10.1016/B978-0-12-380888-2.00003-0
13. Bakkenist CJ, Kastan MB (2003) DNA damage activates ATM through intermolecular autophosphorylation and dimer dissociation. Nature 421:499–506. https://doi.org/10.1038/nature01368

14. Izumi T, Maller JL (1995) Phosphorylation and activation of the Xenopus Cdc25 phosphatase in the absence of Cdc2 and Cdk2 kinase activity. Mol Biol Cell 6:215–226. https://doi.org/10.1091/mbc.6.2.215
15. Blackford AN, Jackson SP (2017) ATM, ATR, and DNA-PK: the trinity at the heart of the DNA damage response. Mol Cell 66:801–817. https://doi.org/10.1016/j.molcel.2017.05.015
16. Bartek J, Lukas J (2003) Chk1 and Chk2 kinases in checkpoint control and cancer. Cancer Cell 3:421–429. https://doi.org/10.1016/s1535-6108(03)00110-7
17. Bucher N, Britten CD (2008) G2 checkpoint abrogation and checkpoint kinase-1 targeting in the treatment of cancer. Br J Cancer 98:523–528. https://doi.org/10.1038/sj.bjc.6604208
18. Byun TS, Pacek M, Yee MC, Walter JC, Cimprich KA (2005) Functional uncoupling of MCM helicase and DNA polymerase activities activates the ATR-dependent checkpoint. Genes Dev 19:1040–1052. https://doi.org/10.1101/gad.1301205
19. Katsuragi Y, Sagata N (2004) Regulation of Chk1 kinase by autoinhibition and ATR-mediated phosphorylation. Mol Biol Cell 15:1680–1689. https://doi.org/10.1091/mbc.e03-12-0874
20. Liu Q et al (2000) Chk1 is an essential kinase that is regulated by Atr and required for the G(2)/M DNA damage checkpoint. Genes Dev 14:1448–1459
21. Wang H, Wang H, Powell SN, Iliakis G, Wang Y (2004) ATR affecting cell radiosensitivity is dependent on homologous recombination repair but independent of nonhomologous end joining. Cancer Res 64:7139–7143. https://doi.org/10.1158/0008-5472.CAN-04-1289
22. Andreassen PR, D'Andrea AD, Taniguchi T (2004) ATR couples FANCD2 monoubiquitination to the DNA-damage response. Genes Dev 18:1958–1963. https://doi.org/10.1101/gad.1196104
23. Wu X, Shell SM, Liu Y, Zou Y (2007) ATR-dependent checkpoint modulates XPA nuclear import in response to UV irradiation. Oncogene 26:757–764. https://doi.org/10.1038/sj.onc.1209828
24. Karnitz LM, Zou L (2015) Molecular pathways: targeting ATR in cancer therapy. Clin Cancer Res 21:4780–4785. https://doi.org/10.1158/1078-0432.CCR-15-0479
25. Pelosi E, Simonsick E, Forabosco A, Garcia-Ortiz JE, Schlessinger D (2015) Dynamics of the ovarian reserve and impact of genetic and epidemiological factors on age of menopause. Biol Reprod 92:130. https://doi.org/10.1095/biolreprod.114.127381
26. Carroll J, Marangos P (2013) The DNA damage response in mammalian oocytes. Front Genet 4:117. https://doi.org/10.3389/fgene.2013.00117
27. Tease C (1983) X-ray-induced chromosome aberrations in dictyate oocytes of young and old female mice. Mutat Res 119:191–194
28. Jacquet P, Adriaens I, Buset J, Neefs M, Vankerkom J (2005) Cytogenetic studies in mouse oocytes irradiated in vitro at different stages of maturation, by use of an early preantral follicle culture system. Mutat Res 583:168–177. https://doi.org/10.1016/j.mrgentox.2005.03.008
29. Hartwell LH, Weinert TA (1989) Checkpoints: controls that ensure the order of cell cycle events. Science 246:629–634. https://doi.org/10.1126/science.2683079
30. Petronczki M, Siomos MF, Nasmyth K (2003) Un menage a quatre: the molecular biology of chromosome segregation in meiosis. Cell 112:423–440. https://doi.org/10.1016/s0092-8674(03)00083-7
31. Bolcun-Filas E, Handel MA (2018) Meiosis: the chromosomal foundation of reproduction. Biol Reprod 99:112–126. https://doi.org/10.1093/biolre/ioy021
32. Zickler D, Kleckner N (2015) Recombination, pairing, and synapsis of homologs during meiosis. Cold Spring Harb Perspect Biol 7:a016626. https://doi.org/10.1101/cshperspect.a016626
33. Lam I, Keeney S (2014) Mechanism and regulation of meiotic recombination initiation. Cold Spring Harb Perspect Biol 7:a016634. https://doi.org/10.1101/cshperspect.a016634
34. Boateng KA, Bellani MA, Gregoretti IV, Pratto F, Camerini-Otero RD (2013) Homologous pairing preceding SPO11-mediated double-strand breaks in mice. Dev Cell 24:196–205. https://doi.org/10.1016/j.devcel.2012.12.002

35. Baudat F, Manova K, Yuen JP, Jasin M, Keeney S (2000) Chromosome synapsis defects and sexually dimorphic meiotic progression in mice lacking Spo11. Mol Cell 6:989–998. https://doi.org/10.1016/s1097-2765(00)00098-8
36. Cole F, Keeney S, Jasin M (2010) Evolutionary conservation of meiotic DSB proteins: more than just Spo11. Genes Dev 24:1201–1207. https://doi.org/10.1101/gad.1944710
37. Keeney S (2008) Spo11 and the formation of DNA double-strand breaks in meiosis. Genome Dyn Stab 2:81–123. https://doi.org/10.1007/7050_2007_026
38. Joyce EF, McKim KS (2010) Chromosome axis defects induce a checkpoint-mediated delay and interchromosomal effect on crossing over during drosophila meiosis. PLoS Genet 6:e1001059. https://doi.org/10.1371/journal.pgen.1001059
39. Hughes SE, Miller DE, Miller AL, Hawley RS (2018) Female meiosis: synapsis, recombination, and segregation in Drosophila melanogaster. Genetics 208:875–908. https://doi.org/10.1534/genetics.117.300081
40. Di Giacomo M et al (2005) Distinct DNA-damage-dependent and -independent responses drive the loss of oocytes in recombination-defective mouse mutants. Proc Natl Acad Sci U S A 102:737–742. https://doi.org/10.1073/pnas.0406212102
41. Rinaldi VD, Bolcun-Filas E, Kogo H, Kurahashi H, Schimenti JC (2017) The DNA damage checkpoint eliminates mouse oocytes with chromosome synapsis failure. Mol Cell 67:1026–1036.e1022. https://doi.org/10.1016/j.molcel.2017.07.027
42. Lukaszewicz A, Lange J, Keeney S, Jasin M (2018) Control of meiotic double-strand-break formation by ATM: local and global views. Cell Cycle 17:1155–1172. https://doi.org/10.1080/15384101.2018.1464847
43. Kurzbauer MT et al (2021) ATM controls meiotic DNA double-strand break formation and recombination and affects synaptonemal complex organization in plants. Plant Cell 33:1633–1656. https://doi.org/10.1093/plcell/koab045
44. Paiano J et al (2020) ATM and PRDM9 regulate SPO11-bound recombination intermediates during meiosis. Nat Commun 11:857. https://doi.org/10.1038/s41467-020-14654-w
45. Pacheco S et al (2015) The ATM signaling cascade promotes recombination-dependent pachytene arrest in mouse spermatocytes. PLoS Genet 11:e1005017. https://doi.org/10.1371/journal.pgen.1005017
46. Natan E, Joerger A (2011) Structure and kinetic stability of the p63 tetramerization domain. J Mol Biol 415:503–513. https://doi.org/10.1016/j.jmb.2011.11.007
47. Yu H et al (2010) Deletion of Puma protects hematopoietic stem cells and confers long-term survival in response to high-dose gamma-irradiation. Blood 115:3472–3480. https://doi.org/10.1182/blood-2009-10-248278
48. Collins JK, Jones KT (2016) DNA damage responses in mammalian oocytes. Reproduction 152:R15–R22. https://doi.org/10.1530/REP-16-0069
49. Roufayel R, Younes K, Al-Sabi A, Murshid N (2022) BH3-only proteins Noxa and Puma are key regulators of induced apoptosis. Life 12:256
50. Allan LA, Clarke PR (2009) Apoptosis and autophagy: regulation of caspase-9 by phosphorylation. FEBS J 276:6063–6073. https://doi.org/10.1111/j.1742-4658.2009.07330.x
51. Suh EK et al (2006) p63 protects the female germ line during meiotic arrest. Nature 444:624–628. https://doi.org/10.1038/nature05337
52. Luan Y, Yu SY, Abazarikia A, Dong R, Kim SY (2022) TAp63 determines the fate of oocytes against DNA damage. Sci Adv 8:eade1846. https://doi.org/10.1126/sciadv.ade1846
53. Kerr JB et al (2012) DNA damage-induced primordial follicle oocyte apoptosis and loss of fertility require TAp63-mediated induction of Puma and Noxa. Mol Cell 48:343–352. https://doi.org/10.1016/j.molcel.2012.08.017
54. Stringer JM, Winship A, Liew SH, Hutt K (2018) The capacity of oocytes for DNA repair. Cell Mol Life Sci 75:2777–2792. https://doi.org/10.1007/s00018-018-2833-9
55. Cheng L, Collyer T, Hardy CF (1999) Cell cycle regulation of DNA replication initiator factor Dbf4p. Mol Cell Biol 19:4270–4278. https://doi.org/10.1128/MCB.19.6.4270

56. Grafi G (1998) Cell cycle regulation of DNA replication: the endoreduplication perspective. Exp Cell Res 244:372–378. https://doi.org/10.1006/excr.1998.4213
57. Harper JV, Brooks G (2005) The mammalian cell cycle: an overview. Methods Mol Biol 296:113–153. https://doi.org/10.1385/1-59259-857-9:113
58. Campos A, Clemente-Blanco A (2020) Cell cycle and DNA repair regulation in the damage response: protein phosphatases take over the reins. Int J Mol Sci 21:446. https://doi.org/10.3390/ijms21020446
59. Houtgraaf JH, Versmissen J, van der Giessen WJ (2006) A concise review of DNA damage checkpoints and repair in mammalian cells. Cardiovasc Revasc Med 7:165–172. https://doi.org/10.1016/j.carrev.2006.02.002
60. Solc P, Schultz RM, Motlik J (2010) Prophase I arrest and progression to metaphase I in mouse oocytes: comparison of resumption of meiosis and recovery from G2-arrest in somatic cells. Mol Hum Reprod 16:654–664. https://doi.org/10.1093/molehr/gaq034
61. Mayer A et al (2016) DNA damage response during mouse oocyte maturation. Cell Cycle 15:546–558. https://doi.org/10.1080/15384101.2015.1128592
62. Marangos P et al (2015) DNA damage-induced metaphase I arrest is mediated by the spindle assembly checkpoint and maternal age. Nat Commun 6:8706. https://doi.org/10.1038/ncomms9706
63. Marangos P, Carroll J (2012) Oocytes progress beyond prophase in the presence of DNA damage. Curr Biol 22:989–994. https://doi.org/10.1016/j.cub.2012.03.063
64. Leem J, Kim JS, Oh JS (2018) WIP1 phosphatase suppresses the DNA damage response during G2/prophase arrest in mouse oocytes. Biol Reprod 99:798–805. https://doi.org/10.1093/biolre/ioy108
65. Macurek L et al (2013) Downregulation of Wip1 phosphatase modulates the cellular threshold of DNA damage signaling in mitosis. Cell Cycle 12:251–262. https://doi.org/10.4161/cc.23057
66. Subramanian GN et al (2020) Oocytes mount a noncanonical DNA damage response involving APC-Cdh1-mediated proteolysis. J Cell Biol 219:e201907213. https://doi.org/10.1083/jcb.201907213
67. Leem J, Oh JS (2022) MDC1 is essential for G2/M transition and spindle assembly in mouse oocytes. Cell Mol Life Sci 79:200. https://doi.org/10.1007/s00018-022-04241-1
68. Sun MH et al (2015) Cumulus cells block oocyte meiotic resumption via gap junctions in cumulus oocyte complexes subjected to DNA double-Strand breaks. PLoS One 10:e0143223. https://doi.org/10.1371/journal.pone.0143223
69. Yuen WS, Merriman JA, O'Bryan MK, Jones KT (2012) DNA double strand breaks but not interstrand crosslinks prevent progress through meiosis in fully grown mouse oocytes. PLoS One 7:e43875. https://doi.org/10.1371/journal.pone.0043875
70. Collins JK, Lane SI, Merriman JA, Jones KT (2015) DNA damage induces a meiotic arrest in mouse oocytes mediated by the spindle assembly checkpoint. Nat Commun 6:8553. https://doi.org/10.1038/ncomms9553
71. Bharadwaj R, Yu H (2004) The spindle checkpoint, aneuploidy, and cancer. Oncogene 23:2016–2027. https://doi.org/10.1038/sj.onc.1207374
72. Jallepalli PV, Lengauer C (2001) Chromosome segregation and cancer: cutting through the mystery. Nat Rev Cancer 1:109–117. https://doi.org/10.1038/35101065
73. Pablo Lara-Gonzalez JP, Desai A (2021) Spindle assembly checkpoint activation and silencing at kinetochores. Semin Cell Dev Biol 117:86–98. https://doi.org/10.1016/j.semcdb.2021.06.009
74. Simon IR, Lane YY, Jones KT (2012) Timing of anaphase-promoting complex activation in mouse oocytes is predicted by microtubule-kinetochore attachment but not by bivalent alignment or tension. Development 139:1947–1955. https://doi.org/10.1242/dev.077040
75. Lane SIR et al (2017) DNA damage induces a kinetochore-based ATM/ATR-independent SAC arrest unique to the first meiotic division in mouse oocytes. Development 144:3475–3486. https://doi.org/10.1242/dev.153965

76. Peters J-M (2006) The anaphase promoting complex/cyclosome: a machine designed to destroy. Nat Rev Mol Cell Biol 7:644–656. https://doi.org/10.1038/nrm1988
77. Blengini CS, Ik Jung G, Aboelenain M, Schindler K (2022) A field guide to Aurora kinase inhibitors: an oocyte perspective. Reproduction 164:V5–V7. https://doi.org/10.1530/REP-22-0292
78. Remillard-Labrosse G et al (2020) Human oocytes harboring damaged DNA can complete meiosis I. Fertil Steril 113:1080–1089.e1082. https://doi.org/10.1016/j.fertnstert.2019.12.029
79. Kyogoku H, Kitajima TS (2017) Large cytoplasm is linked to the error-prone nature of oocytes. Dev Cell 41:287–298.e284. https://doi.org/10.1016/j.devcel.2017.04.009
80. Blengini CS, Nguyen AL, Aboelenain M, Schindler K (2021) Age-dependent integrity of the meiotic spindle assembly checkpoint in females requires Aurora kinase B. Aging Cell 20:e13489. https://doi.org/10.1111/acel.13489
81. Ohkura H (2015) Meiosis: an overview of key differences from mitosis. Cold Spring Harb Perspect Biol 7:a015859. https://doi.org/10.1101/cshperspect.a015859
82. Strom L, Sjogren C (2007) Chromosome segregation and double-strand break repair—a complex connection. Curr Opin Cell Biol 19:344–349. https://doi.org/10.1016/j.ceb.2007.04.003
83. Blackford AN, Stucki M (2020) How cells respond to DNA breaks in mitosis. Trends Biochem Sci 45:321–331. https://doi.org/10.1016/j.tibs.2019.12.010
84. Giunta S, Jackson SP (2011) Give me a break, but not in mitosis: the mitotic DNA damage response marks DNA double-strand breaks with early signaling events. Cell Cycle 10:1215–1221. https://doi.org/10.4161/cc.10.8.15334
85. Xu N et al (2010) Akt/PKB suppresses DNA damage processing and checkpoint activation in late G2. J Cell Biol 190:297–305. https://doi.org/10.1083/jcb.201003004
86. Peterson SE et al (2011) Cdk1 uncouples CtIP-dependent resection and Rad51 filament formation during M-phase double-strand break repair. J Cell Biol 194:705–720. https://doi.org/10.1083/jcb.201103103
87. Giunta S, Belotserkovskaya R, Jackson SP (2010) DNA damage signaling in response to double-strand breaks during mitosis. J Cell Biol 190:197–207. https://doi.org/10.1083/jcb.200911156
88. Nelson G, Buhmann M, von Zglinicki T (2009) DNA damage foci in mitosis are devoid of 53BP1. Cell Cycle 8:379–3383. https://doi.org/10.4161/cc.8.20.9857
89. Leimbacher PA et al (2019) MDC1 interacts with TOPBP1 to maintain chromosomal stability during mitosis. Mol Cell 74:571–583.e578. https://doi.org/10.1016/j.molcel.2019.02.014
90. van Vugt MA et al (2010) A mitotic phosphorylation feedback network connects Cdk1, Plk1, 53BP1, and Chk2 to inactivate the G(2)/M DNA damage checkpoint. PLoS Biol 8:e1000287. https://doi.org/10.1371/journal.pbio.1000287
91. Benada J, Burdoa K, Lidak T, von Morgen P, Macurek L (2015) Polo-like kinase 1 inhibits DNA damage response during mitosis. Cell Cycle 14:219–231. https://doi.org/10.4161/15384101.2014.977067
92. Bakhoum SF, Kabeche L, Compton DA, Powell SN, Bastians H (2017) Mitotic DNA damage response: at the crossroads of structural and numerical cancer chromosome instabilities. Trends Cancer 3:225–234. https://doi.org/10.1016/j.trecan.2017.02.001
93. Orthwein A et al (2014) Mitosis inhibits DNA double-strand break repair to guard against telomere fusions. Science 344:189–193. https://doi.org/10.1126/science.1248024
94. De Marco Zompit M et al (2022) The CIP2A-TOPBP1 complex safeguards chromosomal stability during mitosis. Nat Commun 13:4143. https://doi.org/10.1038/s41467-022-31865-5
95. Leem J, Kim JS, Oh JS (2023) Oocytes can repair DNA damage during meiosis via a microtubule-dependent recruitment of CIP2A-MDC1-TOPBP1 complex from spindle pole to chromosomes. Nucleic Acids Res 51:4899–4913. https://doi.org/10.1093/nar/gkad213
96. Lee C, Leem J, Oh JS (2023) Selective utilization of non-homologous end-joining and homologous recombination for DNA repair during meiotic maturation in mouse oocytes. Cell Prolif 56:e13384. https://doi.org/10.1111/cpr.13384

97. Martin JH, Bromfield EG, Aitken RJ, Lord T, Nixon B (2018) Double strand break DNA repair occurs via non-homologous end-joining in mouse MII oocytes. Sci Rep 8:9685. https://doi.org/10.1038/s41598-018-27892-2
98. Hefferin ML, Tomkinson AE (2005) Mechanism of DNA double-strand break repair by non-homologous end joining. DNA Repair (Amst) 4:639–648. https://doi.org/10.1016/j.dnarep.2004.12.005
99. Taleei R, Girard PM, Sankaranarayanan K, Nikjoo H (2013) The non-homologous end-joining (NHEJ) mathematical model for the repair of double-strand breaks: II. Application to damage induced by ultrasoft X rays and low-energy electrons. Radiat Res 179:540–548. https://doi.org/10.1667/RR3124.1
100. Jasin M, Rothstein R (2013) Repair of strand breaks by homologous recombination. Cold Spring Harb Perspect Biol 5:a012740. https://doi.org/10.1101/cshperspect.a012740
101. Xu Y, Xu D (2020) Repair pathway choice for double-strand breaks. Essays Biochem 64:765–777. https://doi.org/10.1042/EBC20200007
102. Shrivastav M, De Haro LP, Nickoloff JA (2007) Regulation of DNA double-strand break repair pathway choice. Cell Res 18:134–147. https://doi.org/10.1038/cr.2007.111
103. Burma S, Chen BP, Chen DJ (2006) Role of non-homologous end joining (NHEJ) in maintaining genomic integrity. DNA Repair 5:1042–1048. https://doi.org/10.1016/j.dnarep.2006.05.026
104. McVey M, Lee SE (2008) MMEJ repair of double-strand breaks (director's cut): deleted sequences and alternative endings. Trends Genet 24:529–538. https://doi.org/10.1016/j.tig.2008.08.007
105. Lin Y, Lukacsovich T, Waldman AS (1999) Multiple pathways for repair of DNA double-strand breaks in mammalian chromosomes. Mol Cell Biol 19:8353–8360. https://doi.org/10.1128/MCB.19.12.8353
106. Shibata A (2017) Regulation of repair pathway choice at two-ended DNA double-strand breaks. Mutat Res 803–805:51–55. https://doi.org/10.1016/j.mrfmmm.2017.07.011
107. Fei Zhao WK, Kloeber JA, Lou Z (2020) DNA end resection and its role in DNA replication and DSB repair choice in mammalian cells. Exp Mol Med 52:1705–1714. https://doi.org/10.1038/s12276-020-00519-1
108. Li X, Heyer W-D (2008) Homologous recombination in DNA repair and DNA damage tolerance. Cell Res 18:99–113. https://doi.org/10.1038/cr.2008.1
109. Mazón G, Mimitou EP, Symington LS (2010) SnapShot: homologous recombination in DNA double-strand break repair. Cell 142:646, 646.e1. https://doi.org/10.1016/j.cell.2010.08.006
110. Symington LS, Gautier J (2011) Double-Strand break end resection and repair pathway choice. Annu Rev Genet 45:247–271. https://doi.org/10.1146/annurev-genet-110410-132435
111. Dynan WS, Yoo S (1998) Interaction of Ku protein and DNA-dependent protein kinase catalytic subunit with nucleic acids. Nucleic Acids Res 26:1551–1559. https://doi.org/10.1093/nar/26.7.1551
112. Pierce AJ, Hu P, Han M, Ellis N, Jasin M (2001) Ku DNA end-binding protein modulates homologous repair of double-strand breaks in mammalian cells. Genes Dev 15:3237–3242. https://doi.org/10.1101/gad.946401
113. Dion V, Gasser SM (2013) Chromatin movement in the maintenance of genome stability. Cell 152:1355–1364. https://doi.org/10.1016/j.cell.2013.02.010
114. Hauer MH, Gass er SM (2017) Chromatin and nucleosome dynamics in DNA damage and repair. Genes Dev 31:2204–2221. https://doi.org/10.1101/gad.307702.117
115. Dion V, Kalck V, Horigome C, Towbin BD, Gasser SM (2012) Increased mobility of double-strand breaks requires Mec1, Rad9 and the homologous recombination machinery. Nat Cell Biol 14:502–509. https://doi.org/10.1038/ncb2465
116. Mine-Hattab J, Rothstein R (2012) Increased chromosome mobility facilitates homology search during recombination. Nat Cell Biol 14:510–517. https://doi.org/10.1038/ncb2472
117. Foley EA, Kapoor TM (2013) Microtubule attachment and spindle assembly checkpoint signalling at the kinetochore. Nat Rev Mol Cell Biol 14:25–37. https://doi.org/10.1038/nrm3494

118. Matkovic J et al (2022) Kinetochore- and chromosome-driven transition of microtubules into bundles promotes spindle assembly. Nat Commun 13:7307. https://doi.org/10.1038/s41467-022-34957-4
119. Shokrollahi M, Mekhail K (2021) Interphase microtubules in nuclear organization and genome maintenance. Trends Cell Biol 31:721–731. https://doi.org/10.1016/j.tcb.2021.03.014
120. Galletta BJ, Mooren OL, Cooper JA (2010) Actin dynamics and endocytosis in yeast and mammals. Curr Opin Biotechnol 21:604–610. https://doi.org/10.1016/j.copbio.2010.06.006
121. Azoury J et al (2008) Spindle positioning in mouse oocytes relies on a dynamic meshwork of actin filaments. Curr Biol 18:1514–1519. https://doi.org/10.1016/j.cub.2008.08.044
122. Lesca C et al (2005) DNA damage induce gamma-tubulin-RAD51 nuclear complexes in mammalian cells. Oncogene 24:5165–5172. https://doi.org/10.1038/sj.onc.1208723
123. Petsalaki E, Zachos G (2020) DNA damage response proteins regulating mitotic cell division: double agents preserving genome stability. FEBS J 287:1700–1721. https://doi.org/10.1111/febs.15240
124. Roeles J, Tsiavaliaris G (2019) Actin-microtubule interplay coordinates spindle assembly in human oocytes. Nat Commun 10:4651. https://doi.org/10.1038/s41467-019-12674-9
125. Belin BJ, Lee T, Mullins RD (2015) DNA damage induces nuclear actin filament assembly by Formin −2 and spire-(1/2) that promotes efficient DNA repair [corrected]. elife 4:e07735. https://doi.org/10.7554/eLife.07735
126. Brinkley BR (1990) Centromeres and kinetochores: integrated domains on eukaryotic chromosomes. Curr Opin Cell Biol 2:446–452. https://doi.org/10.1016/0955-0674(90)90126-y
127. Cleveland DW, Mao Y, Sullivan KF (2003) Centromeres and kinetochores: from epigenetics to mitotic checkpoint signaling. Cell 112:407–421. https://doi.org/10.1016/s0092-8674(03)00115-6
128. McKinley KL, Cheeseman IM (2016) The molecular basis for centromere identity and function. Nat Rev Mol Cell Biol 17:16–29. https://doi.org/10.1038/nrm.2015.5
129. Garavis M et al (2015) The structure of an endogenous drosophila centromere reveals the prevalence of tandemly repeated sequences able to form i-motifs. Sci Rep 5:13307. https://doi.org/10.1038/srep13307
130. Nassar R, Thompson L, Fouquerel E (2023) Molecular mechanisms protecting centromeres from self-sabotage and implications for cancer therapy. NAR Cancer 5:zcad019. https://doi.org/10.1093/narcan/zcad019
131. Barra V, Fachinetti D (2018) The dark side of centromeres: types, causes and consequences of structural abnormalities implicating centromeric DNA. Nat Commun 9:4340. https://doi.org/10.1038/s41467-018-06545-y
132. Vig BK, Paweletz N (1988) Kinetochores, centromeres, spindles and the induction of aneuploidy. Mutat Res 201:259–269. https://doi.org/10.1016/0027-5107(88)90015-2
133. El Yakoubi W et al (2017) Mps1 kinase-dependent Sgo2 centromere localisation mediates cohesin protection in mouse oocyte meiosis I. Nat Commun 8:694. https://doi.org/10.1038/s41467-017-00774-3
134. Lagirand-Cantaloube J et al (2017) Loss of centromere cohesion in aneuploid human oocytes correlates with decreased kinetochore localization of the sac proteins Bub1 and Bubr1. Sci Rep 7:44001. https://doi.org/10.1038/srep44001
135. Lister LM et al (2010) Age-related meiotic segregation errors in mammalian oocytes are preceded by depletion of cohesin and Sgo2. Curr Biol 20:1511–1521. https://doi.org/10.1016/j.cub.2010.08.023
136. Yin L, Mihajlovic AI, Yang G, FitzHarris G (2023) Kinetochore deterioration concommitant with centromere weakening during aging in mouse oocyte meiosis-I. FASEB J 37:e22922. https://doi.org/10.1096/fj.202300062R
137. Dunkley S, Mogessie B (2023) Actin limits egg aneuploidies associated with female reproductive aging. Sci Adv 9:eadc9161. https://doi.org/10.1126/sciadv.adc9161

138. Chiang T, Duncan FE, Schindler K, Schultz RM, Lampson MA (2010) Evidence that weakened centromere cohesion is a leading cause of age-related aneuploidy in oocytes. Curr Biol 20:1522–1528. https://doi.org/10.1016/j.cub.2010.06.069
139. Greaney J, Wei Z, Homer H (2018) Regulation of chromosome segregation in oocytes and the cellular basis for female meiotic errors. Hum Reprod Update 24:135–161. https://doi.org/10.1093/humupd/dmx035
140. Zielinska AP et al (2019) Meiotic kinetochores fragment into multiple lobes upon Cohesin loss in aging eggs. Curr Biol 29:3749–3765.e3747. https://doi.org/10.1016/j.cub.2019.09.006
141. Nikalayevich E et al (2022) Aurora B/C-dependent phosphorylation promotes Rec8 cleavage in mammalian oocytes. Curr Biol 32:2281–2290.e2284. https://doi.org/10.1016/j.cub.2022.03.041
142. Tachibana-Konwalski K et al (2010) Rec8-containing cohesin maintains bivalents without turnover during the growing phase of mouse oocytes. Genes Dev 24:2505–2516. https://doi.org/10.1101/gad.605910
143. Mikwar M, MacFarlane AJ, Marchetti F (2020) Mechanisms of oocyte aneuploidy associated with advanced maternal age. Mutat Res Rev Mutat Res 785:108320. https://doi.org/10.1016/j.mrrev.2020.108320
144. de Sousa Lopes S, Roelen BA (2010) An overview on the diversity of cellular organelles during the germ cell cycle. Histol Histopathol 25:267–276. https://doi.org/10.14670/HH-25.267
145. van der Reest J, Nardini Cecchino G, Haigis MC, Kordowitzki P (2021) Mitochondria: their relevance during oocyte ageing. Ageing Res Rev 70:101378. https://doi.org/10.1016/j.arr.2021.101378
146. Yang Z, Wei ML, Dong XY (2021) Effects of Yu Linzhu on ovarian function and oocyte mitochondria in natural aging mice. Aging (Albany NY) 13:23328–23337. https://doi.org/10.18632/aging.203626
147. Zhang D, Keilty D, Zhang ZF, Chian RC (2017) Mitochondria in oocyte aging: current understanding. Facts Views Vis Obgyn 9:29–38
148. Menezo Y, Clement P, Dale B, Elder K (2022) Modulating oxidative stress and epigenetic homeostasis in preimplantation IVF embryos. Zygote 30:149–158. https://doi.org/10.1017/S0967199421000356
149. Matthews ML et al (2012) Cancer, fertility preservation, and future pregnancy: a comprehensive review. Obstet Gynecol Int 2012:953937. https://doi.org/10.1155/2012/953937
150. Kerr JB et al (2012) Cisplatin-induced primordial follicle oocyte killing and loss of fertility are not prevented by imatinib. Nat Med 18:1170–1172.; ; author reply 1172–1174. https://doi.org/10.1038/nm.2889
151. Ojo AS, Lipscombe C, Araoye MO, Akinyemi O (2022) Global uptake of fertility preservation by women undergoing cancer treatment: an unmet need in low to high-income countries. Cancer Epidemiol 79:102189. https://doi.org/10.1016/j.canep.2022.102189
152. Liu N et al (2023) L-Proline improves the cytoplasmic maturation of mouse oocyte by regulating glutathione-related redox homeostasis. Theriogenology 195:159–167. https://doi.org/10.1016/j.theriogenology.2022.10.023
153. Lord T, Nixon B, Jones KT, Aitken RJ (2013) Melatonin prevents postovulatory oocyte aging in the mouse and extends the window for optimal fertilization in vitro. Biol Reprod 88:67. https://doi.org/10.1095/biolreprod.112.106450
154. Nikmard F et al (2022) The boosting effects of melatonin on the expression of related genes to oocyte maturation and antioxidant pathways: a polycystic ovary syndrome- mouse model. J Ovarian Res 15:11. https://doi.org/10.1186/s13048-022-00946-w
155. Zhang H et al (2022) Melatonin improves the quality of maternally aged oocytes by maintaining intercellular communication and antioxidant metabolite supply. Redox Biol 49:102215. https://doi.org/10.1016/j.redox.2021.102215
156. Asadi E, Jahanshahi M, Golalipour MJ (2012) Effect of vitamin e on oocytes apoptosis in nicotine-treated mice. Iran J Basic Med Sci 15:880–884

157. Liu MJ et al (2018) Resveratrol improves in vitro maturation of oocytes in aged mice and humans. Fertil Steril 109:900–907. https://doi.org/10.1016/j.fertnstert.2018.01.020
158. Romero S et al (2020) Coenzyme Q10 improves the in vitro maturation of oocytes exposed to the intrafollicular environment of patients on fertility treatment. JBRA Assist Reprod 24:283–288. https://doi.org/10.5935/1518-0557.20200003
159. Leem J, Bai GY, Kim JS, Oh JS (2021) Increased WIP1 expression with aging suppresses the capacity of oocytes to respond to and repair DNA damage. Front Cell Dev Biol 9:810928. https://doi.org/10.3389/fcell.2021.810928

Canonical and Non-canonical Roles of the Nucleolus in Relation to Nucleolar Function in Oocyte Meiosis

Benjamin Wetherall and Suzanne Madgwick

Abstract The nucleolus is the most prominent nuclear organelle in almost all cells including mammalian oocytes and embryos. It is primarily responsible for ribosome biogenesis. However, non-canonical functions have been described and the list of potential nucleolar roles is expanding. Given that in oocytes ribosome-biogenesis and gene transcription shut down, non-canonical roles of the nucleolus are of particular interest in this cell type. In addition, the mature oocyte provides an ideal model in which to study these roles separable from ribosome biogenesis. However, the nucleolus is still relatively mysterious in these cells. Here we first discuss roles of the nucleolus in mitosis where they are relevant to the growing oocyte. We then focus on the role of the nucleolus as ribosome-biogenesis comes to a halt and transcription is silenced at the end of prophase I. Transcription remains silent in oocytes until activation of the embryonic genome, during which time oocytes and early embryos rely on a stockpile of maternal transcripts and proteins. Traditional literature suggested that the 'silent' nucleolus is not more than a passive dense mass, holding nucleolar proteins until they are again required to rebuild an organelle capable of ribosome biogenesis. However, the nucleolus now seems far from obsolete. Chromatin dispersed throughout the nucleoplasm undergoes remarkable rearrangements during the transitions between active-to-silenced and silenced-to-active transcription, appearing to use the nucleolus as a scaffold. Furthermore, a series of enucleolation experiments and other studies indicate the existence of a novel and essential role for the oocyte nucleolus following fertilisation. Importantly, a role that cannot be compensated for by a somatic cell nucleolus.

Keywords Nucleolus · Meiosis · Oocyte · Germinal vesicle breakdown (GVBD) · Chromatin · Embryogenesis

---

B. Wetherall · S. Madgwick (✉)
Biosciences Institute, Faculty of Medical Sciences, Newcastle University, Newcastle upon Tyne, UK
e-mail: Ben.Wetherall@newcastle.ac.uk; Suzanne.Madgwick@newcastle.ac.uk

B. Mogessie (ed.), *Cellular Architecture and Dynamics in Female Meiosis*,
https://doi.org/10.1007/978-3-031-97173-0_6

Introduction

In humans and many model organisms such as the mouse, the nucleolus is by far the most visually remarkable feature of an oocyte. Often so prominent it could be mistaken for the nucleus. In a prophase-arrested oocyte the (usually) single, well-defined nucleolar sphere is clearly contrasted against the nucleoplasm. In comparison the nuclear envelope is rarely visible, the boundary delineated instead by a much subtler change in texture marking the transition from nucleoplasm to cytoplasm (Fig. 1ai,ii).

In the field of reproductive health, the oocyte is for obvious reasons a subject of close study. They are routinely collected, fertilised, and frozen or reimplanted as part of a range of therapeutic approaches to address fertility challenges. But not all oocytes are equal. While an oocyte can reach a stage that appears by light microscopy to be fully competent, that is often not the case. The reason can essentially be reduced to two factors, either the oocyte has been collected prematurely and is not 'ready' yet, or the oocyte is of poor quality (some aspect of development occurred aberrantly and cannot be corrected regardless of more time in the ovary). Consequently, a major focus of oocyte research is to identify 'good' and 'mature' oocytes, while understanding the failings that lead to 'poor quality' oocytes.

Current in vitro fertilisation techniques ideally involve the collection and fertilisation of multiple metaphase II (MII) stage oocytes (also referred to as eggs). This is partly because of the challenges a highly artificial process poses on handling these few highly valuable cells. But importantly, also partly because of the unavoidable range in the quality of the oocytes or eggs collected. In the best-case scenario, a clinician has choice. However, a factor that limits the ability to select the most competent choice is the availability of relevant information. In addition, if we can further characterise the developmental deficits that produce poor quality oocytes, we increase the possibility that treatment regimens can be developed to rescue some aberrant oocytes where there are no naturally competent options.

A great deal of research has been invested in identifying exactly what makes a competent oocyte. That is, an oocyte capable of maturing to MII, being fertilised, and producing a viable embryo. In this context, the nucleolus of the oocyte is frequently discussed given the strength of the correlation between morphological changes surrounding the oocyte's nucleolus and the oocyte's developmental stage or competency.

Briefly, a growing 'immature' oocyte is transcriptionally active and synthesising ribosomes. However, prior to ovulation, transcription and ribosome biogenesis shut down. This is one of, if not *the*, last distinguishable transition an oocyte goes through before becoming fully competent. As the oocyte transitions from a state of transcriptionally active to transcriptionally silent, the chromatin in the nucleus reorganises from an apparently cloudy dispersion throughout the nucleoplasm (Non-Surround-Nucleolus; NSN, Fig. 1bi) to a densely packed sphere around the nucleolus (Surround-Nucleolus; SN, Fig. 1bii). Therefore, in situations where chromatin may be imaged (for example by using intercalating fluorescent dyes and fluorescence microscopy), the NSN or SN status of the oocyte presents an incredibly useful and visually distinct morphology with which to gauge the developmental

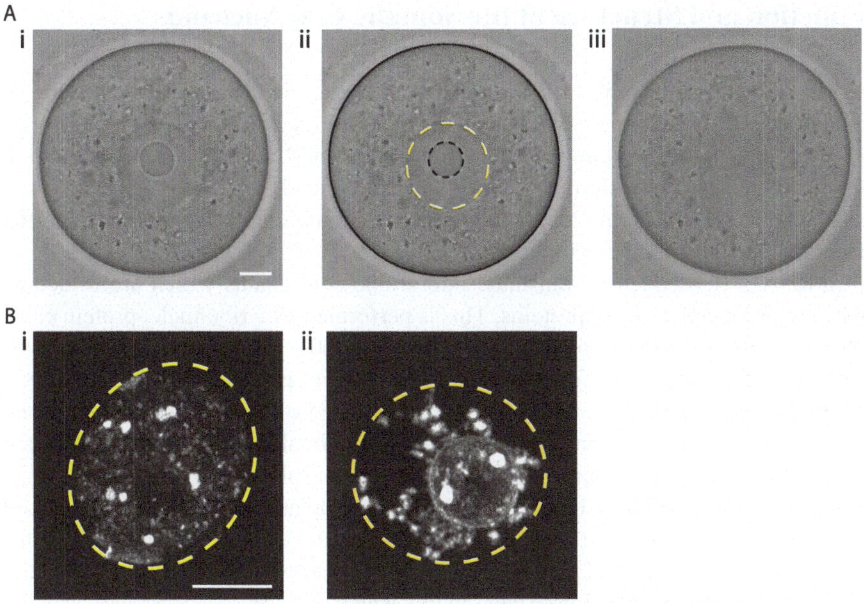

Fig. 1 Observation of mouse oocyte intracellular compartments by transmitted light. (**a, i**) Transmitted-light image of a live and competent-presenting GV-stage mouse oocyte captured using the transmitted-light detector of a confocal microscope (40× magnification oil-immersion lens). Focal plane thickness is approximately 3 μm, the scale bar (white) represents 10 μm. (**a, ii**) Duplicate image annotated. A solid black line, dashed yellow line and dashed black line mark the oocytes cell membrane, nuclear envelope, and nucleolus surface respectively. (**a, iii**) The oocyte presented in (i) 90 min later following GVBD. The nuclear envelope has broken down and nucleolar material has decondensed. The oocyte is now in early prometaphase. (**b**) Representative images showing chromatin organisation in prophase mouse oocytes at the NSN-stage (i) and the SN-stage (ii) by super-resolution fluorescent microscopy. DNA was stained with SiR-DNA (Spirochrome), images are maximum-intensity projections of approximately 60 0.27 μm slices. Dashed yellow lines mark the position of the nuclear envelope. Scale bar = 10 μm. These images are examples generated through studies associated with a MRC Career Development Award to S.M. [MR/T010789/1]. Oocytes were collected from a 10-week-old CD1 mouse (Charles River) handled in accordance with ethics approved by the UK Home Office Animals Scientific Procedures Act 1986. Oocytes were collected from ovaries punctured with a sterile needle and stripped of their cumulus cells mechanically using a pipette. Handling, imaging, and culture took place at 37 °C in medium M2 (Sigma) supplemented with 30 nM 3-isobutyl-1-methylxanthine (Sigma) to maintain prophase I arrest

stage of the oocyte. This transition is extremely well documented, but largely just as a tool or marker to separate oocytes into different stages of development for the study of other organelles, pathways, and proteins. The nucleolus itself remains a mysterious organelle in oocytes.

In this chapter, we will first introduce the role of the mitotic nucleolus before exploring the 'atypical' nucleolus and chromatin structure in mammalian oocytes, examining literature surrounding the NSN-SN transition, and possible non-canonical roles of the nucleolus in this unique cell type.

Function and Structure of the Somatic Cell Nucleolus

Other than a small number of specialised genes carried within mitochondria, a eukaryotic cell's genomic DNA is contained within the nucleus. The vast majority of genes encoded by genomic DNA are transcribed by RNA Polymerase II in situ in the nucleoplasm. They are then capped at 5′ and 3′ ends, and intronic (along with some exonic) DNA is removed by splicing. These mature mRNAs are then exported from the nucleus into the cytoplasm [1].

mRNA is subsequently translated into amino acid chains which are folded as they are produced, to form proteins. This is performed by a ribonucleoprotein complex, the ribosome. Ribosomes comprise a large 60S subunit and a smaller 40S subunit, both of which are formed by a combination of protein and RNA. The RNA content of each subunit can be further divided into 28S and 5.8S (components of the 60S), and 18S and 5S (components of the 40S). Notably 28S, 5.8S and 18S are initially transcribed together as 47S precursor rRNA. Ribosomes exist either free in the cytosol, or membrane-bound in the rough endoplasmic reticulum (ER). Each location translates different mRNAs depending on the environment best suited for the stable formation of that particular protein [2]. However, not all genes within the nucleus are transcribed and translated in this way.

Regions of DNA containing the genes which encode structural RNAs of the ribosome itself (rRNAs) are referred to as ribosomal DNA (rDNA). rDNA is highly repetitive. In humans, rDNA comprises approximately 300 repeats each of the 5S and 47S loci. The 5S locus is located on chromosome 1, while the 47S loci are spread across chromosomes 13, 14, 15, 21, and 22 in Nucleolar Organiser Regions (NORs). The extensive repetition is thought to facilitate the large demand for rRNA in cells producing new ribosomes [3–5].

In somatic cell nuclei, dense protein and RNA-containing sub-compartments form around NORs in telophase, merging to form a small number of large bodies occluding all DNA other than rDNA [6]. These bodies are nucleoli, responsible for the early stages of ribosome biogenesis. The formation of the nucleolar structure is intrinsically tied to the resumption of RNA Polymerase I (RNA PolI) transcription after mitosis. RNA PolI is the polymerase responsible for the transcription of the 47S pre-rRNA at NORs, while RNA Polymerase III (RNA PolIII) is responsible for (among other things) the transcription of the 5S subunit. Nucleoli reduced in both number and size have been observed when NOR reactivation is supressed [7].

The inner-most structures of the nucleolus are Fibrillar Centres (FCs), surrounded by Dense Fibrillar Components (DFCs), named after their appearance when initially identified by electron microscopy [8]. FC-DFC bilayers form around two to three sites of rDNA transcription each, with multiple FC-DFC structures (several dozen in human cells) suspended within a third outer nucleolus compartment, the Granular Component (GC) [9, 10] (see Fig. 2 for detail). RNA PolI transcription of rRNA takes place within the FC and through the FC-DFC boundary, while pre-rRNA processing factors such as fibrillarin form multiple clusters within the DFC. Importantly, transcription itself appears to be the biochemical factor driving the establishment and maintenance of these compartments [10, 11], since the

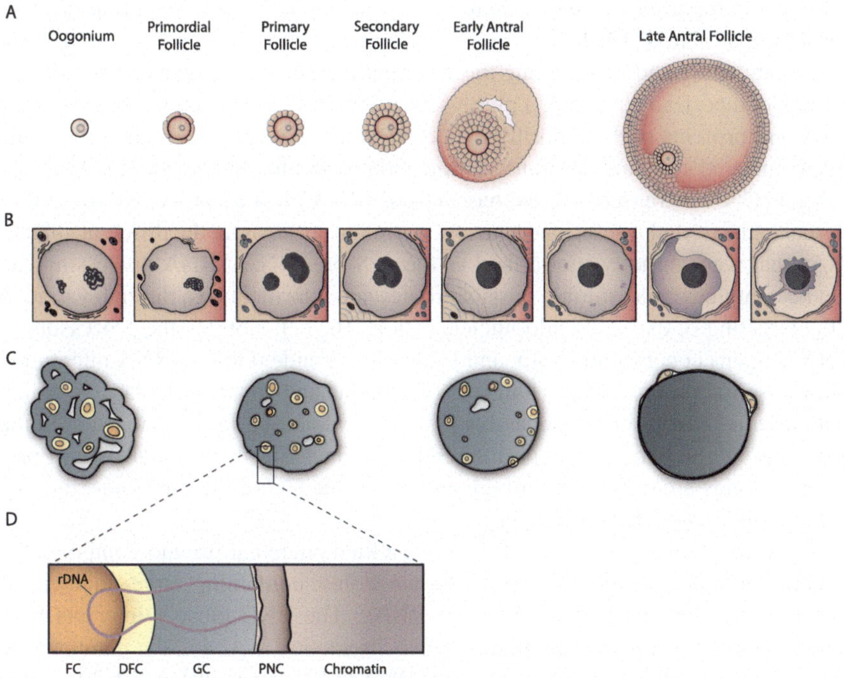

Fig. 2 Nucleolar reorganisation through oocyte maturation. (**a**) Schematic of ovarian follicle development. Oocytes originate from oogonia, produced by early mitotic divisions in foetal development. Oogonia divide asymmetrically to produce an oocyte, which upon surrounding by flat granulosa cells represents the first follicular stage, a primordial follicle. As the oocyte grows through the primary and secondary follicle stages the surrounding granulosa cells become cuboidal in shape and greater in number. The oocyte develops its encapsulating Zona Pellucida. In early antral follicles, a fluid-filled cavity called the antrum begins to form adjacent to the oocyte. Finally, as the follicle approaches the point of ovulation, the antrum greatly expands forming a late-antral follicle. (**b**) As the oocyte develops during folliculogenesis, the appearance of the nucleolus changes. In oogonia and primordial follicles the nucleolus is highly reticulated, and typically 1–3 in number. By the time the oocyte reaches the antral stages, its nucleoli have largely coalesced into one body (although two nucleoli at this point are not uncommon), taking on a denser and rounder shape. This is thought to be driven by cytoplasmic forces. In the late-antral stage chromatin that has previously been decondensed and dispersed through the nucleoplasm (save for some denser foci of heterochromatin) begins to condense around the nucleolus. The majority of DNA is packaged tightly around the nucleolus, although patches of condensed DNA remain in association with the nuclear membrane. This marks a fully developed oocyte ready for ovulation. (**c**) The internal structure of the nucleolus changes as the oocyte develops. In oogonia and primordial follicles, the nucleolus is similar to that of a somatic cell. FC-DFC units are dispersed throughout a reticulated GC. In primordial follicles, the nucleolus appears denser and rounder although still somewhat vacuolar. By the early antral follicle stage, FC-DFC units have begun to reposition, clustering at the cortex of the nucleolus. In late-antral follicle oocytes, the nucleolus has taken on a highly dense, homogenous structure. Residual FC-DFC bodies are sometimes, but not consistently, observed on the surface of the nucleolus. This massive structural change marks the shutdown of ribosome-biogenesis. (**d**) Cross-sectional diagram of the nucleolar compartments. rDNA from the perinucleolar compartment (PNC) extends through the granular component (GC) and into the fibrillar center (FC)/dense fibrillar component (DFC) interface, where rRNA is transcribed

multilayer compartmentalised structure of the nucleolus is lost when transcription of rDNA is inhibited [12, 13].

In establishing and maintaining these compartments, the relative contribution of two model theories is debated; a phase-separation-like interaction between nascent rRNA and proteins such as fibrillarin and NPM1, versus physical tethering of the large connected structures of transcription and processing machinery [11, 14, 15].

After rDNA is transcribed, the nascent pre-rRNA must be processed to remove internal and external transcribed spacers, as well as to separate the different subunit RNAs that were initially transcribed as the 47S pre-rRNA. rRNA processing can then be completed via multiple possible routes and intermediates, through variation in the targeting of exo-and-endo-nucleic action [16, 17]. Notably the processing of rRNA happens concurrently with, and is largely dependent upon, rRNA integration with ribosomal proteins [18]. Hundreds of proteins and small nucleolar RNAs that facilitate this process in the nucleolus have been identified in somatic cells. In the final ribosome biogenesis role performed by the nucleolus, ribosomal subunit precursor particles are trafficked through and out of the GC of the nucleolus, into the nucleoplasm and out to the cytoplasm.

The perinucleolar compartment (PNC) is a final nucleolar pseudo-component; a dynamic structure on the surface of the nucleolus, containing large numbers of RNA-binding proteins and RNA PolIII RNAs. The PNC closely interacts with nucleolus-peripheral heterochromatic and euchromatic DNA, and the maintenance of this compartment is dependent on active RNA PolIII transcription [19, 20]. Some of this perinucleolar chromatin is rDNA, strands of which penetrate the nucleolus to be transcribed by RNA PolI. However, much of this DNA is in inactive chromosomal regions and coding regions for non-ribosomal genes, including developmental and olfactory genes [21].

In addition to a highly active role in ribosome-biogenesis (in a typical human cell ribosomal proteins constitute 4–6% of total protein mass; see [22] for review), given that the nucleolus is so protein dense and lacks a distinct membrane, a longstanding theory is that it can be used by cells to sequester or concentrate proteins as part of their regulation. Indeed, the first detailed nucleolar proteome identified a large fraction of proteins that are not strictly associated with ribosome biogenesis [23, 24]. Some of these additional functions are discussed below.

Changes in Nucleolus Structure in Response to Changing Environments

The structure of the nucleolus is dynamic and liable to change, particularly in response to stress.

Common cellular stressors such as nutrient deprivation or DNA damage trigger different responses dependent on the magnitude of the challenge, ranging from temporary cell-cycle arrest to apoptosis. Given that ribosome biogenesis is a highly energy demanding process tightly coupled to cell growth, it is particularly important that this activity responds to environmental change. Therefore, regardless of whether the cell experiences challenge through reduced resources (for example, nutrient depravation), or nucleolar demand is itself reduced (for example, through damage-induced cell-cycle arrest), the modulation of ribosome biogenesis is crucial for maintaining cellular homeostasis [25]. Importantly, changes in the rate of ribosome biogenesis are coupled with nucleolar structural changes.

Examples of nucleolar structure modulators include genotoxic stressors such as UV radiation-induced DNA lesions, topoisomerase poisons, and rRNA transcription and processing machinery inhibitors. All affect the ability of the nucleolus to transcribe rDNA, and induce a stark structural reorganisation of the nucleolus [26–28]. Here, FC and GC components collapse in a process known as nucleolar segregation, and many nucleolar proteins translocate to new structures on the surface of the nucleoli called nucleolar caps [29]. Indeed, the structure of the nucleolus appears unable to be maintained without active transcription. This is consistent with a model where the different nucleolar compartments are defined by and dependant on rRNA transcription, processing, and assembly into ribosomal subunits. Importantly, if a cell goes on to survive the insult that induced the stress response, nucleolar structure is restored concurrent with ribosome biogenesis resumption [30].

A second example of changes in nucleolar structure includes experiments in cells challenged with reduced pH environments or heat shock. In this situation, in human cells, a non-coding RNA is transcribed by RNA PolI from an intergenic spacer region in rDNA. This particular RNA captures and immobilises proteins that hold a 'nucleolar detention sequence' [31]. Captured proteins are then detectable in discrete 'detention centre' regions of structurally remodelled nucleoli as transcription stalls [32]. By this storage method, the cell can rapidly, but reversibly, inhibit specific pathways in response to stress. Alternatively (or in addition), it has been suggested this response could actually facilitate efficient biochemical interactions by bringing proteins into close proximity [33]. Notably, it has been demonstrated that the structural change forming 'detention centres' is not simply a result of transcriptional silencing [32]. However, given that transcriptional silencing is the next step in this stress response, the co-dependency of transcriptional activity and nucleolar structure maintenance makes it difficult to untangle events that drive one response over another. In a similar process, an RNA PolI transcript-driven nucleolar sequestration of proteins has also been observed in *Drosophila*, where curiously one of the proteins sequestered is itself an rDNA transcriptional repressor [34].

It is important to consider these known 'non-biogenesis' roles of the somatic cell nucleolus when thinking about the role of the nucleolus in oocytes that spend significant time in a transcriptionally inactive state.

Additional Functions of the Somatic Nucleolus

The literature reports further 'other roles' of the nucleolus, particularly in relation to stress responses. However, we will not discuss these in this chapter. Many of those roles involve the nucleolus shutting down ribosome biogenesis and it could therefore be argued that this more modulation of the canonical role of the nucleolus in response to different challenging stimuli (as opposed to new or different functions of the nucleolus).

Importantly though, the role of the somatic cell nucleolus does extend beyond ribosome biogenesis and there are several examples of the nucleolus functioning as a reservoir or storage compartment outside of a stress response. This organelle is well suited to this role; highly dynamic in protein content with no physical barrier preventing the diffusion or shuttling of proteins in either direction [35]. Indeed, many proteins hold a nucleolar localisation sequence (NoLS), a positively charged motif that is known to facilitate interaction with negatively charged components of the nucleolus [36, 37].

Some examples of protein sequestration not linked to stress responses include Yeast Cdc14 (a mitotic exit promoter) sequestered in the nucleolus until anaphase [38]; a pool of Human PP1gamma (thought to localise to chromatin to aid condensation in anaphase) stored in the nucleolus in interphase [39]; and RNP telomerase sequestered until S phase [40, 41]. In addition, some nuclear proteins have been found to move into the nucleolus if they have been misfolded. Here their aggregation is prevented and they can be re-folded correctly [42]. Certainly, in somatic cells, it seems clear that the nucleolus has influence over its environment by the timely removal and release of cell-cycle licensing factors (as above), and by removal of damaged proteins. It is therefore possible that these and similar roles may be equally (if not more) important in the oocyte. The oocyte is an incredibly long-lived cell and for a significant period of oocyte maturation and early embryogenesis transcription is shut down. This unique situation also makes the oocytes a useful tool with which to study the role of the nucleolus. Non-ribosome biogenesis nucleolar roles can be studied independently of transcription, but without the need for chemical inhibitors. However, the nucleolus is understudied in oocyte meiosis and structurally perhaps even more dynamic than those in somatic cells. We will discuss this after a brief introduction to the generation of oocytes from their mitotic germline mother cells.

The Formation of an Oocyte and the Changing Role of the Nucleolus

The formation of oocytes begins in the foetal ovary with the mitotic proliferation of vast numbers of germline stem cells (millions in humans, thousands in rodents). Germline stem cells then enter meiosis, ending their stemness, to produce a highly specialised cell population: primordial oocytes. Primordial oocytes are unable to replicate themselves, but instead aim to divide to produce a competent egg with a unique assortment of paternal and maternal genes.

On entry into meiosis, primordial oocytes pass through three stages of meiosis I prophase (leptotene, zygotene, and pachytene) before arresting at the last stage;

diplotene. At diplotene, each oocyte becomes surrounded by a monolayer of somatic cells (pre-granulosa cells), and together this structure is termed the primordial follicle. These events all take place before birth, and the oocyte spends the rest of its life following the fate of the follicle. Primordial follicles are non-growing and provide a finite pool from which all preovulatory follicles will eventually be derived.

During a female's reproductive lifespan, small numbers of primordial follicles are activated in waves to exit the ovarian reserve and begin growing; metabolic activity increases, and the surrounding pre-granulosa cells proliferate and differentiate. Indeed, the oocyte increases in volume over 100-fold while the primordial follicle evolves through secondary-, small antral-, large antral-, and preovulatory follicular stages. This process takes approximately 4 months in humans and culminates in a much shorter phase of meiosis resumption and ovulation (24 h). However, it should be noted that <0.1% of primordial oocytes reach this stage. Most oocytes die either before they are surrounded by pre-granulosa cells (attrition) or later in development as follicles degenerate (atresia). See Telfer et al. [43] for review.

Acquisition of meiotic competence is not only reliant on changes in oocyte growth and follicular development but on germinal vesicle (GV) morphology, chromatin organisation, and related to this, transcriptional activity. The GV of the oocyte is akin to the nucleus of a mitotic cell, containing the cells chromatin and nucleolus. The nucleolus however is atypical for several reasons, often instead referred to as a nucleolus-like body (NLB) or nucleolus precursor body (NPB). Note that like in mitosis, the nucleolar material frequently exists as multiple bodies at earlier stages of the oocyte development or growth. However unlike in mitosis, nucleoli usually merge in oocytes towards the end of prophase. Indeed, studies of the ultrastructure of the nucleus in mammalian oogonia and primordial oocytes typically show 1–3 reticulated nucleoli surrounded by uncondensed chromatin [44–48]. While in contrast, most late GV stage oocytes have a single, highly compacted, rounded nucleolus. The merging of many bodies is thought to be directed by cytoplasmic forces that reorganise nuclear condensates such that multiple nucleoli collide and coalesce [49, 50]. Thereby, once prepared for ovulation, GV oocytes are largely (but not exclusively) characterised by a single smoothed nucleolus which is easily identifiable as a prominent dark dot by light microscopy in *transparent* oocytes. The dynamics of merging nucleoli can be seen exaggerated in enucleolated oocytes. In this setting, re-injected nucleolar material is rapidly translocated into the nucleus, where first many miniature NLBs/NPBs form, but then fuse and form a single large nucleolus [51].

Importantly, it must be acknowledged that while oocyte nucleoli are clearly identifiable in humans, mice, and rats, they are not readily visible in all mammals. This includes many livestock species, which have dark or opaque oocytes. Cattle particularly are an extremely valuable food source, but with high rates of infertility [52, 53]. The inability to track nucleoli morphology as a non-invasive method of oocyte and embryo health determination hampers the selection of viable cells this in these species. Appropriate criteria for selecting bovine oocytes and embryos are currently lacking, and this is concerning given our increasing reliance on IVF and embryo transfer for livestock reproduction [54]. The reason for this species-specific difference is unknown, for review see [55].

Unfortunately, due to the difficulty of obtaining sufficient nucleolar material of useful purity, the oocyte NLB/NPB proteome is far from perfect. A significant difficulty is that while nucleoli can be removed, they readily dissolve when kept in manipulation medium making them difficult to handle [56]. In addition, they likely contain many different proteins in very low relative abundance. This dense structure is also largely impenetrable to antibodies by normal protocols. Components identified so far are well summarised by Fulka et al. [57].

Throughout the oocytes' growth phase nucleolar material is highly active, producing large numbers of ribosomes to support rapid growth. Accordingly, their internal structures resemble that of the nucleoli in mitotic cells, containing three distinct compartments—the fibrillar centre, dense fibrillar component, and granular component (see Fig. 2d) [58]. At this time, not surprisingly, the major RNA species found in oocytes is rRNA [59].

However, in completing oocyte growth, and as nucleoli centre and merge in developmentally advanced oocytes, ribosome biogenesis shuts down. This facilitates transcriptional silencing in the oocyte, and results in the disassembly of the fine ultrastructure of an active nucleolus, now a single dense mass with only a few remaining ribosomal particles present at the surface [60–62]. Alongside this change in nucleoli structure and transcriptional silencing [63, 64], oocyte chromatin morphology alters dramatically as nucleoli become enclosed in condensing heterochromatin. This reorganisation is distinct from chromatin morphology in somatic cells and takes place as the oocyte is prepared for ovulation, prior to GV breakdown (GVBD; akin to nuclear envelope breakdown in mitosis). Following GVBD, nucleolar material disperses (Fig. 1aiii), and chromatin arranges and segregates through meiosis I and II divisions. Nucleoli later reform in the one-cell embryo following successful fertilisation.

Changes in Chromatin Morphology and the Unknown Role of the Nucleolus in Oocytes and Embryos

In the actively transcribing and rapidly growing oocyte, chromatin within the nucleus is uncondensed and dispersed across the GV. In this configuration, chromatin morphology is often described as 'non-surrounded nucleolus': NSN. Then, as the oocyte is prepared for ovulation, chromatin is reorganised and envelops the dense fibrillar mass of the nucleolus with morphology described as 'surrounded nucleolus': SN (Fig. 1b). It should be noted though that the NSN-SN terminology is not always used, and many other descriptors have been used to classify chromatin arrangements in a wide variety of different species [65]. This can make the literature more difficult to reconcile. A second point of confusion is introduced by the terminology used to describe the oocyte at this stage; sometimes referred to as a 'mature

GV stage oocyte', but also described as a 'mature oocyte' (a term more frequently used to describe fertilisation competency and the later MII arrest stage). However, what is clear and widely accepted is that across species, chromatin contained in the GV achieves a high degree of condensation and compaction before the resumption of meiosis in healthy oocytes.

The change in chromatin distribution and associated transcriptional silencing is also accompanied by a collapse of nucleolar ultrastructure and a change in the biochemical composition of the nucleolus. While both NSN and SN oocyte nucleoli contain RNA and nuclear proteins, SN nucleoli have been described as impoverished for 18S and 28S ribosomal RNAs. Without a ribosome biogenesis role, this led to the suggestion that nucleoli of fully-grown mammalian oocytes serve only to store nucleolar proteins, but not rRNA which will only become important again when ribosome biogenesis is reinstated in the early embryo [66].

The theory above is supported by experiments that enucleolated GV stage oocytes, first in pig [67] and then in mouse [68]. Fulka et al. [67] reported that nucleoli can be removed relatively simply from oocytes. Remarkably, enucleolated oocytes matured to MII with the same frequency as controls. Nucleoli are therefore not essential for the onset of meiosis or oocyte maturation. Indeed, when enucleolated oocytes were then fertilised, pronuclei of the resultant embryos did not contain visible nucleoli (no new nucleolar material was synthesised), but cells still managed one or two divisions before development arrested. Critically, in this same setting reinjection of nucleolar material after fertilisation rescued some developmental potential. It is worth noting that following oocyte enucleolation, the paternally-derived pronucleus also lacked a nucleolus. However, this was similarly rescued by reintroduction of nucleolar material to the zygote, an excellent demonstration that nucleolar material in the male pronucleus is maternally derived [68].

Together these experiments support the idea that the oocyte nucleolus is 'simply a material repository'. However, this is more complicated. First, somatic cell or embryonic stem cell (ESC) nucleoli cannot substitute for the original oocyte nucleolar material in rescue experiments [68]. In these investigations no nucleoli formed despite ESCs being pluripotent and ordinarily extremely efficient as donor cells in somatic cell nuclear transfer [69, 70]: ESC-reconstructed embryos were unable to develop without oocyte-specific nucleolar material [68].

Second, Ogushi and Saitou [71] studied the potential of enucleolated oocytes in more detail. Enucleolated oocytes were reinjected with an oocyte nucleolus at MII stage, followed by fertilisation. These cells formed pronuclei with nucleoli and developed to full term. However, in contrast, where the nucleolus was reinjected at the one-cell pronucleus (PN or zygote) stage, while most still divided and formed two-cell embryos, the rate of blastocyst formation was substantially reduced. If the nucleolus was only necessary again once ribosome biogenesis reinitiates, then theoretically the timing of this reintroduction shouldn't matter as long as it takes place before genome activation (two to four cell in mouse embryos). Interestingly, zygotes without nucleoli had disordered chromatin, not only in the female pronucleus but in the male pronucleus. These experiments suggest a function of the nucleolus in oocytes beyond our current understanding, and perhaps a role in early embryogenesis. The existence of an oocyte-specific nucleolar protein role unrelated to ribosome biogenesis is plausible [56], but not yet fully identified. We will revisit enucleolation experiments further down.

One aspect of oocyte nucleolus and chromatin morphology that is widely conserved in mammals is the link between NSN and SN conformations and transcriptional silencing. The compacted layer of chromatin surrounding the nucleolus is often referred to as the karyosphere or karyosome, and morphology has been noted and studied for some time. The difference between SN and NSN nucleolar chromatin state was first evidenced in mouse in 1995 [72], and then in human in 2002 [73]. These morphology changes were then correlated with differential transcriptional activity [74–76]. SN-type oocytes are reported to be silent while NSN-type oocytes are reported to be transcriptionally active and synthesisze all classes of RNA [74, 77, 78]. Longo et al. [62] further cemented the link between karyosphere morphology and transcriptional activity with an EM study of nucleolar material in the developing oocyte. Demonstrating the transition of NORs (rDNA transcription sites) from being present throughout the nucleolus initially, but more exclusively towards the edge of the nucleolus as the oocyte prepares to resume meiosis. NSN oocytes have NORs on the surface of the nucleolus, while in SN oocytes NOR staining is absent. However, the link between the NSN–SN transition and transcriptional silencing is perhaps not as tightly coupled as first thought. This will be discussed in the next section.

NSN and SN Karyosphere Morphology Informs Competency

Successful development across the oocyte-to-embryo transition requires the establishment of a full cohort of maternal mRNAs ahead of transcriptional silencing [79]. Transcription is then silenced globally and does not significantly resume until early embryo stages. During this period the oocyte relies on its reservoir of stored maternal mRNA. GV oocytes that fail to undergo global transcriptional silencing have significantly reduced rates of oocyte maturation, fertilisation and embryo development [74, 78, 80, 81]. Some factors contributing to global transcriptional silencing have been identified [80, 82–86], and both transcriptional silencing and developmental competence are closely associated with extensive chromatin modification (including changes in histone methylation and acetylation, and DNA methylation) [80, 87–89].

Given that the NSN–SN transition occurs at the end of the growth stage, concomitant with transcriptional silencing, NSN and SN morphology strongly indicates not only maturity but oocyte competency. Specifically, while oocytes with either chromatin arrangement are able to resume meiosis, they do not have comparable developmental potential [81, 90–92]; embryos derived from NSN oocytes often halt development at the two-cell stage, while embryos from SN oocytes are capable of developing to the blastocyst stage and full term. Whether nucleoli play an *active* role in this transition is unknown, but their function as a scaffold for DNA is clear.

(It is also perhaps worth noting that the developmental arrest stage in NSN fertilised oocytes is the same as the arrest point in fertilised enucleolated oocytes.)

While transcriptional silencing and chromatin condensation are closely linked temporally (NSN-SN transitioning takes place as transcription gradually ceases) [93], these two events are in fact separable and appear to be driven by distinct pathways. For example, an H3K4 methyltransferase, MLL2, controls most of the promoter-specific chromatin modifications (H3K4me3) during oogenesis and early development. In mice with a conditional knockout and hypomorph model of MLL2, oocytes proceed to a normal SN chromatin configuration, but remain transcriptionally active. SN configuration did not depend on transcriptional repression [80]. Alternatively, in NPM2 (a nuclear protein highly expressed in oocytes) knockout mice, chromatin in full size oocytes is diffuse with no condensation around multiple dispersed nucleoli. This demonstrates that NPM2 is essential for the organisation of the nucleolus and the compaction of chromatin. However, in this knockout model, female meiosis progressed with no notable defects in MII oocytes despite disruption of nucleolar structure. There was no detectable alteration in 18S and 28S ribosomal RNA (rRNA) levels or changes in the absolute levels of protein synthesis [94]. Ultimately MLL2 or NPM2 loss are lethal, no viable embryos were produced from either knockout model. Therefore, while this supports that both correct chromatin configuration and transcriptional silencing are essential, this is evidence that they are not causally linked.

It also seems possible that transcription shuts down gradually while chromatin remodelling is a relatively discrete event at the end of that period. By FISH analysis, Shishova et al. [66] probed for 18S and 28S ribosomal RNAs. While FISH signals were clearly seen within nucleoli of pre-antral oocytes, under those same hybridisation conditions 18S rRNA and 28S rRNA could only be detected inside the nucleoli of 10% of NSN oocytes. Furthermore, only 40% of NSN oocytes presented with peripheral 18S and 28S rRNA signatures. This suggests in 50–90% of the NSN oocytes, transcription was either at the end stages of, or already silenced before the SN transition. Expectedly, no FISH signals were detected in SN oocytes.

Switching Transcription Back on in the Early Embryo, a Return to Mitotic-Type Nucleolar and Chromatin Morphology

Through meiosis I and meiosis II oocytes rely on maternal transcripts. This is possible because mRNAs are extremely stable during the oocytes growth phase, permitting their accumulation in preparation for a period without transcript production [95–97]. The absence of global transcription lasts until the embryonic genome activates, after fertilisation when a new embryo is formed. Interestingly, maternal transcripts begin a programme of degradation almost immediately following GVBD which marks the resumption of meiosis. Initially this degradation is gradual but triggers a transition from mRNA stability to instability even at this early stage. Importantly however, a substantial amount of maternal mRNA still persists in one-cell embryos with the vast majority of mRNAs degraded after this stage [98–103].

Nucleolar material, that decondensed following GVBD, is retained in the oocyte through meiosis I and II divisions. This allows ready recondensation and nucleoli formation in zygotes (or PN stage embryos), which have intranuclear organelles in pronuclei that morphologically appear like those in late GV stage oocytes. At this stage, it is generally accepted that RNA PolI is not active. However, as embryos develop further, these structures gradually transform into the tripartite nucleoli typical of somatic cells. Concomitantly with this transformation, ribosomal RNA production is reinitiated [104]. Curiously, differences between the female and male pronuclei can be detected following fertilisation when comparing the organisation of chromatin in zygotes. In female PNs, centromeres surround the surface of nucleoli while in the male pronucleus, they remained associated but in less organised clusters [105].

Interestingly, while transcription is silenced at the same time point in all mammalian oocyte species tested (prior to GVBD and ovulation), the duration of the period without transcription varies. This is due to differences in the timing of the resumption of transcription, now governed by the embryo (reviewed in references [106–108]). Details concerning the molecular control over the resumption of transcription are not yet fully understood [109], but thought to be organised in two waves [106, 110]. In mice, a minor wave of transcription is detected in the zygote [111], followed by an acceleration of transcription in two-cell embryos [112, 113]. In humans, while a minor level of transcription is also reported in zygotes (specifically genes related to protein transport and signalling [102]), the major wave of transcription occurs later, in four-to-eight-cell embryos [114]. Waves of transcriptional activation in cattle largely follow human-like timings [115–117].

As previously mentioned, using enucleolation experiments it was demonstrated that nucleolar material is exclusively of maternal origin, stored by the oocyte and essential for embryonic development [68]. Embryos originating from enucleolated oocytes fail to replace nucleoli material and arrest development at the two-cell stage. A simple explanation for this is that embryos developing from enucleolated oocytes lack a store of material necessary to activate transcription by RNA PolI. They thereby are consequently unable to initiate ribosome biogenesis. However, oocyte nucleoli are essential prior to this time point.

In 2010, Ogushi and Saitou [71] narrowed down the window during which nucleoli are required, returning nucleolar material to enucleolated mouse oocytes at different time points. Here, the reintroduction of the nucleolus just 8 h after fertilisation was too late to rescue the arrest in development. In contrast, when nucleoli were transplanted into enucleolated oocytes just prior to fertilisation, embryos developed normally. This suggests that nucleoli are essential during the first few hours after fertilisation. Given that there is no RNA PolI activity at this time [118–120], and that embryos originating from enucleolated oocytes are able to synthesise proteins [68], this suggests a nucleolus function or factor unrelated to ribosome biogenesis, but essential to re-establish cell plasticity after fertilisation. One that is not present in somatic nucleoli, but essential for embryonic development, perhaps for an embryo-specific process involved in reprograming.

Evidence from inter-species nuclear transfer experiments suggests this factor is conserved. Mouse nucleoli injected into previously enucleolated porcine oocytes allow development beyond the enucleolation-associated developmental block [121]. Interestingly, in the reverse situation, two porcine nucleoli were needed per one enucleolated mouse egg to overcome the developmental block suggesting the 'dose' of this conserved factor might be species specific [122].

Indeed, it is possible that storage of this unknown factor is the primary essential role of oocyte nucleoli once transcription has shut down; oocytes may not need a 'store' of material for ribosome biogenesis. Kyogoku et al. [123] removed nucleoli from late mouse zygotes (10 h after sperm injection). Not surprisingly nucleoli were not visible in two-cell embryos but formed at the four-cell stage and embryos developed to term. While it cannot be ruled out that some proteins necessary for ribosome biogenesis remained in the nucleoplasm of enucleolated zygotes, this raises the question of the origin of the nucleolar proteins. In addition, many nucleolar proteins are lost during waves of degradation following meiosis resumption and again following fertilisation (reviewed in [57]). Early embryos may have capacity to synthesise nucleolar proteins de novo.

It seems highly likely that the majority of oocyte nucleoli content is identical to somatic cell content, it has of course been derived from a mitotic cell. However, strong evidence as described above suggests there is an additional factor contained in this organelle essential for early embryonic development. OCT4 (also known as POU5F1) was implicated in this role; a pluripotency factor expressed in stem and germ cells [124, 125] that localises to nucleoli in mouse oocytes [126]. It was suggested that OCT4 might regulate the maternal to embryonic transition in embryos [127]. However, when *Oct4* conditional knockout mice were generated with oocytes severely depleted of OCT4 they remained fully fertile [128]. Nevertheless, OCT4 might not be the only pluripotency-associated factor stored in the nucleolus capable of establishing cell plasticity [104].

Another important nucleolar protein is nucleoplasmin 2 (NPM2). Ogushi et al. [129], injected a high dose of *Npm2* mRNA into MII oocytes previously enucleolated. When fertilised, embryos formed nucleolar-like structures in their pronuclei. Chromatin appeared to condense normally around these structures, and remarkably 48% of embryos reached the blastocyst stage. This demonstrated that NPM2 alone was sufficient to reconstitute the nucleolar structure in enucleolated embryos and suggests that this structure is an essential platform for chromatin organisation in early embryonic development. Notably NPM2 had previously been identified as a histone chaperone for sperm chromatin decondensation in Xenopus [130, 131]. It seems plausible that a crucial nucleolar role that is missing in enucleated oocytes is appropriate chromatin organisation as pronuclei form in the zygote, ahead of a role in transcription.

The observation that chromatin condensed around the NPM2-derived structures in oocytes was important. Like fully grown competent mouse oocytes, heterochromatin reorganises in a sphere around nucleoli as they form in zygotes [132, 133]. It is not clear, however, why this close nucleoli–chromatin arrangement is so faithful or how essential it is. For example, in NPM2 null mouse embryos an absence of a histone H3 mark has been noted [94] and defects in chromatin remodelling are apparent [129]. However, where Ogushi and Saitou [71] investigated the formation of heterochromatin in embryos derived from enucleolated oocytes, while they did observe heterochromatin architectural defects, these defects were so substantial, that it is difficult to interpret changes in histone modification patterns in a nucleolus free environment.

Furthermore, in the study by Kyogoku et al. [123], when removing the nucleolus from late zygotes, nucleoli did not fully reform until the four-cell stage. However, in this study, pups were generated from enucleolated zygotes. This could suggest that a nucleolar role as a scaffold was not essential in late zygotes and two-cell embryos. Yet, in the example images provided, nucleolar material removal does not appear 100% complete and PN chromatin still looks condensed either around or in close proximity to a body of residual nucleolar material.

Given the timings of the experiments descried above, it seems possible that any necessity for nucleoli to organise chromatin architecture is a fleeting event immediately post fertilisation.

Changing Chromatin Landscape in Early Embryos

It is tempting to speculate that, like in somatic cells, oocyte and zygote nucleoli might act as epigenetic organising centres, bringing together particular regions or preventing long-distance DNA interactions [134, 135]. Indeed, certain sequences are known to associate with nucleoli in oocytes, and histone modifications related to both active and repressive chromatin structures seem to follow the large-scale chromatin movements that occur during the NSN to SN transition [93]. Certainly, both correct epigenetic programming and the nucleolus are essential for development. Yet their link and reliance on one another is still to be concluded. Understanding this is important given that early embryonic development is mainly regulated by epigenetic mechanisms. The epigenome of early embryos undergoes significant changes, especially on histone H3, regulating embryonic genome activation, X-chromosome inactivation, and the chromatin reorganisation required for proper development (See Wilkinson et al. [136] for review).

The current trend towards delayed childbirth and our reliance on in-vitro fertilisation techniques highlight a need for better understanding of epigenetics in oocytes and early embryos. Epigenetic markers in oocytes are significantly altered in ageing in oocytes [137–139], and the epigenome is particularly vulnerable to stress and change [140], likely considerable during in vitro culture.

Conclusions

There are many unknowns regarding the content and role of the nucleolus in oocytes. In fully grown oocytes, there is an undeniable connection between the NSN and SN morphology of the karyosphere and oocyte competency, yet how directly coupled transcriptional silencing is to this transition is debatable. They are closely coupled in time but are not simultaneous and not necessarily coordinated by the same pathway. From the GV stage, and for several divisions, the oocyte and the early embryo have no need for ribosome biogenesis. Indeed, this would likely be highly detrimental. A prominent theory therefore, was that an oocyte's nucleolus is simply a storage facility of material with which to reform embryonic nucleoli, reinitiating ribosome biogenesis as the embryo switches over to its own programme of transcription. However, a growing body of literature now instead points to this dense, fibrillar, structureless mass at the centre of the oocyte as carrying a factor critical immediately after fertilisation. The presence of this factor (or structure) is essential two to three divisions before ribosome biogenesis is reinstated, which could suggest that it is not involved in reactivating ribosome biogenesis, or in the switch from the reliance on maternal to embryonic transcripts. Interestingly, however, the defect induced by the loss of this early factor does not present until that later stage. It is likely that there is still a connection. Nevertheless this factor is not a component of somatic cell nucleoli or ESC nucleoli.

Like the oocyte nucleolus, the zygote nucleolus is necessary for 'normal' chromatin structure, wrapping around the nucleolus in a tight sphere. In both cases, the nucleolus certainly acts as a scaffold and chromatin condenses around this structure. However, it is not fully clear how critical this structure is for normal embryonic development. Chromatin structure is demonstrably lost in oocytes that don't have a nucleolus and don't develop beyond two-cell stage, but we cannot fully conclude from these experiments which loss was the insult. The presence of the nucleolus and chromatin structure are so intrinsically linked. It is possible that the early zygotic role of a nucleolus is in fact for an aspect of chromatin modelling that requires this scaffold. Separating the role of the nucleolus and the need for correct chromatin scaffolding in the newly fertilised zygote would be technically very challenging.

A role for the oocyte's nucleolus as a retention site is also possible and underexplored. Enucleation experiments do not tell us anything about the importance of compartmentalising or detaining proteins that are misfolded or damaged (and would otherwise be detrimental to cell health). The action of removing the nucleolus would relieve the oocyte of any potentially damaging agents within. Sequestering proteins to the nucleolus is not uncommon in somatic cells and this function represents an important regulatory mechanism (for a review, see [140, 142] which may be even more critical in long lived cells.

References

1. Licatalosi DD, Darnell RB (2010) RNA processing and its regulation: global insights into biological networks. Nat Rev Genet 11:75–87. https://doi.org/10.1038/nrg2673
2. Nyathi Y, Wilkinson BM, Pool MR (2013) Co-translational targeting and translocation of proteins to the endoplasmic reticulum. Biochim Biophys Acta 1833:2392–2402. https://doi.org/10.1016/j.bbamcr.2013.02.021
3. Gonzalez I, Sylvester J (1995) Complete sequence of the 43-kb human ribosomal DNA repeat: analysis of the intergenic spacer. Genomics 27:320–328. https://doi.org/10.1006/geno.1995.1049
4. McStay B (2016) Nucleolar organizer regions: genomic 'dark matter' requiring illumination. Genes Dev 30:1598–1610. https://doi.org/10.1101/gad.283838.116
5. Warmerdam DO, Wolthuis RMF (2019) Keeping ribosomal DNA intact: a repeating challenge. Chromosom Res 27:57–72. https://doi.org/10.1007/s10577-018-9594-z
6. Savino TM, Gébrane-Younès J, De Mey J, Sibarita JB, Hernandez-Verdun D (2001) Nucleolar assembly of the rRNA processing machinery in living cells. J Cell Biol 153:1097–1110. https://doi.org/10.1083/jcb.153.5.1097
7. Booth DG, Takagi M, Sanchez-Pulido L, Petfalski E, Vargiu G, Samejima K, Imamoto N, Ponting CP, Tollervey D, Earnshaw WC, Vagnarelli P (2014) Ki-67 is a PP1-interacting protein that organises the mitotic chromosome periphery. elife 3:e01641. https://doi.org/10.7554/eLife.01641
8. Hadjiolov AA (1985) The nucleolus and ribosome biogenesis. Cell biology monographs. Springer, Vienna. https://doi.org/10.1007/978-3-7091-8742-5
9. Cheutin T, O'Donohue M-F, Beorchia A, Vandelaer M, Kaplan H, Deféver B, Ploton D, Thiry M (2002) Three-dimensional organization of active rRNA genes within the nucleolus. J Cell Sci 115:3297–3307. https://doi.org/10.1242/jcs.115.16.3297
10. Yao R-W, Xu G, Wang Y, Shan L, Luan P-F, Wang Y, Wu M, Yang L-Z, Xing Y-H, Yang L, Chen L-L (2019) Nascent pre-rRNA sorting via phase separation drives the assembly of dense fibrillar components in the human nucleolus. Mol Cell 76:767–783.e11. https://doi.org/10.1016/j.molcel.2019.08.014
11. Lafontaine DLJ, Riback JA, Bascetin R, Brangwynne CP (2021) The nucleolus as a multiphase liquid condensate. Nat Rev Mol Cell Biol 22:165–182. https://doi.org/10.1038/s41580-020-0272-6
12. Boulon S, Westman BJ, Hutten S, Boisvert F-M, Lamond AI (2010) The nucleolus under stress. Mol Cell 40:216–227. https://doi.org/10.1016/j.molcel.2010.09.024
13. Daniel L, Cerutti E, Donnio L-M, Nonnekens J, Carrat C, Zahova S, Mari P-O, Giglia-Mari G (2018) Mechanistic insights in transcription-coupled nucleotide excision repair of ribosomal DNA. Proc Natl Acad Sci U S A 115:E6770–E6779. https://doi.org/10.1073/pnas.1716581115
14. Miller OL, Beatty BR (1969) Visualization of nucleolar genes. Science 164:955–957. https://doi.org/10.1126/science.164.3882.955
15. Tartakoff A, DiMario P, Hurt E, McStay B, Panse VG, Tollervey D (2022) The dual nature of the nucleolus. Genes Dev 36:765–769. https://doi.org/10.1101/gad.349748.122
16. Aubert M, O'Donohue M-F, Lebaron S, Gleizes P-E (2018) Pre-ribosomal RNA processing in human cells: from mechanisms to congenital diseases. Biomolecules 8:123. https://doi.org/10.3390/biom8040123
17. Henras AK, Plisson-Chastang C, O'Donohue MF, Chakraborty A, Gleizes PE (2015) An overview of pre-ribosomal RNA processing in eukaryotes. Wiley Interdiscip Rev RNA 6:225–242. https://doi.org/10.1002/wrna.1269
18. O'Donohue M-F, Choesmel V, Faubladier M, Fichant G, Gleizes P-E (2010) Functional dichotomy of ribosomal proteins during the synthesis of mammalian 40S ribosomal subunits. J Cell Biol 190:853–866. https://doi.org/10.1083/jcb.201005117

19. Pollock C, Huang S (2010) The Perinucleolar compartment. Cold Spring Harb Perspect Biol 2:a000679. https://doi.org/10.1101/cshperspect.a000679
20. Schöfer C, Weipoltshammer K (2018) Nucleolus and chromatin. Histochem Cell Biol 150:209–225. https://doi.org/10.1007/s00418-018-1696-3
21. Németh A, Conesa A, Santoyo-Lopez J, Medina I, Montaner D, Péterfia B, Solovei I, Cremer T, Dopazo J, Längst G (2010) Initial genomics of the human nucleolus. PLoS Genet 6:e1000889. https://doi.org/10.1371/journal.pgen.1000889
22. Shore D, Albert B (2022) Ribosome biogenesis and the cellular energy economy. Curr Biol 32:R611–R617. https://doi.org/10.1016/j.cub.2022.04.083
23. Andersen JS, Lam YW, Leung AKL, Ong S-E, Lyon CE, Lamond AI, Mann M (2005) Nucleolar proteome dynamics. Nature 433:77–83. https://doi.org/10.1038/nature03207
24. Andersen JS, Lyon CE, Fox AH, Leung AKL, Lam YW, Steen H, Mann M, Lamond AI (2002) Directed proteomic analysis of the human nucleolus. Curr Biol 12:1–11. https://doi.org/10.1016/S0960-9822(01)00650-9
25. Sengupta S, Peterson TR, Sabatini DM (2010) Regulation of the mTOR complex 1 pathway by nutrients, growth factors, and stress. Mol Cell 40:310–322. https://doi.org/10.1016/j.molcel.2010.09.026
26. Al-Baker EA, Oshin M, Hutchison CJ, Kill IR (2005) Analysis of UV-induced damage and repair in young and senescent human dermal fibroblasts using the comet assay. Mech Ageing Dev 126:664–672. https://doi.org/10.1016/j.mad.2004.12.002
27. Govoni M, Farabegoli F, Pession A, Novello F (1994) Inhibition of topoisomerase II activity and its effect on nucleolar structure and function. Exp Cell Res 211:36–41. https://doi.org/10.1006/excr.1994.1055
28. Louvet E, Junéra HR, Le Panse S, Hernandez-Verdun D (2005) Dynamics and compartmentation of the nucleolar processing machinery. Exp Cell Res 304:457–470. https://doi.org/10.1016/j.yexcr.2004.11.018
29. Shav-Tal Y, Blechman J, Darzacq X, Montagna C, Dye BT, Patton JG, Singer RH, Zipori D (2005) Dynamic sorting of nuclear components into distinct nucleolar caps during transcriptional inhibition. Mol Biol Cell 16:2395–2413. https://doi.org/10.1091/mbc.e04-11-0992
30. Musawi S, Donnio L-M, Zhao Z, Magnani C, Rassinoux P, Binda O, Huang J, Jacquier A, Coudert L, Lomonte P, Martinat C, Schaeffer L, Mottet D, Côté J, Mari P-O, Giglia-Mari G (2023) Nucleolar reorganization after cellular stress is orchestrated by SMN shuttling between nuclear compartments. Nat Commun 14:7384. https://doi.org/10.1038/s41467-023-42390-4
31. Audas TE, Jacob MD, Lee S (2012) Immobilization of proteins in the nucleolus by ribosomal intergenic spacer noncoding RNA. Mol Cell 45:147–157. https://doi.org/10.1016/j.molcel.2011.12.012
32. Jacob MD, Audas TE, Uniacke J, Trinkle-Mulcahy L, Lee S (2013) Environmental cues induce a long noncoding RNA–dependent remodeling of the nucleolus. Mol Biol Cell 24:2943–2953. https://doi.org/10.1091/mbc.E13-04-0223
33. Wang M, Bokros M, Theodoridis PR, Lee S (2019) Nucleolar sequestration: remodeling nucleoli into amyloid bodies. Front Genet 10:1179. https://doi.org/10.3389/fgene.2019.01179
34. Mikhaleva EA, Leinsoo TA, Ishizu H, Gvozdev VA, Klenov MS (2019) The nucleolar transcriptome regulates Piwi shuttling between the nucleolus and the nucleoplasm. Chromosom Res 27:141–152. https://doi.org/10.1007/s10577-018-9595-y
35. Phair RD, Misteli T (2000) High mobility of proteins in the mammalian cell nucleus. Nature 404:604–609. https://doi.org/10.1038/35007077
36. Martin RM, Ter-Avetisyan G, Herce HD, Ludwig AK, Lättig-Tünnemann G, Cardoso MC (2015) Principles of protein targeting to the nucleolus. Nucleus 6:314–325. https://doi.org/10.1080/19491034.2015.1079680
37. Musinova YR, Lisitsyna OM, Golyshev SA, Tuzhikov AI, Polyakov VY, Sheval EV (2011) Nucleolar localization/retention signal is responsible for transient accumulation of histone H2B in the nucleolus through electrostatic interactions. Biochim Biophys Acta 1813:27–38. https://doi.org/10.1016/j.bbamcr.2010.11.003

38. Shou W, Seol JH, Shevchenko A, Baskerville C, Moazed D, Chen ZW, Jang J, Shevchenko A, Charbonneau H, Deshaies RJ (1999) Exit from mitosis is triggered by Tem1-dependent release of the protein phosphatase Cdc14 from nucleolar RENT complex. Cell 97:233–244. https://doi.org/10.1016/s0092-8674(00)80733-3
39. Trinkle-Mulcahy L, Chusainow J, Lam YW, Swift S, Lamond A (2007) Visualization of intracellular PP1 targeting through transiently and stably expressed fluorescent protein fusions. In: Moorhead G (ed) Protein phosphatase protocols, methods in molecular biology. Springer, New York, pp 133–154. https://doi.org/10.1385/1-59745-267-X:133
40. Klump BM, Perez GI, Patrick EM, Adams-Boone K, Cohen SB, Han L, Yu K, Schmidt JC (2023) TCAB1 prevents nucleolar accumulation of the telomerase RNA to facilitate telomerase assembly. Cell Rep 42:112577. https://doi.org/10.1016/j.celrep.2023.112577
41. Wong JMY, Kusdra L, Collins K (2002) Subnuclear shuttling of human telomerase induced by transformation and DNA damage. Nat Cell Biol 4:731–736. https://doi.org/10.1038/ncb846
42. Frottin F, Schueder F, Tiwary S, Gupta R, Körner R, Schlichthaerle T, Cox J, Jungmann R, Hartl FU, Hipp MS (2019) The nucleolus functions as a phase-separated protein quality control compartment. Science 365:342–347. https://doi.org/10.1126/science.aaw9157
43. Telfer EE, Grosbois J, Odey YL, Rosario R, Anderson RA (2023) Making a good egg: human oocyte health, aging, and in vitro development. Physiol Rev 103:2623–2677. https://doi.org/10.1152/physrev.00032.2022
44. Ariel O, Yao S, Kearney MT, Joudrey SD, Al-Bagdadi F (2016) Morphometric and ultrastructure studies of primordial follicles and expression of Estrogen-dependent genes in the ovaries of domestic cats. J Microsc Ultrastruct 4:46–54. https://doi.org/10.1016/j.jmau.2015.10.003
45. Baker TG, Franchi LL (1967) The structure of the chromosomes in human primordial oocytes. Chromosoma 22:358–377. https://doi.org/10.1007/BF00319880
46. Henry Sathananthan A, Wood C, Leeton JF (1982) Ultrastructural evaluation of 8–16 cell human embryos cultured in vitro. Micron 1969(13):193–203. https://doi.org/10.1016/0047-7206(82)90084-X
47. Hyttel P, Tosti E, Raffaele B (2011) Oocyte maturation and fertilization: a long history for a short event. Bentham
48. Paulini F, Silva RC, de Paula Rôlo JLJ, Lucci CM (2014) Ultrastructural changes in oocytes during folliculogenesis in domestic mammals. J Ovarian Res 7:102. https://doi.org/10.1186/s13048-014-0102-6
49. Al Jord A, Letort G, Chanet S, Tsai F-C, Antoniewski C, Eichmuller A, Da Silva C, Huynh J-R, Gov NS, Voituriez R, Terret M-É, Verlhac M-H (2022) Cytoplasmic forces functionally reorganize nuclear condensates in oocytes. Nat Commun 13:5070. https://doi.org/10.1038/s41467-022-32675-5
50. Colin A, Letort G, Razin N, Almonacid M, Ahmed W, Betz T, Terret M-E, Gov NS, Voituriez R, Gueroui Z, Verlhac M-H (2020) Active diffusion in oocytes nonspecifically centers large objects during prophase I and meiosis I. J Cell Biol 219:e201908195. https://doi.org/10.1083/jcb.201908195
51. Fulka J, Langerova A, Loi P, Martinkova S, Fulka H (2011) Transplantation of nucleoli into human zygotes: not as simple as expected? J Assist Reprod Genet 28:385–389. https://doi.org/10.1007/s10815-011-9565-1
52. Hornak M, Kubicek D, Broz P, Hulinska P, Hanzalova K, Griffin D, Machatkova M, Rubes J (2016) Aneuploidy detection and mtDNA quantification in bovine embryos with different cleavage onset using a next-generation sequencing-based protocol. Cytogenet Genome Res 150:60–67. https://doi.org/10.1159/000452923
53. Turner KJ, Silvestri G, Black DH, Dobson G, Smith C, Handyside AH, Sinclair KD, Griffin DK (2019) Karyomapping for simultaneous genomic evaluation and aneuploidy screening of preimplantation bovine embryos: the first live-born calves. Theriogenology 125:249–258. https://doi.org/10.1016/j.theriogenology.2018.11.014
54. Ferré LB, Kjelland ME, Strøbech LB, Hyttel P, Mermillod P, Ross PJ (2020) Review: recent advances in bovine in vitro embryo production: reproductive biotechnology history and methods. Animal 14:991–1004. https://doi.org/10.1017/S1751731119002775

55. Fulka H, Loi P, Palazzese L, Benc M, Fulka J (2022) Nucleus reprogramming/remodeling through selective enucleation (SE) of immature oocytes and zygotes: a nucleolus point of view. J Reprod Dev 68:165–172. https://doi.org/10.1262/jrd.2022-004
56. Fulka H, Langerova A (2019) Nucleoli in embryos: a central structural platform for embryonic chromatin remodeling? Chromosom Res 27:129–140. https://doi.org/10.1007/s10577-018-9590-3
57. Fulka H, Rychtarova J, Loi P (2020) The nucleolus-like and precursor bodies of mammalian oocytes and embryos and their possible role in post-fertilization centromere remodelling. Biochem Soc Trans 48:581–593. https://doi.org/10.1042/BST20190847
58. Fair T, Hyttel P, Lonergan P, Boland MP (2001) Immunolocalization of nucleolar proteins during bovine oocyte growth, meiotic maturation, and fertilization. Biol Reprod 64:1516–1525. https://doi.org/10.1095/biolreprod64.5.1516
59. Bachvarova R (1974) Incorporation of tritiated adenosine into mouse ovum RNA. Dev Biol 40:52–58. https://doi.org/10.1016/0012-1606(74)90106-7
60. Antoine N, Lepoint A, Baeckeland E, Goessens G (1988) Ultrastructural cytochemistry of the nucleolus in rat oocytes at the end of the folliculogenesis. Histochemistry 89:221–226. https://doi.org/10.1007/BF00493143
61. Biggiogera M, Martin TE, Gordon J, Amalric F, Fakan S (1994) Physiologically inactive nucleoli contain nucleoplasmic ribonucleoproteins: immunoelectron microscopy of mouse spermatids and early embryos. Exp Cell Res 213:55–63. https://doi.org/10.1006/excr.1994.1172
62. Longo F, Garagna S, Merico V, Orlandini G, Gatti R, Scandroglio R, Redi CA, Zuccotti M (2003) Nuclear localization of NORs and centromeres in mouse oocytes during folliculogenesis. Mol Reprod Dev 66:279–290. https://doi.org/10.1002/mrd.10354
63. Bogolyubov DS (2018) Chapter one—Karyosphere (Karyosome): a peculiar structure of the oocyte nucleus. In: Galluzzi L (ed) International review of cell and molecular biology. Academic Press, pp 1–48. https://doi.org/10.1016/bs.ircmb.2017.12.001
64. Crozet N (1989) Nucleolar structure and RNA synthesis in mammalian oocytes. J Reprod Fertil Suppl 38:9–16
65. Luciano AM, Franciosi F, Dieci C, Lodde V (2014) Changes in large-scale chromatin structure and function during oogenesis: a journey in company with follicular cells. Anim Reprod Sci 149:3–10. https://doi.org/10.1016/j.anireprosci.2014.06.026
66. Shishova KV, Lavrentyeva EA, Dobrucki JW, Zatsepina OV (2015) Nucleolus-like bodies of fully-grown mouse oocytes contain key nucleolar proteins but are impoverished for rRNA. Dev Biol 397:267–281. https://doi.org/10.1016/j.ydbio.2014.11.022
67. Fulka J, Moor RM, Loi P, Fulka J (2003) Enucleolation of porcine oocytes. Theriogenology 59:1879–1885. https://doi.org/10.1016/S0093-691X(02)01226-8
68. Ogushi S, Palmieri C, Fulka H, Saitou M, Miyano T, Fulka J (2008) The maternal nucleolus is essential for early embryonic development in mammals. Science 319:613–616. https://doi.org/10.1126/science.1151276
69. Wakayama T, Rodriguez I, Perry AC, Yanagimachi R, Mombaerts P (1999) Mice cloned from embryonic stem cells. Proc Natl Acad Sci U S A 96:14984–14989. https://doi.org/10.1073/pnas.96.26.14984
70. Wakayama T, Tabar V, Rodriguez I, Perry AC, Studer L, Mombaerts P (2001) Differentiation of embryonic stem cell lines generated from adult somatic cells by nuclear transfer. Science 292:740–743. https://doi.org/10.1126/science.1059399
71. Ogushi S, Saitou M (2010) The nucleolus in the mouse oocyte is required for the early step of both female and male pronucleus organization. J Reprod Dev 56:495–501. https://doi.org/10.1262/jrd.09-184H
72. Zuccotti M, Piccinelli A, Giorgi Rossi P, Garagna S, Redi CA (1995) Chromatin organization during mouse oocyte growth. Mol Reprod Dev 41:479–485. https://doi.org/10.1002/mrd.1080410410
73. Combelles CMH, Cekleniak NA, Racowsky C, Albertini DF (2002) Assessment of nuclear and cytoplasmic maturation in in-vitro matured human oocytes. Hum Reprod 17:1006–1016. https://doi.org/10.1093/humrep/17.4.1006

74. Bouniol-Baly C, Hamraoui L, Guibert J, Beaujean N, Szöllösi MS, Debey P (1999) Differential transcriptional activity associated with chromatin configuration in fully grown mouse germinal vesicle oocytes. Biol Reprod 60:580–587. https://doi.org/10.1095/biolreprod60.3.580
75. Christians E, Boiani M, Garagna S, Dessy C, Redi CA, Renard JP, Zuccotti M (1999) Gene expression and chromatin organization during mouse oocyte growth. Dev Biol 207:76–85. https://doi.org/10.1006/dbio.1998.9157
76. Miyara F, Migne C, Dumont-Hassan M, Le Meur A, Cohen-Bacrie P, Aubriot F-X, Glissant A, Nathan C, Douard S, Stanovici A, Debey P (2003) Chromatin configuration and transcriptional control in human and mouse oocytes. Mol Reprod Dev 64:458–470. https://doi.org/10.1002/mrd.10233
77. De La Fuente R, Eppig JJ (2001) Transcriptional activity of the mouse oocyte genome: companion granulosa cells modulate transcription and chromatin remodeling. Dev Biol 229:224–236. https://doi.org/10.1006/dbio.2000.9947
78. Dundr M, Misteli T. Biogenesis of nuclear bodies. Cold Spring Harb Perspect Biol. 2010 Dec;2(12):a000711. https://doi.org/10.1101/cshperspect.a000711. Epub 2010 Nov 10. PMID: 21068152; PMCID: PMC2982170. Emmott and Hiscox 2009:
79. Emmott E, Hiscox JA. Nucleolar targeting: the hub of the matter. EMBO Rep. 2009 Mar;10(3):231-8. https://doi.org/10.1038/embor.2009.14. Epub 2009 Feb 20. PMID: 19229283; PMCID: PMC2658561. Hadjiolov 1985
80. Liu H, Aoki F (2002) Transcriptional activity associated with meiotic competence in fully grown mouse GV oocytes. Zygote (Cambridge, England) 10:327–332. https://doi.org/10.1017/s0967199402004069
81. Li L, Zheng P, Dean J (2010) Maternal control of early mouse development. Development (Cambridge, England) 137:859–870. https://doi.org/10.1242/dev.039487
82. Andreu-Vieyra CV, Chen R, Agno JE, Glaser S, Anastassiadis K, Stewart AF, Matzuk MM (2010) MLL2 is required in oocytes for bulk histone 3 lysine 4 trimethylation and transcriptional silencing. PLoS Biol 8:e1000453. https://doi.org/10.1371/journal.pbio.1000453
83. Zuccotti M, Ponce RH, Boiani M, Guizzardi S, Govoni P, Scandroglio R, Garagna S, Redi CA (2002) The analysis of chromatin organisation allows selection of mouse antral oocytes competent for development to blastocyst. Zygote (Cambridge, England) 10:73–78. https://doi.org/10.1017/s0967199402002101
84. Chousal JN, Cho K, Ramaiah M, Skarbrevik D, Mora-Castilla S, Stumpo DF, Lykke-Andersen J, Laurent LC, Blackshear PF, Wilkinson MF, Cook-Andersen H (2018) Global transcriptional silencing and developmental competence in the oocyte mediated by the mRNA decay activator ZFP36L2. Dev Cell 44:392–402.e7. https://doi.org/10.1016/j.devcel.2018.01.006
85. Liu Y-J, Nakamura T, Nakano T (2012) Essential role of DPPA3 for chromatin condensation in mouse oocytogenesis. Biol Reprod 86:40. https://doi.org/10.1095/biolreprod.111.095018
86. Lowther KM, Mehlmann LM (2015) Embryonic poly(a)-binding protein is required during early stages of mouse oocyte development for chromatin organization, transcriptional silencing, and meiotic competence. Biol Reprod 93:43. https://doi.org/10.1095/biolreprod.115.131359
87. Medvedev S, Pan H, Schultz RM (2011) Absence of MSY2 in mouse oocytes perturbs oocyte growth and maturation, RNA stability, and the transcriptome. Biol Reprod 85:575–583. https://doi.org/10.1095/biolreprod.111.091710
88. Xia M, He H, Wang Y, Liu M, Zhou T, Lin M, Zhou Z, Huo R, Zhou Q, Sha J (2012) PCBP1 is required for maintenance of the transcriptionally silent state in fully grown mouse oocytes. Cell Cycle Georget Tex 11:2833–2842. https://doi.org/10.4161/cc.21169
89. De La Fuente R (2006) Chromatin modifications in the germinal vesicle (GV) of mammalian oocytes. Dev Biol 292:1–12. https://doi.org/10.1016/j.ydbio.2006.01.008
90. Kageyama S, Liu H, Kaneko N, Ooga M, Nagata M, Aoki F (2007) Alterations in epigenetic modifications during oocyte growth in mice. Reproduction (Cambridge, England) 133:85–94. https://doi.org/10.1530/REP-06-0025

91. Zuccotti M, Bellone M, Longo F, Redi CA, Garagna S (2011) Fully-mature antral mouse oocytes are transcriptionally silent but their heterochromatin maintains a transcriptional permissive histone acetylation profile. J Assist Reprod Genet 28:1193–1196. https://doi.org/10.1007/s10815-011-9562-4
92. Inoue A, Nakajima R, Nagata M, Aoki F (2008) Contribution of the oocyte nucleus and cytoplasm to the determination of meiotic and developmental competence in mice. Hum Reprod 23:1377–1384. https://doi.org/10.1093/humrep/den096
93. Zuccotti M, Giorgi Rossi P, Martinez A, Garagna S, Forabosco A, Redi CA (1998) Meiotic and developmental competence of mouse antral oocytes. Biol Reprod 58:700–704. https://doi.org/10.1095/biolreprod58.3.700
94. Zuccotti M, Merico V, Sacchi L, Bellone M, Brink TC, Bellazzi R, Stefanelli M, Redi CA, Garagna S, Adjaye J (2008) Maternal Oct-4 is a potential key regulator of the developmental competence of mouse oocytes. BMC Dev Biol 8:97. https://doi.org/10.1186/1471-213X-8-97
95. Bonnet-Garnier A, Feuerstein P, Chebrout M, Fleurot R, Jan H-U, Debey P, Beaujean N (2012) Genome organization and epigenetic marks in mouse germinal vesicle oocytes. Int J Dev Biol 56:877–887. https://doi.org/10.1387/ijdb.120149ab
96. Burns KH, Viveiros MM, Ren Y, Wang P, DeMayo FJ, Frail DE, Eppig JJ, Matzuk MM (2003) Roles of NPM2 in chromatin and nucleolar organization in oocytes and embryos. Science 300:633–636. https://doi.org/10.1126/science.1081813
97. Ma J, Flemr M, Strnad H, Svoboda P, Schultz RM (2013) Maternally recruited DCP1A and DCP2 contribute to messenger RNA degradation during oocyte maturation and genome activation in mouse. Biol Reprod 88:11. https://doi.org/10.1095/biolreprod.112.105312
98. Paynton BV, Bachvarova R (1994) Polyadenylation and deadenylation of maternal mRNAs during oocyte growth and maturation in the mouse. Mol Reprod Dev 37:172–180. https://doi.org/10.1002/mrd.1080370208
99. Yu C, Ji S-Y, Sha Q-Q, Dang Y, Zhou J-J, Zhang Y-L, Liu Y, Wang Z-W, Hu B, Sun Q-Y, Sun S-C, Tang F, Fan H-Y (2016) BTG4 is a meiotic cell cycle-coupled maternal-zygotic-transition licensing factor in oocytes. Nat Struct Mol Biol 23:387–394. https://doi.org/10.1038/nsmb.3204
100. Abe K-I, Funaya S, Tsukioka D, Kawamura M, Suzuki Y, Suzuki MG, Schultz RM, Aoki F (2018) Minor zygotic gene activation is essential for mouse preimplantation development. Proc Natl Acad Sci U S A 115:E6780–E6788. https://doi.org/10.1073/pnas.1804309115
101. Pikó L, Clegg KB (1982) Quantitative changes in total RNA, total poly(A), and ribosomes in early mouse embryos. Dev Biol 89:362–378. https://doi.org/10.1016/0012-1606(82)90325-6
102. Sha Q-Q, Zheng W, Wu Y-W, Li S, Guo L, Zhang S, Lin G, Ou X-H, Fan H-Y (2020a) Dynamics and clinical relevance of maternal mRNA clearance during the oocyte-to-embryo transition in humans. Nat Commun 11:4917. https://doi.org/10.1038/s41467-020-18680-6
103. Sha Q-Q, Zhu Y-Z, Li S, Jiang Y, Chen L, Sun X-H, Shen L, Ou X-H, Fan H-Y (2020b) Characterization of zygotic genome activation-dependent maternal mRNA clearance in mouse. Nucleic Acids Res 48:879–894. https://doi.org/10.1093/nar/gkz1111
104. Xue Z, Huang K, Cai C, Cai L, Jiang C, Feng Y, Liu Z, Zeng Q, Cheng L, Sun YE, Liu J, Horvath S, Fan G (2013) Genetic programs in human and mouse early embryos revealed by single-cell RNA sequencing. Nature 500:593–597. https://doi.org/10.1038/nature12364
105. Alizadeh Z, Kageyama S, Aoki F (2005) Degradation of maternal mRNA in mouse embryos: selective degradation of specific mRNAs after fertilization. Mol Reprod Dev 72:281–290. https://doi.org/10.1002/mrd.20340
106. Fulka H, Aoki F (2016) Nucleolus precursor bodies and ribosome biogenesis in early mammalian embryos: old theories and new discoveries. Biol Reprod 94(143):1–8. https://doi.org/10.1095/biolreprod.115.136093
107. Aguirre-Lavin T, Adenot P, Bonnet-Garnier A, Lehmann G, Fleurot R, Boulesteix C, Debey P, Beaujean N (2012) 3D-FISH analysis of embryonic nuclei in mouse highlights several abrupt changes of nuclear organization during preimplantation development. BMC Dev Biol 12:30. https://doi.org/10.1186/1471-213X-12-30

108. Lee MT, Bonneau AR, Giraldez AJ (2014) Zygotic genome activation during the maternal-to-zygotic transition. Annu Rev Cell Dev Biol 30:581–613. https://doi.org/10.1146/annurev-cellbio-100913-013027
109. Schultz RM, Stein P, Svoboda P (2018) The oocyte-to-embryo transition in mouse: past, present, and future†. Biol Reprod 99:160–174. https://doi.org/10.1093/biolre/ioy013
110. Tadros W, Lipshitz HD (2009) The maternal-to-zygotic transition: a play in two acts. Development 136:3033–3042. https://doi.org/10.1242/dev.033183
111. Anger M, Radonova L, Horakova A, Sekach D, Charousova M (2021) Impact of global transcriptional silencing on cell cycle regulation and chromosome segregation in early mammalian embryos. Int J Mol Sci 22:9073. https://doi.org/10.3390/ijms22169073
112. Jukam D, Shariati SAM, Skotheim JM (2017) Zygotic genome activation in vertebrates. Dev Cell 42:316–332. https://doi.org/10.1016/j.devcel.2017.07.026
113. Aoki F, Worrad DM, Schultz RM (1997) Regulation of transcriptional activity during the first and second cell cycles in the preimplantation mouse embryo. Dev Biol 181:296–307. https://doi.org/10.1006/dbio.1996.8466
114. Hamatani T, Carter MG, Sharov AA, Ko MSH (2004) Dynamics of global gene expression changes during mouse preimplantation development. Dev Cell 6:117–131. https://doi.org/10.1016/S1534-5807(03)00373-3
115. Latham KE, Garrels JI, Chang C, Solter D (1991) Quantitative analysis of protein synthesis in mouse embryos. I. Extensive reprogramming at the one- and two-cell stages. Development 112:921–932. https://doi.org/10.1242/dev.112.4.921
116. Vassena R, Boué S, González-Roca E, Aran B, Auer H, Veiga A, Belmonte JCI (2011) Waves of early transcriptional activation and pluripotency program initiation during human preimplantation development. Development 138:3699–3709. https://doi.org/10.1242/dev.064741
117. Graf A, Krebs S, Zakhartchenko V, Schwalb B, Blum H, Wolf E (2014) Fine mapping of genome activation in bovine embryos by RNA sequencing. Proc Natl Acad Sci U S A 111:4139–4144. https://doi.org/10.1073/pnas.1321569111
118. Memili E, First NL (2000) Zygotic and embryonic gene expression in cow: a review of timing and mechanisms of early gene expression as compared with other species. Zygote 8:87–96. https://doi.org/10.1017/S0967199400000861
119. Misirlioglu M, Page GP, Sagirkaya H, Kaya A, Parrish JJ, First NL, Memili E (2006) Dynamics of global transcriptome in bovine matured oocytes and preimplantation embryos. Proc Natl Acad Sci U S A 103:18905–18910. https://doi.org/10.1073/pnas.0608247103
120. Engel W, Zenzes MT, Schmid M (1977) Activation of mouse ribosomal RNA genes at the 2-cell stage. Hum Genet 38:57–63. https://doi.org/10.1007/BF00295808
121. Knowland J, Graham C (1972) RNA synthesis at the two-cell stage of mouse development. J Embryol Exp Morphol 27:167–176
122. Zatsepina O, Baly C, Chebrout M, Debey P (2003) The step-wise assembly of a functional nucleolus in preimplantation mouse embryos involves the Cajal (coiled) body. Dev Biol 253:66–83. https://doi.org/10.1006/dbio.2002.0865
123. Morovic M, Strejcek F, Nakagawa S, Deshmukh RS, Murin M, Benc M, Fulka H, Kyogoku H, Pendovski L, Fulka J, Laurincik J (2017) Mouse oocytes nucleoli rescue embryonic development of porcine enucleolated oocytes. Zygote 25:675–685. https://doi.org/10.1017/S0967199417000491
124. Benc M, Martinkova S, Rychtarova J, Fulka J, Bartkova A, Fulka H, Laurincik J (2020) Assessing the effect of interspecies oocyte nucleolar material dosage on embryonic development. Theriogenology 155:17–24. https://doi.org/10.1016/j.theriogenology.2020.06.001
125. Kyogoku H, Fulka J, Wakayama T, Miyano T (2014) De novo formation of nucleoli in developing mouse embryos originating from enucleolated zygotes. Development (Cambridge, England) 141:2255–2259. https://doi.org/10.1242/dev.106948
126. Niwa H, Miyazaki J, Smith AG (2000) Quantitative expression of Oct-3/4 defines differentiation, dedifferentiation or self-renewal of ES cells. Nat Genet 24:372–376. https://doi.org/10.1038/74199

127. Pesce M, Wang X, Wolgemuth DJ, Schöler H (1998) Differential expression of the Oct-4 transcription factor during mouse germ cell differentiation. Mech Dev 71:89–98. https://doi.org/10.1016/s0925-4773(98)00002-1
128. Parfenov VN, Pochukalina GN, Davis DS, Reinbold R, Schöler HR, Murti KG (2003) Nuclear distribution of Oct-4 transcription factor in transcriptionally active and inactive mouse oocytes and its relation to RNA polymerase II and splicing factors. J Cell Biochem 89:720–732. https://doi.org/10.1002/jcb.10545
129. Foygel K, Choi B, Jun S, Leong DE, Lee A, Wong CC, Zuo E, Eckart M, Reijo Pera RA, Wong WH, Yao MWM (2008) A novel and critical role for Oct4 as a regulator of the maternal-embryonic transition. PLoS One 3:e4109. https://doi.org/10.1371/journal.pone.0004109
130. Wu G, Schöler HR (2014) Role of Oct4 in the early embryo development. Cell Regen (London, England) 3:7. https://doi.org/10.1186/2045-9769-3-7
131. Ogushi S, Yamagata K, Obuse C, Furuta K, Wakayama T, Matzuk MM, Saitou M (2017) Reconstitution of the oocyte nucleolus in mice through a single nucleolar protein, NPM2. J Cell Sci 130:2416–2429. https://doi.org/10.1242/jcs.195875
132. Laskey RA, Honda BM, Mills AD, Finch JT (1978) Nucleosomes are assembled by an acidic protein which binds histones and transfers them to DNA. Nature 275:416–420. https://doi.org/10.1038/275416a0
133. Philpott A, Leno GH, Laskey RA (1991) Sperm decondensation in Xenopus egg cytoplasm is mediated by nucleoplasmin. Cell 65:569–578. https://doi.org/10.1016/0092-8674(91)90089-H
134. Fulka H, Langerova A (2014) The maternal nucleolus plays a key role in centromere satellite maintenance during the oocyte to embryo transition. Development 141:1694–1704. https://doi.org/10.1242/dev.105940
135. Probst AV, Almouzni G (2011) Heterochromatin establishment in the context of genome-wide epigenetic reprogramming. Trends Genet 27:177–185. https://doi.org/10.1016/j.tig.2011.02.002
136. Quinodoz SA, Ollikainen N, Tabak B, Palla A, Schmidt JM, Detmar E, Lai MM, Shishkin AA, Bhat P, Takei Y, Trinh V, Aznauryan E, Russell P, Cheng C, Jovanovic M, Chow A, Cai L, McDonel P, Garber M, Guttman M (2018) Higher-order inter-chromosomal hubs shape 3D genome organization in the nucleus. Cell 174:744–757.e24. https://doi.org/10.1016/j.cell.2018.05.024
137. Vertii A, Ou J, Yu J, Yan A, Pagès H, Liu H, Zhu LJ, Kaufman PD (2019) Two contrasting classes of nucleolus-associated domains in mouse fibroblast heterochromatin. Genome Res 29:1235–1249. https://doi.org/10.1101/gr.247072.118
138. Wilkinson AL, Zorzan I, Rugg-Gunn PJ (2023) Epigenetic regulation of early human embryo development. Cell Stem Cell 30:1569–1584. https://doi.org/10.1016/j.stem.2023.09.010
139. Manosalva I, González A (2010) Aging changes the chromatin configuration and histone methylation of mouse oocytes at germinal vesicle stage. Theriogenology 74:1539–1547. https://doi.org/10.1016/j.theriogenology.2010.06.024
140. Petri T, Dankert D, Demond H, Wennemuth G, Horsthemke B, Grümmer R (2020) In vitro postovulatory oocyte aging affects H3K9 trimethylation in two-cell embryos after IVF. Ann Anat 227:151424. https://doi.org/10.1016/j.aanat.2019.151424
141. Shao G-B, Wang J, Zhang L-P, Wu C-Y, Jin J, Sang J-R, Lu H-Y, Gong A-H, Du F-Y, Peng W-X (2015) Aging alters histone H3 lysine 4 methylation in mouse germinal vesicle stage oocytes. Reprod Fertil Dev 27:419–426. https://doi.org/10.1071/RD13293
142. Bar-Sadeh B, Rudnizky S, Pnueli L, Bentley GR, Stöger R, Kaplan A, Melamed P (2020) Unravelling the role of epigenetics in reproductive adaptations to early-life environment. Nat Rev Endocrinol 16:519–533. https://doi.org/10.1038/s41574-020-0370-8

Breaking Trends: Large Animal Models to Study Spindle Assembly and Chromosome Segregation in Human Oocytes

Ainhoa Larreategui Aparicio, Thaiz Emerencia, and Marta de Ruijter-Villani

Abstract Human oocytes are extremely prone to chromosome segregation errors. Although the mouse model has been extremely helpful in the study of oocyte meiosis, recent research has highlighted several important differences between murine and human oocytes in the mechanisms regulating spindle assembly and chromosome segregation. These limitations have led researchers to look for alternative animal models that would more closely recapitulate human meiosis. Oocytes from large animals such as cows, pigs, sheep, and horses have proved to share several features with human oocytes making them a useful addition to studies on human oocyte meiosis. In particular, the length of meiosis, the mechanisms of microtubule nucleation, the oocyte size and the aneuploidy rate in large animal oocytes closely resemble their human counterpart. Therefore, understanding the molecular mechanisms driving correct chromosome segregation in large animal oocytes can provide valuable information that can be translated to human meiosis.

Keywords Oocyte · Meiosis · Spindle · Chromosome segregation · Large animals

---

A. L. Aparicio
Department of Clinical Sciences, Faculty of Veterinary Medicine, Utrecht University, Utrecht, The Netherlands

Hubrecht Institute, Utrecht, The Netherlands

T. Emerencia
Department of Clinical Sciences, Faculty of Veterinary Medicine, Utrecht University, Utrecht, The Netherlands

M. de Ruijter-Villani (✉)
Department of Clinical Sciences, Faculty of Veterinary Medicine, Utrecht University, Utrecht, The Netherlands

Hubrecht Institute, Utrecht, The Netherlands

Division Woman and Baby, University Medical Centrum Utrecht, Utrecht, The Netherlands
e-mail: m.villani@uu.nl

B. Mogessie (ed.), *Cellular Architecture and Dynamics in Female Meiosis*,
https://doi.org/10.1007/978-3-031-97173-0_7

Introduction

Human oocytes very often show an incorrect number of chromosomes, a condition known as aneuploidy. When an aneuploid oocyte is fertilized, this will give rise to an aneuploid embryo. Aneuploidy in embryos is the major cause of implantation failure, miscarriage and congenital birth defects in human reproduction. Approximately 20% of oocytes of young women between 20 and 30 years old is aneuploid [1]. The proportion of aneuploid oocytes further increases to 50% in women older than 35 years [1]. Oocyte aneuploidy originates from errors in the segregation of the chromosomes during the meiotic divisions. Meiotic divisions consist of two successive divisions, meiosis I and II, which reduce the genetic content by half.

Although extremely common and with severe consequences for human reproductive health, the mechanisms that drive chromosome segregation errors in human oocytes are not completely understood. During the last decade several studies on human oocytes have led to a better understanding of human oocyte meiosis. However, access to human oocytes is limited by availability and ethical concerns and although non-human primates might represent the closest animal model, some of the aforementioned limitations also apply.

The use of non-primate animal models is not only ethically more acceptable, but also allows the use of a larger number of oocytes under more controlled conditions. The model organism most commonly employed for studies on mammalian meiosis has so far been the mouse model. The use of mice for these studies has been meaningful as the basic processes of oocyte meiosis are relatively conserved among mammals. Moreover, the extensive use of the murine model in scientific research has fuelled the development of a large variety of remarkably powerful molecular tools which allow sophisticated genetic manipulations, further extending its advantages as a model to study early mammalian development. Nevertheless, recent research has highlighted several important differences between murine and human oocytes in the mechanisms regulating spindle assembly and chromosome segregation [2]. These limitations in the translational potential of murine oocyte research have led researchers to look for alternative animal models that would more closely recapitulate human meiosis. Several research groups worldwide have started to use large animals such as cows, pigs, sheep and horses to study mechanisms underlying spindle assembly and chromosome segregation. The oocytes of these species have proved to share several features with human oocytes making them a useful addition to studies on human and murine oocytes (Table 1). In particular, the length of meiosis, the size, and the aneuploidy rate in large animal oocytes closely resemble their human counterpart. Moreover, there are several analogies between women and large animals in other aspects of their reproductive biology. Large animals are monoovulatory (with the exception of the pig), with a long follicular phase and a similar time course to ovulation; they exhibit prolonged meiotic oocyte arrest and (some of them) have a long time interval to reproductive senescence.

Table 1 Cross-species comparison of reproductive cycles and oocyte meiosis parameters

	Human	Murine	Bovine	Equine	Porcine	Ovine	References
Puberty onset	10–15 years	27–30 days	14 months	14 months	200 days	30 weeks	(Ahima, Dushay et al. 1997; Nah, Park et al. 2011; Hu, Yan 2022; Tummaruk, Tantasuparuk et al. 2009)
Ovulation type	Monovulatory	Polyovulatory	Monovulatory	Monovulatory	Polyovulatory	Monovulatory	–
Reproductive cycle length (days)	28	4–5	21	21	21	19	[3]; (Brown, Fabre Nys et al. 2014)
Follicular phase length (days)	14	10–17	11	14	7	4	[3]; (Pedersen 1970)
Mature oocyte size (μm)	150–180	90–100	150–180	150–180	120	–	(Miyano, Ogushi et al. 2007)
Meiotic resumption (hours)	4–8[a]	4–6	6–8	8–16	27	6–7	[4–6]; (Miyano, Ogushi et al. 2007)
Meiotic division duration (hours)	20–24[a]	8	15–16	16–18	15	17–18	[2, 4–7]
Aneuploidy rate in MII oocytes	<19 years: ~30% 20–30 years: ~20% >33 years: 50%	2–3 months: 1.7% 16 months: 23.1%	20–50% (IVM)	<15 years: 15% >16 years: 55%	<15 months: 12% 69 months: 9% 92 months: 10%	–	[2, 8–10]

[a] Data from immature GV oocytes retrieved from women undergoing controlled ovarian stimulation

In this review, we discuss the use of large animals as models to study oocyte meiosis and in particular the processes of spindle assembly and of chromosome segregation. We will highlight the limitations and the advantages of these models, with particular attention to those aspects that are highly translational to their human counterpart.

Fetal Oocyte Development and Meiotic Recombination

During early female embryogenesis, primordial germ cells (PGCs) proliferate and migrate to the developing ovary. PGCs migration takes place from week 5 in human pregnancy [11], between days 17 and 21 in sheep [12], between days 18 and 31 in cattle [13], between 20 and 30 days in horses [14] and between days 7 to 11 in mouse pregnancies [15] (Fig. 1). The number of PGCs decreases drastically when they initiate meiosis and undergo a single round of DNA duplication, becoming the primordial oocytes [16]. The onset of meiosis varies between different mammalian species, although it always occurs during fetal life (Fig. 1). Meiosis is initiated in

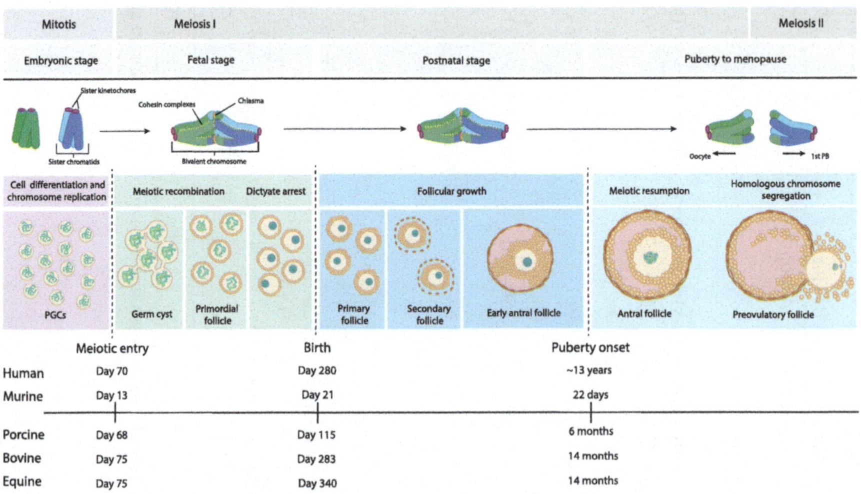

Fig. 1 Schematic representation of oocyte development and folliculogenesis and cross-species comparison of meiotic entry, birth and puberty onset. During early embryonic development primordial germ cells (PGCs) migrate to the developing ovary. These germ cells divide and form a group of interconnected cells called oogonium. The number of PGCs decreases drastically when they initiate meiosis and undergo a single round of DNA duplication. The timing of meiotic entry varies among different species but always occurs during fetal life. The early stages of meiosis involve the formation of bivalents, where homologous chromosomes pair up and undergo genetic recombination. Once recombination is complete, meiosis is arrested and oocytes enter a state called dictyate arrest. The oocytes associate with pre-granulosa cells to form primordial follicles. Folliculogenesis is characterized by the development of the follicles from primordial to antral stage. Only after puberty antral follicles are able to develop further to pre-ovulatory follicles and their oocytes are able to resume meiosis and complete the first meiotic division

human females from day 70 (Garcia, Dietrich et al. 1987), in sheep from day 52 [17], in cattle from day 75 [18], in horses from day 75 [19], and in mice from day 13 of pregnancy [20] (Jones 2007; Garcia, Dietrich et al. 1987). The first stage of meiosis, also termed prophase, is characterized by structural chromosome rearrangements. The replicated chromosomes, composed of two sister chromatids held together by cohesion complexes, are paired with their homologues (Fig. 1). The two sister chromatids of each homologous chromosome become linked as a pair by the synaptonemal complex. This meiosis-specific protein complex holds the homologous chromosomes in close proximity while they undergo a process of genetic recombination [21]. During this process DNA double-strand breaks (DSBs) are induced in homologues chromosome and subsequently repaired leading to crossover or non-crossover products. On average only one to four DSBs per homologous chromosome pair are repaired as crossover, where the strands of genomic DNA of the two homologues is exchanged. Repair of DNA DSBs between non-sister chromatids results in the appearance of chiasmata. The cohesin complexes that are located in the chromosome arms distally to the chiasmata hold the homologues chromosomes together in a typical configuration also referred to as a bivalent [22] (Baudat, Imai et al. 2013). Once recombination is complete, meiosis is arrested and oocytes enter a state called dictyate arrest. At the same time oocytes become associated to pre-granulosa cells to form a structure called the primordial follicle. In women, sheep, and cattle follicular growth starts pre-natally, while in rodents, the first signs of follicle growth are observed in the post-natal period [16, 23] (Kerr, Myers et al. 2013; Kallen 2022; Tingen, Kim et al. 2009). During the pre-natal period and childhood follicles are able to grow and differentiate to preantral and antral follicles, but degenerate at various stages of their development before they reach pre-ovulatory sizes.

Puberty Onset and Reproductive Cycle

Puberty onset is characterized by the activation of the hypothalamic-pituitary-ovarian axis, which, in turn, stimulates follicular growth and ovarian steroids production. Puberty occurs around 22 days in mice, 6 months in pigs, 14 months in cows and mares, and 10–15 years in girls (Fig. 1) (Reiland 1978; Day, Nogueira 2013; Lea, England 2019). From this moment on females are reproductively mature and exhibit reproductive cycles. Reproductive cycles are composed by a follicular phase and a luteal phase (see Table 1 for details in the different species). The follicular phase is characterized by the growth of a group of follicles also called a "cohort." This phase terminates with the ovulation of the dominant pre-ovulatory follicle (in mono-ovulatory species), secondary to a surge of Luteinizing Hormone (LH) from the pituitary gland. The luteal phase is characterized by the presence on the ovary of a corpus luteum, an endocrine structure which develops from the remnants of the ovulated follicle. The follicular phase of women, cows, and horses is extremely similar with a length of approximately 14 days of the 28-day menstrual cycle in women, about 11 days of the 21-day oestrous cycle in cows, and about 14 days of the 21 day oestrous cycle in horses [3].

Acquisition of Developmental Competence and Meiotic Resumption

Primordial follicles remain in a resting state until they are recruited and start to grow. During follicular growth, the oocyte also increases in size as it accumulates mRNAs and organelles, acquires the ability to resume and complete meiosis, and secretes a specialized extracellular matrix called zona pellucida [24, 25]. During this growing phase, the oocyte is intimately connected with granulosa cells through gap junctions. Through these heterologous gap junctions granulosa cells support oocyte growth and prevent its premature meiotic resumption by direct transfer of metabolic substrates and meiosis-arresting signals [26, 27]. The pituitary LH surge induces the resumption of meiosis in the oocyte within the pre-ovulatory follicle [28]. Meiotic resumption is morphologically characterized by the nuclear envelope breakdown (NEBD) also called germinal-vesicle breakdown in oocytes (GVBD). During NEBD the chromosomes fully condense and are released into the cytoplasm. After NEBD, the meiotic spindle assembles and aligns the bivalent chromosomes. During anaphase of the first meiotic division (anaphase I) separase releases the cohesin complexes on the chromosome arms of the homologous chromosomes and allows the two homologues to be segregated by the spindle microtubules (Wassmann 2022). Only one set of homologous chromosomes is retained in the oocyte, while the other is extruded in the first polar body. The oocyte then directly enters the second meiotic division and the cell cycle gets further arrested at metaphase II (MII) until fertilization. At fertilization, the oocyte fuses with the sperm cell triggering anaphase II and the second polar body formation. During anaphase I, the cohesin complexes present on the centromeric region of the chromosomes, which hold the sister of each homologue together, are protected from the separase activity through the action of the Shugoshin-PP2A complex. During anaphase II, instead, centromeric cohesion is no longer protected and is cleaved by separase allowing sister chromosomes segregation [29].

Meiosis resumption is dependent on the activation of the maturation-promoting factor (MPF) heterodimer comprised of the regulatory subunit cyclin B (CCNB) and the catalytic subunit cyclin-dependent kinase 1 (CDK1) [28]. Cyclin B continues to accumulate in the oocyte cytoplasm until metaphase I, and then is degraded by the anaphase-promoting complex-cyclosome (APC/C)-dependent. After the first polar body extrusion (PBE), MPF reactivation by cyclin B1 re-accumulation is required to allow the oocytes to progress directly into metaphase II without a defined interphase [28] (Strączyńska, Papis et al. 2022; Adhikari, Carroll 2018). The time necessary to resume meiosis and the duration of the meiotic division in vitro is different in oocytes from different mammalian species (Table 1). The time necessary to resume meiosis is relatively short in human, mouse, cow and sheep oocytes, ranging between 4 and 8 h [4, 5, 7, 30, 31]. By contrast, horse and pig oocytes need more time to resume meiosis, respectively 8–16 and 27 h [6, 31]. The duration of the meiotic divisions, calculated as the time necessary to reach the MII stage from NEBD, also varies dramatically between different mammalians. Mouse oocytes undergo meiotic divisions in only 8 h, while large animal and human oocytes require respectively 15–18 and 20–24 h. This is probably the consequence of the different strategies used to assemble the meiotic spindles in the different species [2, 5, 7, 30]. Interestingly, the in vitro maturation of large animal and mouse oocytes shows a relatively high efficiency, with 80% of immature oocytes able to reach the MII stage. Instead, only 60% of immature human oocytes are able to reach the MII stage in vitro [32].

Actin-Driven Chromosome Clustering

Human oocytes show an extreme delay in the onset of microtubule nucleation after NEBD, with a time between NEBD and the onset of microtubule nucleation from kinetochores of 2–4 h [7, 33]. Although shorter, a delay in the onset of microtubule nucleation after NEBD is also visible in porcine oocytes [34], while in mouse oocytes no delay is visible and, instead, microtubule nucleation starts before NEBD [35]. To avoid the dispersion of chromosomes in the large oocyte cytoplasm during this long gap phase before the onset of spindle assembly, human and porcine oocytes cluster their chromosomes through their actin cytoskeleton [34] (Fig. 2). Dynamic actin cables are formed by Formin-2 and Spire from the disassembling nuclear envelope and these cables migrate toward the center of the nucleus transporting the chromosomes along with them. The actin-dependent clustering of the chromosomes is achieved through two different mechanisms. The first mechanism is through the interaction of actin with the kinetochores, which is mainly responsible for the movement and clustering of the small chromosomes. The large chromosomes seem to mainly interact with actin through their chromosome arms [34]. Actin-driven chromosome clustering is important to facilitate the capture of the chromosomes by the microtubules. This clustering mechanism has not been studied in other mammalian species, however the observations that in bovine oocytes chromosomes cluster before microtubule nucleation onset and that in mice oocytes chromosomes can cluster in absence of microtubules, suggest that this might be a conserved mechanism across species [2, 36].

Kinetics and Mechanisms of Spindle Assembly

Correct chromosome segregation during oocyte meiosis is dependent on the formation of a functional bipolar spindle. Mammalian oocytes lack canonical centrosomes, the main microtubule organizing centers (MTOC) that orchestrate spindle assembly in mitotic and male meiotic cells. The centrosomes, formed by two centrioles and pericentrosomal material, are degraded during oogenesis before the onset of the first meiotic division [37–39]. Consequently, mammalian oocytes use alternative mechanisms to assemble and stabilize their spindles. Interestingly, the

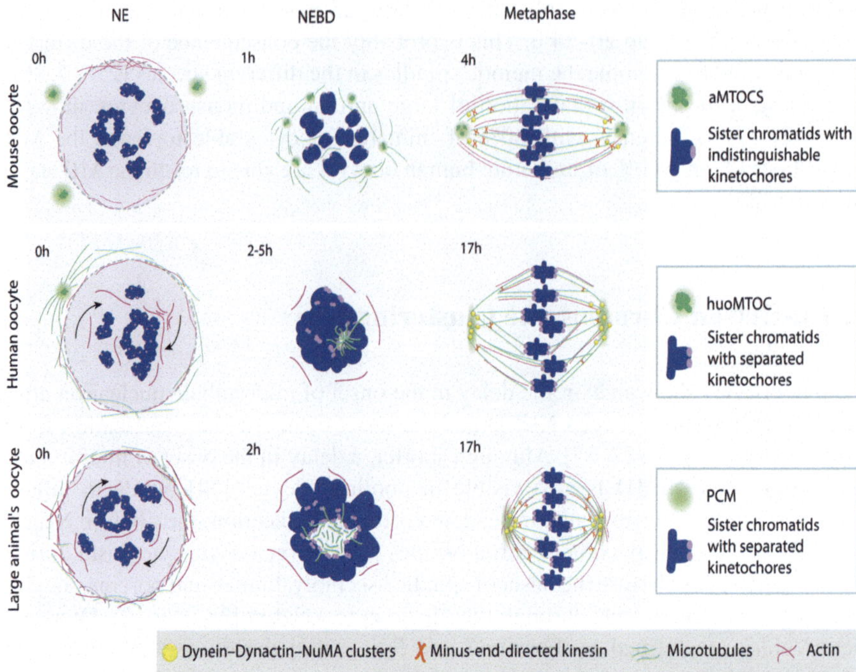

Fig. 2 Meiotic spindle assembly in mice, human and large animal oocytes. Schematic representation of the different mechanisms used by the different species to assemble and stabilize a meiotic acentrosomal spindle. Time from nuclear envelope break down (NEBD) is indicated. Before NEBD, acentrosomal Microtubule Organizing Centers (aMTOCs), in mice, and human Microtubule Organizing Centers (huoMTOC), in human oocytes, initiate microtubule nucleation, however, while in mice this process continue after NEBD, the huoMTOC and its associated microtubules disappear at NEBD. After NEBD actin cables forms in human and pig oocytes and cluster chromosomes, facilitating their subsequent capture by the microtubules. A few hours after NEBD the microtubules starts to be nucleated from the kinetochores in human and large animal oocytes, and then the microtubules form an aster within the clustered chromosomes. In murine oocytes, the process of microtubule nucleation is much faster due to the presence of aMTOCs, which quickly form a multipolar spindle. Once the spindle is formed, mouse oocytes clusters their aMTOCs to form two defined spindle poles. Unlike in human and large animals' oocytes where the clustering of the microtubules minus ends in the two spindle poles is achieved by the combined action of NuMA and dynein-dynactin. Several minus ends directed kinesins acts as a spindle stabilizing factor. While in mouse oocytes, the distance between sister kinetochores is very small, in human and bovine oocytes sister kinetochores are often visible as separated entities. The distance between sister kinetochores represent a challenge for the correct alignment of the bivalents in metaphase I

mechanisms used differ between mammalian species, highlighting species specific adaptions to achieve the formation of an acentriolar bipolar spindle.

In human oocytes, spindle assembly during meiosis I is an extremely long process, requiring approximately 16 h [7]. Approximately 2–4 h after NEBD microtubules starts to be nucleated from the kinetochores [7, 33]. Within 1 h from the onset of nucleation, the microtubules form an aster within the clustered chromosomes

(Fig. 2). From this stage, an additional 8–10 h are required before a bipolar spindle, with chromosomes congressed in the metaphase plate, is assembled [7, 34]. Large animal oocytes, such as cows, pigs, horses and sheep, also assemble their meiotic spindle in an extremely slow fashion. Similarly to humans, cow oocytes nucleate microtubules from the kinetochores. These kinetochore-derived microtubules form an aster within the clustered chromosomes [2, 33, 40]. The time between NEBD and bipolar spindle formation is 8–10 h [2, 6, 31, 40–43]. In stark contrast, bipolar spindle assembly only requires 4 h in mouse oocytes [35]. These differences in spindle assembly kinetics are probably the results of the different mechanisms driving microtubule nucleation and spindle assembly in the different species (Fig. 3). Indeed, while all mammalian oocytes lack canonical centrosomes, some species, such as mice, possess multiple acentriolar microtubule organizing centers (aMTOCs) [35]. These aMTOCs lack centrioles yet contain many pericentrosomal proteins, and efficiently replace centrosome function during spindle assembly [35, 44, 45]. Before NEBD, aMTOCs form from the cytoplasmic microtubule network. Subsequently, they are pulled centripetally toward the nuclear envelope by the motor protein dynein and start nucleating microtubules [35]. After NEBD, aMTOCs remain the predominant nucleation sites of new microtubules and mediate a RanGTP-dependent increase in microtubule mass (Fig. 3). The microtubules then organize to form a multipolar spindle or microtubule-ball which lacks well-defined poles. The aMTOCs then cluster forming two defined spindle poles [35].

A recent study reports the existence of one human oocyte microtubule organizing center (huoMTOC) [33]. The huoMTOC is able to recruit selected centrosomal (CCP110 and DISC1) and microtubule associated (TACC3 and CKAP6) proteins known to act as microtubule regulatory factors [33]. During early prophase, approximately 6 h before NEBD, a single huoMTOC is formed near the cell cortex of human oocytes and progressively migrates toward the nuclear envelope. During late prophase, approximately 2 h before NEBD the huoMTOC starts nucleating microtubules causing an asymmetrical microtubule distribution around the nuclear envelope. Within 1 h after NEBD the huoMTOC and its associated microtubules disappear and no microtubule nucleation is observable for the first 2–4 h after NEBD. After this time CCP110, DISC1, TACC3 and CKAP6 re-localize to the kinetochores where microtubule nucleation is resumed (Fig. 3). Unlike the mouse, the huoMTOC does not act as a microtubule nucleating center after NEBD, nor relocates at the spindle poles during spindle bipolarization. Therefore the huoMTOC seems to have a function completely different from mice aMTOCs, which consists mainly in enriching selected microtubule regulatory factors in close proximity to the nuclear envelope, so that, after NEBD, they can promptly relocate to the kinetochores to support microtubule nucleation. Interestingly, the huoMTOC seems to be specific to human oocytes, as this particular structure was not detected in mouse or pig oocytes [33]. Several studies reported that large animal oocytes, such as cow, pig, and sheep oocytes, do not contain canonical aMTOCs [31, 46], however, similarly to mouse oocytes, they are able to enrich microtubule regulatory factors in close proximity to the spindle microtubules [45]. This suggests that oocytes of different mammalian species use different strategies to enrich

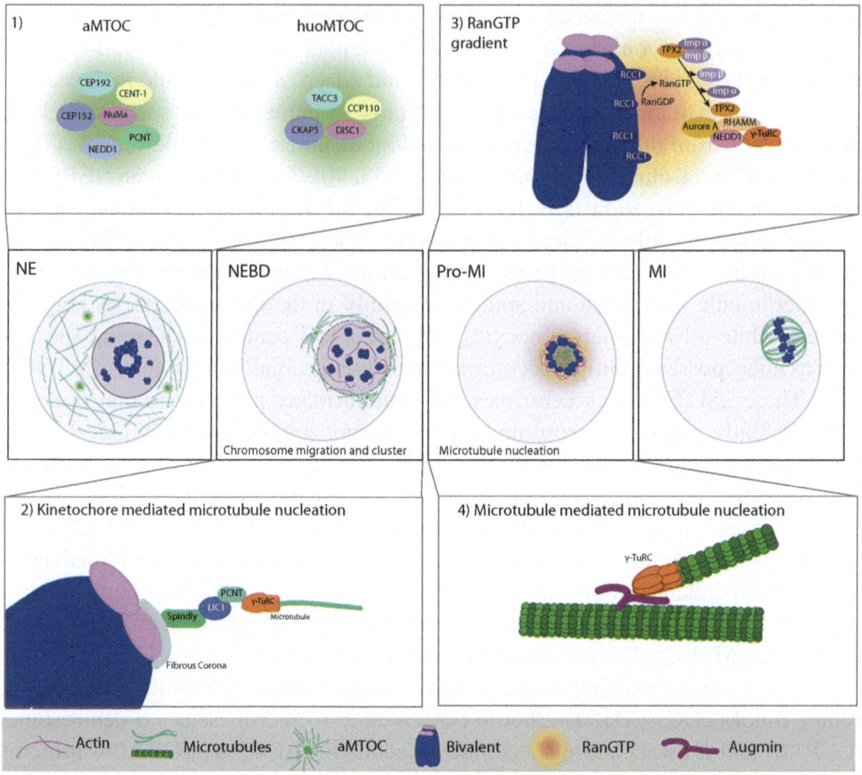

Fig. 3 Microtubule nucleation pathways in mammalian oocytes. (1) In mouse oocytes, multiple aMTOCS are responsible for nucleating microtubules and defining the spindle poles before and after NEBD. The aMTOCs contain pericentrosome proteins such as pericentrin (PCNT), Peri Centriolar Matrix 1 (CENT-1), Nuclear Mitotic Apparatus (NuMA), centrosomal proteins 192 and 152 (CEP192 and CEP152). Human oocytes show only one huoMTOC before GVBD. The huoMTOC recruits selected centrosomal (CCP110 and DISC1) and microtubule associated (TACC3 and CKAP6) proteins. (2) after NEBD microtubules nucleation from kinetochores is driven by the fibrous corona. This process relies on the interaction between the LIC1 subunit of the dynein motor complex with the γ-tubulin-tethering protein pericentrin (PCNT), which, in turn, directly promotes microtubule nucleation. (3) After nuclear envelope break down the RAN-GTP is generated around the chromosomes and releases spindle assembly factors, such as TPX2, from inhibitory complexes containing importin. (4) Following the onset of microtubule nucleation, Augmin nucleates additional microtubules by recruiting γ-tubulin ring complex (γ-TuRC), a potent microtubules nucleator

microtubule regulatory factors in close proximity to the chromosomes, avoiding their dispersion throughout the large oocyte cytoplasm.

A recent study showed that, in human mitotic cells, microtubule nucleation from kinetochores is driven by the fibrous corona, a transient large fibrous meshwork which assembles on the outer kinetochores [40]. This process relies on the LIC1 subunit of the dynein motor complex, a known component of the fibrous corona (Fig. 3). LIC1 interacts with the γ-tubulin-tethering protein pericentrin (PCNT), which, in turn, directly promotes microtubule nucleation [40]. This mechanism seems to be conserved in mammalian oocytes, since the unusually large fibrous coronas of bovine oocytes also drive microtubule nucleation from the kinetochores. Moreover, preventing a key step in fibrous corona expansion in bovine oocytes, besides impairing microtubule nucleation from the kinetochores, caused defects in chromosome congression and spindle assembly [40].

Microtubule nucleation in mammalian oocytes is highly dependent on the RanGTP pathway. The Ran-GTP is generated around chromosomes, by the regulator of chromosome condensation 1 (RCC1), a guanine nucleotide exchange factor of Ran that associates with chromosomes and produces the active Ran-GTP. Ran-GTP, in turn, releases spindle assembly factors, such as TPX2, from inhibitory complexes containing importin (Fig. 3). These spindle assembly factors become then locally active around chromosomes [47, 48] promoting microtubule nucleation either at aMTOCs in mouse oocytes or at the kinetochores in human and large animal oocytes [7, 35]. Inhibition of Ran-GTP function in murine, human and porcine oocytes greatly delays microtubule nucleation and leads to either failure to form the meiotic spindle or spindle assembly defects [34, 35, 49].

Following the onset of microtubule nucleation, the nucleation of additional microtubules is further promoted by the Augmin pathway (Fig. 3). Augmin initiates microtubule nucleation by recruiting γ-tubulin ring complex (γ-TuRC), a potent microtubules nucleator, to the existing microtubules [50]. Interestingly, this nucleation mechanism is accelerated by the presence of Ran-GTP, suggesting that Ran-GTP is also required for microtubule-dependent microtubule nucleation [51]. Augmin contributes to the microtubule mass during initial spindle assembly in mouse embryonic cells, which similarly to mouse oocytes, assemble their spindles using aMTOCs [52]. Although this pathway has been well characterized in *Drosophila melanogaster* and *Xenopous levis* oocytes, it will be necessary to investigate the role of the Augmin pathway in oocytes that lack aMTOCs such as large animals and human oocytes.

Spindle Bipolarization and Stabilization

A crucial step in meiotic spindle assembly is the organization of a stable bipolar spindle which enables correct chromosome congression and segregation. In order for spindle poles to assemble, the minus end of the spindle microtubules need to be bundled together. In mouse oocytes, where the minus ends of the microtubules are anchored to aMTOCs, this process is achieved by the clustering of the aMTOCs (Fig. 2). In human and large animal oocytes that do not possess conventional aMTOCs, the microtubule minus ends binds to the microtubule cross-linking protein nuclear mitotic apparatus (NuMA). Lack of NuMA or dynein–dynactin results in unfocused spindle poles [2]. Mechanistically, NuMA stabilizes the spindle poles of acentrosomal, aMTOCs-free spindles by cross-linking microtubules at the minus ends and engaging with dynein complexes present at the poles for further pole clustering [2]. Interestingly, while bovine and porcine oocytes show a similar NuMA localization pattern as human oocytes, their spindles are more stable than human oocytes [2, 7]. In fact, after reaching the bipolarized state spindles of human oocytes often fail to maintain their bipolarity, resulting in apolar and monopolar spindle intermediates. It is known that these intermediates promote erroneous microtubule-kinetochore interactions, which, in turn, result in chromosome lagging at anaphase and chromosome segregation errors [7]. This instability seems to depend on the lack of the kinesin superfamily protein C1 (KIFC1), a minus-end-directed kinesin that acts as a spindle stabilizing factor, which, in contrast, is present in mouse, bovine, and porcine oocytes (Fig. 2). KIFC1 seems to ensure the coalescence of clustered microtubule minus ends into two spindle poles. Interestingly, the depletion of KIFC1 in bovine oocytes leads to a spindle pole instability similar to the one observed in human oocytes, and introduction of exogenous KIFC1 in human oocytes seems to rescue spindle instability. It is likely that KIFC1 ensures spindle stabilization through the formation of static cross-links along parallel microtubules at the poles and the alignment of antiparallel microtubules in the central region of the spindle [2].

The spindle of human, mouse, and large animal oocytes is permeated by prominent actin filaments [53]. Actin seems to stabilize the meiotic spindle by promoting the formation of kinetochore fibers (k-fibers), as perturbations of spindle actin induce abnormal k-fibers formation, leading to chromosome alignment and segregation defects [53].

It is probable that other proteins enriched at the microtubules and spindle poles of mammalian oocytes are also involved in the process of spindle stabilization and bipolarization [45].

Mechanisms of Chromosome Mis-Segregation

Human oocytes are extremely prone to chromosome segregation errors, with an incidence of mature eggs being aneuploid of 30% in girls younger than 20 years, 20% in adult women, and 50% in women over 33 years [1]. In the majority (70%) of aneuploid human oocytes, only one single chromosome is affected, while in 30% of the cases, a more complex aneuploidy is seen, with two to ten affected chromosomes [1]. The type of segregation error during the first meiotic division varies depending upon maternal age, with non-disjunction (NDJ) being more frequent in girls younger than 20 years and premature separation of sister chromatids (PSSC) and reverse segregation (RS) being more frequent in women older than 33 years [1].

One of the animals that seems a promising natural model to study the mechanisms driving chromosome segregation errors in oocytes of aged females is the mare. Similarly to human oocytes, in the majority (76%) of aneuploid equine oocytes (matured in vitro) only one single chromosome is affected [9]. Moreover, not only the incidence of oocyte aneuploidy in young (15%; younger than 15 years) and aged (55%; older than 15 years) mares is comparable to that reported for women, but the prevalent mechanism of mis-segregation in aged mares is PSSC [9].

The cow also seems a good model to study oocyte aneuploidy. Bovine oocytes show an incidence of aneuploidy (20–50%; in vitro matured) similar to women [8], however, the influence of maternal age on oocyte aneuploidy in this species in unknown.

Surprisingly, a recent study showed no difference between the frequency of aneuploidy in oocytes from young (12%; 15 months) and aged (9%; 69 months and 10%; 92 months) sows [10]. Therefore, although pigs can be useful to study some aspects of oocyte meiosis, the low frequency of oocyte aneuploidy in this species and the fact that it does not increase with maternal age might suggest that this might not be a good model to study how meiotic segregation errors naturally arise.

Although the mechanisms involved in chromosome segregation errors have not yet been studied in details in large animal oocytes, in the following paragraphs, we have attempted to highlight the chromosome segregation errors conserved in cows, horses, and pigs oocytes.

Chromosome Non-disjunction

Chromosome non-disjunction is the most frequent error observed in young women and the mechanisms driving this error are poorly understood both in human and large animal oocytes. Interestingly, large chromosomes (chromosome 1 and 5) seem to be more prone to NDJ and are more often incorrectly segregated in young girls. The reason why larger chromosomes are more prone to NDJ is still unknown. Based on studies in somatic cells, this might be due to the fact that large chromosomes are frequently positioned at the periphery of the nucleus, and therefore, they need more time to congress and achieve biorientation increasing the likeliness of mis-segregation [54]. If this mechanism is conserved in mammalian oocytes remains to be investigated.

Premature Separation of Sister Chromatids (PSSC) and Reverse Segregation (RS)

Reproductive aging in women increases the risk of premature separation of sister chromatids (PSSC) and reverse segregation (RS) [1]. In particular, acrocentric chromosomes (chromosomes 13–15, 21, and 22) seem to be more often affected by these types of errors. Sister-chromatid cohesion has been shown to weaken in human oocytes upon advancing maternal age causing the disruption of the bivalent structure needed for accurate segregation in meiosis I [55, 56]. Localized centromeric and more systemic cohesion weakening drive the increase in PSSC and RS as women age, by, respectively, inducing the deterioration of the bivalents and increasing the distance between the sister kinetochores, which, in turn, predisposes the sister kinetochores to mis-orient and form bipolar attachment in metaphase I [57–60] (Fig. 4). Progressive loss of cohesin first affects chromosomes with shorter arms, because these have less total cohesin. Moreover, when cohesion is weakened, the bivalents rely entirely on their chiasmata to retain their integrity [61]. Since chiasmata distribute evenly across DNA, small acrocentric chromosomes are more likely to be achiasmate and therefore are more prone to being mis-segregated in reproductively aged women [62].

Similarly to women, PSSC in aged mares seems to proceed from the weakening of centromeric cohesion as evidenced by the increased distance between sister kinetochores in MII oocytes. It is interesting to note that the age-related increase in interkinetochore distance observed in mares (1.33 ± 0.40 μm in mares <15 years, compared to 1.96 ± 0.74 μm in mares >15 years) is of a similar magnitude to that identified in women (0.82 ± 0.03 μm in women <35 years, compared to 1.1 ± 0.03 μm in women >35 years) [1, 9]. A recent study on cattle crossover interference, shows that, similarly to human oocytes, small chromosomes are more likely to be achiasmate and that in aged females an increased count of recombination events enhances the likelihood of an oocyte resulting in a successful live birth [63]. Moreover, similar to human oocytes, the sister kinetochores of each homologous pair are physically separated in bovine bivalents during the first meiotic division [40]. This suggests that bovine oocytes could experience similar challenges in correctly aligning their bivalents during the first meiotic division as human oocytes.

Spindle Instability

Another factor increasing the risk of chromosome segregation errors in human oocytes is the instability of their spindles [7] (Fig. 4). Approximately, 80% of human oocytes form multipolar-spindle intermediates during metaphase, which have been shown to favor chromosome segregation errors [7]. While spindle instability is also detected in oocytes from young women, the observation that oocytes from women of advanced age more often show defective spindle morphology, might indicate that reproductive aging exacerbates spindle instability in human oocytes [2, 64]. Advanced maternal age has been shown to impair spindle stability and the ability to correctly align chromosomes in horse oocytes [65, 66] Interestingly the incidence of chromosome misalignment in MII oocytes from aged mares (47.4% in mare older than 15 years vs. 4.5% in mares younger than 15 years) is quite similar to the incidence reported in women (79% in women between 40 and 45 years old vs. 17% in women between 20 and 25 years old) [64, 65].

Permissive Cellular Check Points

The DNA damage response (DDR) induces cellular arrest in response to endogenous or exogenous DNA damage (Jackson, Bartek 2009). In mammalian oocyte meiosis, the DDR plays an important role during the recombination of homologous chromosomes (Keeney, Lange et al. 2014). The Spindle assembly check point (SAC) prevents cells from entering anaphase until all the kinetochores are correctly attached to spindle microtubules (Marston, Wassmann 2017; Touati, Wassmann 2016). Unattached or incorrectly attached kinetochores recruit SAC components and activate the mitotic check point complex (MCC) composed of MAD2, BUBR1 and BUB3 and CDC20 (Fig. 4). When sequestered by the MCC, CDC20 is unable to activate the anaphase-promoting complex/cyclosome (APC/C) and the progression into anaphase is halted (Marston, Wassmann 2017; Touati, Wassmann 2016). When all the kinetochores are correctly attached, CDC20 is released and is able to bind and activate the APC/C. The APC/C:Cdc20 complex degrades cyclin B (CCNB) and securin, leading to cohesin cleavage and anaphase onset (Fig. 4) (Marston, Wassmann 2017; Touati, Wassmann 2016).

Although all the SAC components are present in mammalian oocytes, the SAC seems to be more permissive in these cells allowing anaphase onset even if kinetochores are still unattached or not correctly attached to spindle microtubules (Greaney, Wei et al. 2018). This is believed to be caused by a dilution of the SAC signal within the large oocyte cytoplasm. Up to now all the studies on the SAC function have been performed on mouse oocytes. Although mouse oocytes are reasonably large (90–100 μm) and show reduced SAC stringency, mammalian oocytes with a size more similar to human oocytes (150–180 μm) might be a better model to study this process [67]. Moreover in mouse oocytes a cross-talk between the DDR and SAC efficiently prevents oocyte maturation in presence of DNA damage (Collins, Lane et al. 2015), this cross-talk does not take place in human oocytes, since they can successfully undergo maturation while harboring experimentally induced DNA damage (Remillard-Labrosse, Dean et al. 2020). Similarly to human oocytes, large animal oocytes require a prolonged time to complete meiosis (Table 1), therefore, it is reasonable to expect that they use similar strategies for preventing the APC/C:Cdc20 complex from precociously initiating anaphase. Finally, both in women and mares, advanced maternal age has been shown to induce a dysregulation of key SAC components [66, 68].

Conclusions

A better understanding of the mechanisms driving chromosome segregation in human oocytes is crucial for understanding potential causes of implantation failure or early miscarriages. While the mouse has traditionally served as animal model for studying oocyte meiosis, large animals, such as cows, horses and pigs, with mechanisms of spindle assembly and rate of oocyte aneuploidy closely matching that of human counterparts, provide a new opportunity for studying these mechanisms in an animal where oocyte aneuploidy naturally occur and overcome the current limitation on the use of human oocytes.

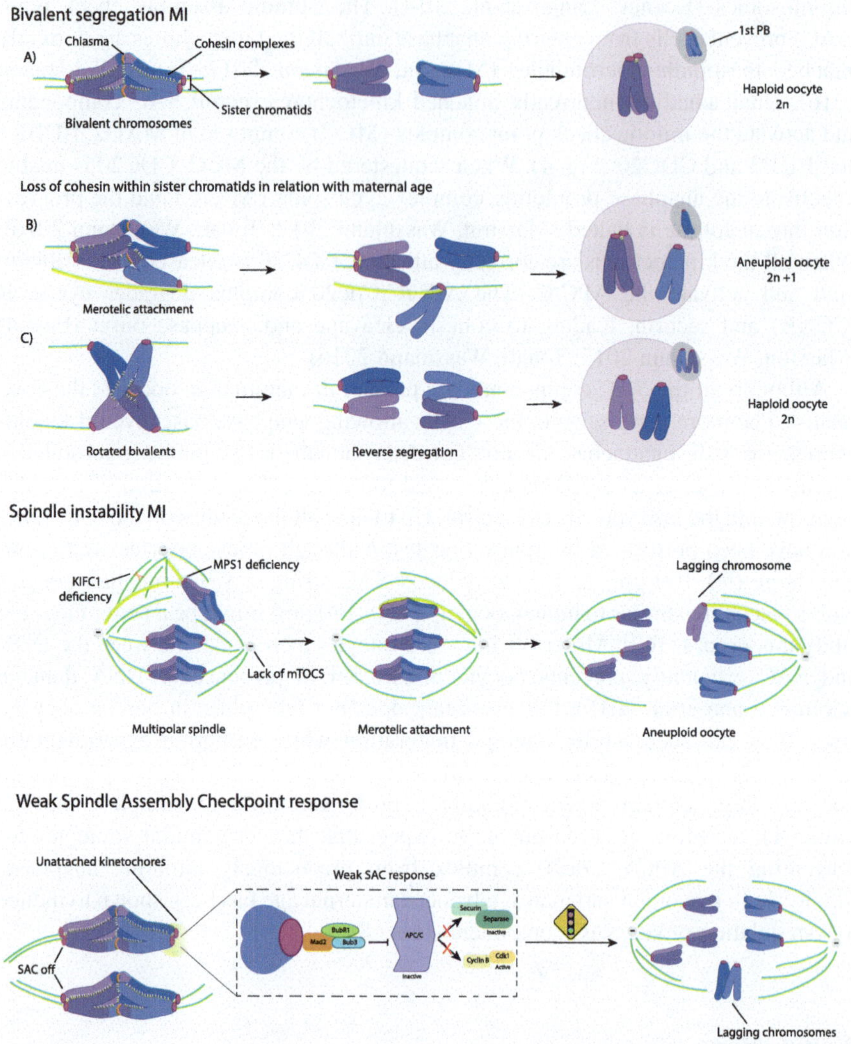

Fig. 4 Schematic representation of the most frequent errors in chromosome segregation during meiosis I in mammalian oocytes. During meiosis I the two homologous chromosomes are held together by cohesin complexes and form a bivalent. During anaphase I the two homologous chromosomes are segregated while the sister chromatids remains together (**a**). Reproductive aging induces loss of cohesin complexes resulting in a greater distance between the sister chromatids. This, in turn, predisposes the sister kinetochores to mis-orient and form incorrect attachment in metaphase I. The most frequent forms of orientation errors are (**b**) merotelic attachments, present when one kinetochore is attached to both spindle poles, and (**c**) bivalent rotation. At anaphase these incorrect attachments will result in premature separation of sister chromatids and reverse segregation, respectively. Although in the second error the number of chromosomes at the end of meiosis I is correct, the lack of centromeric cohesin between the two segregated chromatids will induce a random chromatid segregation at anaphase of meiosis II. (**d**) The absence of centrosomes and aMTOCs in humans and large animals gives rise to transient multipolar spindles, which favors the

The main impediment to using human oocytes is in the lack of access to healthy oocytes that can be used for research purposes only. The oocytes that are generally donated are those that have not been able to mature in vivo of to be fertilized in vitro and are, therefore, not suitable for ICSI or IVF treatments. Moreover the lack of a standardized IVM protocol for human oocytes yielding a maturation rate comparable to those in mice or large animals further limits the ability to study human meiosis in vitro (Yang, Kolben et al. 2021). Additionally, the donated oocytes are often obtained from subfertile patients, which may not accurately represent the healthy human population. This introduces potential biases and limitations in research outcomes, as the findings may not fully reflect the behavior of normal, healthy oocytes during meiosis.

While the human oocytes available for research are very limited in number, this does not apply to oocytes from large animals. A large number of oocytes can be obtained by puncturing the follicles of the animal's ovaries (Ovum-Pick-Up or OPU) or by aspiration of oocytes from ovaries obtained from a slaughterhouse. In the first case the animals do not need to be euthanized after the procedure and OPUs can be repeated safely on the same animal with only minimum discomfort. Both approaches reduce or eliminate the need of experimental animals. The main limiting factor in the use of large animals compared to mice is the restricted range of molecular techniques that can be applied to these species. Powerful tools have been developed to allow the generation of genetically engineered mouse models which enable exploration of the function of specific genes, proteins, and pathways. While, in theory, large animal can also be genetically engineered the prolonged gestational length, the low number of offspring and the strict regulatory compliance limit the applicability of this approach. Up to now much of the research conducted in large animals has been largely observational, however, recent break-throughs in molecular techniques have opened new possibilities for conducting molecular studies also in large animal models. For example the acute and rapid degradation of endogenous proteins by the Trim Away technique enables to knock-down or knock-out specific proteins in oocytes (Clift, So et al. 2018). Moreover the injection of oocytes with in vitro produced mRNA allows the transient expression of proteins tagged with a fluorophore which, in turn, can be then visualized live during the meiotic division. Since oocyte from large animals, such as cows and horses, naturally exhibit similar percentages of aneuploidy as their human counterpart [1, 9] (Kuliev, Zlatopolsky et al. 2011), it is likely that they share also the same mechanisms of chromosome mis-segregations during meiosis. Large animals and human oocytes use similar mechanisms to nucleate microtubules and assemble a meiotic spindle, however, how these mechanisms impact chromosome segregation is not well understood.

Fig. 4 (continued) appearance of merotelic attachments and later incorrect chromosome segregation. In particular, increased spindle instability has been shown to proceed from a KIFC1 deficiency in human oocytes and a MPS1 deficiency in equine oocytes. (**e**) The spindle assembly check point (SAC), which prevents cells from entering anaphase until all the kinetochores are correctly attached to spindle microtubules, is permissive in oocytes. This weak SAC response allows anaphase onset even if kinetochores are still unattached or not correctly attached to spindle microtubules, leading to chromosome segregation errors

Therefore, understanding the general mechanisms driving correct chromosome segregation in large animal oocytes can provide valuable information that can be translated to human meiosis.

References

1. Gruhn JR, Zielinska AP, Shukla V et al (2019) Chromosome errors in human eggs shape natural fertility over reproductive life span. Science 365(6460):1466–1469. https://doi.org/10.1126/science.aav7321
2. So C, Menelaou K, Uraji J et al (2022) Mechanism of spindle pole organization and instability in human oocytes. Science 375(6581). https://doi.org/10.1126/science.abj3944
3. Adams GP, Singh J, Baerwald AR (2012) Large animal models for the study of ovarian follicular dynamics in women. Theriogenology 78(8):1733–1748. https://doi.org/10.1016/j.theriogenology.2012.04.010
4. Yang Q, Zhu L, Wang M et al (2021) Analysis of maturation dynamics and developmental competence of in vitro matured oocytes under time-lapse monitoring. Reprod Biol Endocrinol 19(1):183. https://doi.org/10.1186/s12958-021-00868-0
5. Sirard MA, Florman HM, Leibfried-Rutledge ML et al (1989) Timing of nuclear progression and protein synthesis necessary for meiotic maturation of bovine oocytes. Biol Reprod 40(6):1257–1263. https://doi.org/10.1095/biolreprod40.6.1257
6. Hinrichs K, Schmidt AL, Friedman PP et al (1993) In vitro maturation of horse oocytes: characterization of chromatin configuration using fluorescence microscopy. Biol Reprod 48(2):363–370. https://doi.org/10.1095/biolreprod48.2.363
7. Holubcová Z, Blayney M, Elder K et al (2015) Human oocytes. Error-prone chromosome-mediated spindle assembly favors chromosome segregation defects in human oocytes. Science 348(6239):1143–1147. https://doi.org/10.1126/science.aaa9529
8. Luciano AM, Franciosi F, Lodde V et al (2013) Oocytes isolated from dairy cows with reduced ovarian reserve have a high frequency of aneuploidy and alterations in the localization of progesterone receptor membrane component 1 and aurora kinase B. Biol Reprod 88(3):58. https://doi.org/10.1095/biolreprod.112.106856
9. Rizzo M, Preez ND, Ducheyne KD et al (2020) The horse as a natural model to study reproductive aging-induced aneuploidy and weakened centromeric cohesion in oocytes. Aging 12(21):22220–22232. https://doi.org/10.18632/aging.104159
10. Hornak M, Jeseta M, Musilova P et al (2011) Frequency of aneuploidy related to age in porcine oocytes. PLoS One 6(4):e18892. https://doi.org/10.1371/journal.pone.0018892
11. Mamsen LS, Brochner CB, Byskov AG et al (2012) The migration and loss of human primordial germ stem cells from the hind gut epithelium towards the gonadal ridge. Int J Dev Biol 56(10–12):771–778. https://doi.org/10.1387/ijdb.120202lm
12. Ledda S, Bogliolo L, Bebbere D et al (2010) Characterization, isolation and culture of primordial germ cells in domestic animals: recent progress and insights from the ovine species. Theriogenology 74(4):534–543
13. Wrobel K, Süß F (1998) Identification and temporospatial distribution of bovine primordial germ cells prior to gonadal sexual differentiation. Anat Embryol 197:451–467
14. Curran S, Urven L, Ginther OJ (1997) Distribution of putative primordial germ cells in equine embryos. Equine Vet J Suppl 25:72–76. https://doi.org/10.1111/j.2042-3306.1997.tb05105.x
15. Anderson R, Copeland TK, Schöler H et al (2000) The onset of germ cell migration in the mouse embryo. Mech Dev 91(1–2):61–68
16. Smith P, Wilhelm D, Rodgers RJ (2014) Development of mammalian ovary. J Endocrinol 221(3):145. https://doi.org/10.1530/JOE-14-0062

17. Mauléon P, Mariana JC (1977) Oogenesis and folliculogenesis. In: Cole HH, Cupps PT (eds) Reproduction in domestic animals, 3rd edn. Academy Press, New York, pp 175–201
18. Erickson BH (1966) Development and radio-response of the prenatal bovine ovary. Reproduction 11(1):97–105
19. Deanesly R (1977) Germ cell proliferations in the fetal horse ovary. Cell Tissue Res 185(3):361–371. https://doi.org/10.1007/BF00220296
20. Borum K (1961) Oogenesis in the mouse: a study of the meiotic prophase. Exp Cell Res 24(3):495–507
21. Vallente RU, Cheng EY, Hassold TJ (2006) The synaptonemal complex and meiotic recombination in humans: new approaches to old questions. Chromosoma 115(3):241–249. https://doi.org/10.1007/s00412-006-0058-4
22. Youds JL, Boulton SJ (2011) The choice in meiosis—defining the factors that influence crossover or non-crossover formation. J Cell Sci 124(4):501–513. https://doi.org/10.1242/jcs.074427
23. Peters H, Byskov AG, Grinsted J (1978) Follicular growth in fetal and prepubertal ovaries of humans and other primates. Clin Endocrinol Metab 7(3):469–485. https://doi.org/10.1016/s0300-595x(78)80005-x
24. Gosden R, Krapez J, Briggs D (1997) Growth and development of the mammalian oocyte. BioEssays 19(10):875–882. https://doi.org/10.1002/bies.950191007
25. Wassarman PM, Albertini DF (1994) The mammalian ovum. In: The physiology of reproduction, vol 1. Raven Press, New York, pp 79–122
26. Bornslaeger EA, Wilde MW, Schultz RM (1984) Regulation of mouse oocyte maturation: involvement of cyclic AMP phosphodiesterase and calmodulin. Dev Biol 105(2):488–499. https://doi.org/10.1016/0012-1606(84)90306-3
27. Dekel N, Lawrence TS, Gilula NB et al (1981) Modulation of cell-to-cell communication in the cumulus-oocyte complex and the regulation of oocyte maturation by LH. Dev Biol 86(2):356–362. https://doi.org/10.1016/0012-1606(81)90193-7
28. Li J, Qian W, Sun Q (2019) Cyclins regulating oocyte meiotic cell cycle progression dagger. Biol Reprod 101(5):878–881. https://doi.org/10.1093/biolre/ioz143
29. Ishiguro T, Tanaka K, Sakuno T, Watanabe Y (2010) Shugoshin-PP2A counteracts casein-kinase-1-dependent cleavage of Rec8 by separase. Nat Cell Biol 12(5):500–506
30. De Smedt V, Szollosi D (1991) Cytochalasin D treatment induces meiotic resumption in follicular sheep oocytes. Mol Reprod Dev 29(2):163–171. https://doi.org/10.1002/mrd.1080290212
31. Lee J, Miyano T, Moor RM (2000) Spindle formation and dynamics of gamma-tubulin and nuclear mitotic apparatus protein distribution during meiosis in pig and mouse oocytes. Biol Reprod 62(5):1184–1192. https://doi.org/10.1095/biolreprod62.5.1184
32. Walls ML, Ryan JP, Keelan JA et al (2015) In vitro maturation is associated with increased early embryo arrest without impairing morphokinetic development of useable embryos progressing to blastocysts. Hum Reprod 30(8):1842–1849. https://doi.org/10.1093/humrep/dev125
33. Wu T, Dong J, Fu J et al (2023b) The mechanism of acentrosomal spindle assembly in human oocytes. Obstet Gynecol Surv 78(3):158–159. https://doi.org/10.1097/OGX.0000000000001134
34. Harasimov K, Uraji J, Mönnich EU et al (2023) Actin-driven chromosome clustering facilitates fast and complete chromosome capture in mammalian oocytes. Nat Cell Biol 25(3):439–452. https://doi.org/10.1038/s41556-022-01082-9
35. Schuh M, Ellenberg J (2007) Self-organization of MTOCs replaces centrosome function during acentrosomal spindle assembly in live mouse oocytes. Cell 130(3):484–498. https://doi.org/10.1016/j.cell.2007.06.025
36. Reichmann J, Nijmeijer B, Hossain MJ et al (2018) Dual-spindle formation in zygotes keeps parental genomes apart in early mammalian embryos. Science 361(6398):189–193. https://doi.org/10.1126/science.aar7462
37. Schatten H, Sun Q (2011) Centrosome dynamics during mammalian oocyte maturation with a focus on meiotic spindle formation. Mol Reprod Dev 78(10–11):757–768. https://doi.org/10.1002/mrd.21380

38. Szollosi D, Calarco P, Donahue RP (1972) Absence of centrioles in the first and second meiotic spindles of mouse oocytes. J Cell Sci 11(2):521–541. https://doi.org/10.1242/jcs.11.2.521
39. Szollosi D, Mandelbaum J, Plachot M et al (1986) Ultrastructure of the human preovulatory oocyte. J In Vitro Fert Embryo Transf 3(4):232–242. https://doi.org/10.1007/BF01132810
40. Wu J, Larreategui-Aparicio A, Lambers MLA et al (2023a) Microtubule nucleation from the fibrous corona by LIC1-pericentrin promotes chromosome congression. Curr Biol 33(5):912–925.e6. https://doi.org/10.1016/j.cub.2023.01.010
41. Li G, Liu Y, Bunch TD et al (2005) Asymmetric division of spindle microtubules and microfilaments during bovine meiosis from metaphase I to metaphase III. Mol Reprod Dev 71(2):220–226. https://doi.org/10.1002/mrd.20255
42. Li X, Qin Y, Wilsher S et al (2006) Centrosome changes during meiosis in horse oocytes and first embryonic cell cycle organization following parthenogenesis, fertilization and nuclear transfer. Reproduction 131(4):661–667. https://doi.org/10.1530/rep.1.00795
43. Peng L, He Y, Wang W et al (2023) PAK1 is involved in the spindle assembly during the first meiotic division in porcine oocytes. Int J Mol Sci 24(2):1123. https://doi.org/10.3390/ijms24021123
44. Gueth-Hallonet C, Antony C, Aghion J et al (1993) Gamma-tubulin is present in acentriolar MTOCs during early mouse development. J Cell Sci 105(Pt 1):157–166. https://doi.org/10.1242/jcs.105.1.157
45. So C, Seres KB, Steyer AM et al (2019) A liquid-like spindle domain promotes acentrosomal spindle assembly in mammalian oocytes. Science 364(6447):eaat9557. https://doi.org/10.1126/science.aat9557
46. Kim Y, Lee I, Jo Y et al (2019) Acentriolar microtubule organization centers and Ran-mediated microtubule formation pathways are both required in porcine oocytes. Mol Reprod Dev 86(8):972–983. https://doi.org/10.1002/mrd.23172
47. Bennabi I, Terret M, Verlhac M (2016) Meiotic spindle assembly and chromosome segregation in oocytes. J Cell Biol 215(5):611–619. https://doi.org/10.1083/jcb.201607062
48. Clarke PR, Zhang C (2008) Spatial and temporal coordination of mitosis by Ran GTPase. Nat Rev Mol Cell Biol 9(6):464–477. https://doi.org/10.1038/nrm2410
49. Drutovic D, Duan X, Li R et al (2020) RanGTP and importin beta regulate meiosis I spindle assembly and function in mouse oocytes. EMBO J 39(1):e101689. https://doi.org/10.15252/embj.2019101689
50. Goshima G, Wollman R, Goodwin SS et al (2007) Genes required for mitotic spindle assembly in drosophila S2 cells. Science 316(5823):417–421. https://doi.org/10.1126/science.1141314
51. Petry S, Groen AC, Ishihara K et al (2013) Branching microtubule nucleation in Xenopus egg extracts mediated by augmin and TPX2. Cell 152(4):768–777. https://doi.org/10.1016/j.cell.2012.12.044
52. Watanabe S, Shioi G, Furuta Y et al (2016) Intra-spindle microtubule assembly regulates clustering of microtubule-organizing centers during early mouse development. Cell Rep 15(1):54–60. https://doi.org/10.1016/j.celrep.2016.02.087
53. Mogessie B, Schuh M (2017) Actin protects mammalian eggs against chromosome segregation errors. Science 357(6353):eaal1647. https://doi.org/10.1126/science.aal1647
54. Klaasen SJ, Truong MA, van Jaarsveld RH et al (2022) Nuclear chromosome locations dictate segregation error frequencies. Nature 607(7919):604–609. https://doi.org/10.1038/s41586-022-04938-0
55. Chiang T, Duncan FE, Schindler K et al (2010) Evidence that weakened centromere cohesion is a leading cause of age-related aneuploidy in oocytes. Curr Biol 20(17):1522–1528. https://doi.org/10.1016/j.cub.2010.06.069
56. Duncan FE, Hornick JE, Lampson MA et al (2012) Chromosome cohesion decreases in human eggs with advanced maternal age. Aging Cell 11(6):1121–1124. https://doi.org/10.1111/j.1474-9726.2012.00866.x
57. Angell R (1995) Meiosis I in human oocytes. Cytogenet Cell Genet 69(3–4):266–272. https://doi.org/10.1159/000133977

58. Kouznetsova A, Lister L, Nordenskjöld M et al (2007) Bi-orientation of achiasmatic chromosomes in meiosis I oocytes contributes to aneuploidy in mice. Nat Genet 39(8):966–968. https://doi.org/10.1038/ng2065
59. Patel J, Tan SL, Hartshorne GM et al (2016) Unique geometry of sister kinetochores in human oocytes during meiosis I may explain maternal age-associated increases in chromosomal abnormalities. Biol Open 5(2):178–184. https://doi.org/10.1242/bio.016394
60. Zielinska AP, Holubcova Z, Blayney M et al (2015) Sister kinetochore splitting and precocious disintegration of bivalents could explain the maternal age effect. elife 4:e11389. https://doi.org/10.7554/eLife.11389
61. Kong A, Barnard J, Gudbjartsson DF et al (2004) Recombination rate and reproductive success in humans. Nat Genet 36(11):1203–1206. https://doi.org/10.1038/ng1445
62. Wang S, Hassold T, Hunt P et al (2017) Inefficient crossover maturation underlies elevated aneuploidy in human female meiosis. Cell 168(6):977–989.e17. https://doi.org/10.1016/j.cell.2017.02.002
63. Wang Z, Shen B, Jiang J et al (2016) Effect of sex, age and genetics on crossover interference in cattle. Sci Rep 6:37698. https://doi.org/10.1038/srep37698
64. Battaglia DE, Goodwin P, Klein NA et al (1996) Influence of maternal age on meiotic spindle assembly in oocytes from naturally cycling women. Hum Reprod 11(10):2217–2222. https://doi.org/10.1093/oxfordjournals.humrep.a019080
65. Rizzo M, Ducheyne KD, Deelen C et al (2019) Advanced mare age impairs the ability of in vitro-matured oocytes to correctly align chromosomes on the metaphase plate. Equine Vet J 51(2):252–257. https://doi.org/10.1111/evj.12995
66. Rizzo M, Stout TAE, Cristarella S et al (2021) Compromised MPS1 activity induces multipolar spindle formation in oocytes from aged mares: establishing the horse as a natural animal model to study age-induced oocyte meiotic spindle instability. Front Cell Dev Biol 9:657366. https://doi.org/10.3389/fcell.2021.657366
67. Menezo YJR, Herubel F (2002) Mouse and bovine models for human IVF. Reprod Biomed Online 4(2):170–175. https://doi.org/10.1016/s1472-6483(10)61936-0
68. Steuerwald N, Cohen Herrera RJ et al (2001) Association between spindle assembly checkpoint expression and maternal age in human oocytes. Mol Hum Reprod 7(1):49–55. https://doi.org/10.1093/molehr/7.1.49J,
69. Adhikari D, Carroll J (2018) Oocyte Meiotic Resumption Upon Puberty. In: Encyclopedia of Reproduction. Elsevier, pp 167–171. https://doi.org/10.1016/B978-0-12-801238-3.64444-6
70. Ahima RS, Dushay J, Flier SN, et al (1997) Leptin accelerates the onset of puberty in normal female mice. Journal of Clinical Investigation 99:391–395. https://doi.org/10.1172/JCI119172
71. Baudat F, Imai Y, de Massy B (2013) Meiotic recombination in mammals: localization and regulation. Nat Rev Genet 14:794–806. https://doi.org/10.1038/nrg3573
72. Brown HM, Fabre Nys C, Cognié J, Scaramuzzi RJ (2014) Short oestrous cycles in sheep during anoestrus involve defects in progesterone biosynthesis and luteal neovascularisation. REPRODUCTION 147:357–367. https://doi.org/10.1530/REP-13-0400
73. Clift D, So C, McEwan WA, et al (2018) Acute and rapid degradation of endogenous proteins by Trim-Away. Nat Protoc 13:2149–2175. https://doi.org/10.1038/s41596-018-0028-3
74. Collins JK, Lane SIR, Merriman JA, Jones KT (2015) DNA damage induces a meiotic arrest in mouse oocytes mediated by the spindle assembly checkpoint. Nat Commun 6:8553. https://doi.org/10.1038/ncomms9553
75. Day ML, Nogueira GP (2013) Management of age at puberty in beef heifers to optimize efficiency of beef production. Animal Frontiers 3:6–11. https://doi.org/10.2527/af.2013-0027
76. Garcia M, Dietrich AJJ, Freixa L, et al (1987) Development of the first meiotic prophase stages in human fetal oocytes observed by light and electron microscopy. Hum Genet 77:223–232. https://doi.org/10.1007/BF00284474
77. Greaney J, Wei Z, Homer H (2018) Regulation of chromosome segregation in oocytes and the cellular basis for female meiotic errors. Hum Reprod Update 24:135–161. https://doi.org/10.1093/humupd/dmx035
78. Hu J, Yan P (2022) Effects of Backfat Thickness on Oxidative Stress and Inflammation of Placenta in Large White Pigs. Vet Sci 9:302. https://doi.org/10.3390/vetsci9060302

79. Jackson SP, Bartek J (2009) The DNA-damage response in human biology and disease. Nature 461:1071–1078. https://doi.org/10.1038/nature08467
80. Jones KT (2008) Meiosis in oocytes: predisposition to aneuploidy and its increased incidence with age. Hum Reprod Update 14:143–158. https://doi.org/10.1093/humupd/dmm043
81. Kallen AN (2022) Ovarian Reserve and Early Follicle Development: Prerequisite Knowledge for Understanding Ovarian Tissue Harvesting for Cryopreservation. In: Principles and Practice of Ovarian Tissue Cryopreservation and Transplantation. Elsevier, pp 37–48. https://doi.org/10.1016/B978-0-12-823344-3.00002-9
82. Keeney S, Lange J, Mohibullah N (2014) Self-Organization of Meiotic Recombination Initiation: General Principles and Molecular Pathways. Annu Rev Genet 48:187–214. https://doi.org/10.1146/annurev-genet-120213-092304
83. Kerr JB, Myers M, Anderson RA (2013) The dynamics of the primordial follicle reserve. REPRODUCTION 146:R205–R215. https://doi.org/10.1530/REP-13-0181
84. Kuliev A, Zlatopolsky Z, Kirillova I, et al (2011) Meiosis errors in over 20,000 oocytes studied in the practice of preimplantation aneuploidy testing. Reprod Biomed Online 22:2–8. https://doi.org/10.1016/j.rbmo.2010.08.014
85. Lea R, England GCW (2019) Puberty and Seasonality. In: Veterinary Reproduction and Obstetrics. Elsevier, pp 54–62. https://doi.org/10.1016/B978-0-7020-7233-8.00003-3
86. Marston AL, Wassmann K (2017) Multiple Duties for Spindle Assembly Checkpoint Kinases in Meiosis. Front Cell Dev Biol 5:. https://doi.org/10.3389/fcell.2017.00109
87. Miyano T, Ogushi S, Bui H-T, Lee J (2007) Meiotic Resumption and Spindle Formation of Pig Oocytes. J Mamm Ova Res 24:92–98. https://doi.org/10.1274/jmor.24.92
88. Nah WH, Park MJ, Gye MC (2011) Effects of early prepubertal exposure to bisphenol A on the onset of puberty, ovarian weights, and estrous cycle in female mice. Clin Exp Reprod Med 38:75. https://doi.org/10.5653/cerm.2011.38.2.75
89. Pedersen T (1970) FOLLICLE KINETICS IN THE OVARY OF THE CYCLIC MOUSE. Acta Endocrinol (Copenh) 64:304–323. https://doi.org/10.1530/acta.0.0640304
90. Reiland S (1978) Growth and skeletal development of the pig. Acta Radiol Suppl 358:15–22
91. Rémillard-Labrosse G, Dean NL, Allais A, et al (2020) Human oocytes harboring damaged DNA can complete meiosis I. Fertil Steril 113:1080-1089.e2. https://doi.org/10.1016/j.fertnstert.2019.12.029
92. Strączyńska P, Papis K, Morawiec E, et al (2022) Signaling mechanisms and their regulation during in vivo or in vitro maturation of mammalian oocytes. Reproductive Biology and Endocrinology 20:37. https://doi.org/10.1186/s12958-022-00906-5
93. Tingen C, Kim A, Woodruff TK (2009) The primordial pool of follicles and nest breakdown in mammalian ovaries. Mol Hum Reprod 15:795–803. https://doi.org/10.1093/molehr/gap073
94. Touati SA, Wassmann K (2016) How oocytes try to get it right: spindle checkpoint control in meiosis. Chromosoma 125:321–335. https://doi.org/10.1007/s00412-015-0536-7
95. Tummaruk P, Tantasuparuk W, Techakumphu M, Kunavongkrit A (2009) The association between growth rate, body weight, backfat thickness and age at first observed oestrus in crossbred Landrace×Yorkshire gilts. Anim Reprod Sci 110:108–122. https://doi.org/10.1016/j.anireprosci.2008.01.004
96. Wassmann K (2022) Separase Control and Cohesin Cleavage in Oocytes: Should I Stay or Should I Go? Cells 11:3399. https://doi.org/10.3390/cells11213399
97. Yang H, Kolben T, Meister S, et al (2021) Factors Influencing the In Vitro Maturation (IVM) of Human Oocyte.Biomedicines 9:1904. https://doi.org/10.3390/biomedicines9121904

Aurora Kinases: A Summary of Mouse Genetic Models Used to Distinguish Their Roles in Oocyte Meiosis and Female Fertility

Mayra Romero, Gyu Ik Jung, and Karen Schindler

Abstract The Aurora kinases (AURK) are key regulatory proteins essential for spindle building and chromosome alignment and segregation in mitosis and meiosis. AURKA and AURKB are expressed in all cell types but the third isoform, AURKC, is exclusively expressed in germ cells. Unique expression of all three Aurora kinase isoforms in germ cells raises questions about identifying their specific and redundant functions, oocyte and meiosis-specific roles and generally why germ cells require an additional isoform. Conditional knockout mice were used to address these questions and reveal additional information about regulation of oocyte meiosis. This chapter provides an update on recent findings uncovering more details on the role of Aurora kinases in oocyte meiosis. We dive into their individual functions during establishment of the meiotic spindle and coordinating chromosome alignment but also address newly discovered non-canonical roles in regulating mitochondria, translation, and cytokinesis. Understanding the functional implications of these kinases offers valuable insight into the fundamental processes that control meiosis and fertility.

Keywords Oocyte · Aurora kinase · Meiosis · Aneuploidy · Gamete

Introduction

Female fertility is highly reliant on egg quality. Before oocytes become developmentally competent eggs, they must undergo a multi-step process comprising meiotic initiation and genetic recombination, followed by cell-cycle arrest, cellular growth, meiotic resumption and finally meiosis I completion [2, 3]. Oocyte meiosis begins during fetal development with DNA replication and completion of the

M. Romero · G. I. Jung · K. Schindler (✉)
Department of Genetics, Rutgers, The State University of New Jersey, Piscataway, NJ, USA

Human Genetics Institute of New Jersey, Piscataway, NJ, USA
e-mail: ks804@hginj.rutgers.edu

B. Mogessie (ed.), *Cellular Architecture and Dynamics in Female Meiosis*,
https://doi.org/10.1007/978-3-031-97173-0_8

substages of Prophase I, which involve double strand DNA breaks and their repair [4, 5]. Oocytes remain arrested at the last substage of Prophase I (diplotene/dictyate) until ovulation; ovulation and meiotic resumption are triggered by hormonal surges in sexually mature mammals. Once meiosis resumes, oocytes complete one meiotic division (meiosis I or MI) and arrest at Metaphase II. The second meiotic division (meiosis II or MII) is completed upon fertilization and egg activation. Successful chromosome segregation requires proper completion of early meiotic events, such as synapsis and meiotic recombination, and later the establishment of the bipolar spindle after meiotic resumption [5, 6]. Aneuploidy, the incorrect number of chromosomes in a cell, most commonly results from chromosome segregation errors in meiosis I and is the leading cause of early miscarriage and failure in in vitro fertilization [7]. If a meiotic spindle is not properly assembled during meiosis I, chromosomes may not align or separate correctly, most likely resulting in an aneuploid egg. If this aneuploid egg is fertilized, an aneuploid conceptus forms.

MTOCs and Spindle Assembly

Microtubule organizing centers (MTOCs), located at spindle poles of both mitotic and meiotic cells, are responsible for the nucleation of microtubules forming the spindle [6, 8, 9]. Unlike mitotic cells, oocyte MTOCs lack centrioles, which are responsible for regulating microtubule dynamics and organization required for cell division [10, 11]. These acentriolar MTOCs (aMTOCs) still assemble bipolar spindles. However, oocytes complete meiosis I with high frequency of chromosome segregation errors [12]. Spindle bipolarity in mouse oocytes is established from self-organization of multiple aMTOCs that undergo several reorganization steps [13, 14]. Clusters of aMTOCs are present in Prophase I-arrested oocytes, and once oocytes resume meiosis and the nuclear envelope breaks down, the aMTOCs fragment [14, 15]. The aMTOC fragmentation process occurs in three distinct steps: decondensation, stretching, and fragmentation. Once fragmentation is complete, aMTOCs redistribute and cluster at opposing ends of the bipolar spindle as oocytes reach Metaphase I [6, 13, 16]. Microtubule nucleation from aMTOCs is an integral step in stabilizing bipolar spindles and, once kinetochore-microtubule attachments are established, for ensuring chromosome alignment at the metaphase plate.

Chromosome Alignment and Segregation

To ensure genomic integrity, chromosomes must properly align at Metaphase prior to their segregation and completion of cell division. Processes critical to the fidelity of chromosome segregation include a tension-sensing mechanism at kinetochores, the chromosome passenger complex (CPC), the spindle assembly checkpoint (SAC), and cohesion [17, 18]. In both mitosis and meiosis, chromosomes are aligned at the Metaphase/Metaphase I plates by establishing end-on kinetochore-microtubule (K-MT) attachments [19–21]. The initial process is described as highly dynamic, where K-MT attachments are formed, disrupted, and reformed, a cyclical process which allows for detection of chromosomes that are incorrectly attached. In mitosis, K-MT error correction continues through microtubule-based pulling and pushing forces until sister chromatids are attached to and pulled toward opposite ends of the spindle. Once stably attached, maximum tension is achieved across sister chromatid kinetochores. On the other hand, chromosome congression in oocytes takes place at the same time as the correction of incorrect attachments, demonstrated by the presence of lateral and merotelic attachments even after chromosome biorientation is observed [21]. Incorrect K-MT attachments are rectified by the CPC, which localizes at kinetochores during Metaphase/Metaphase I and destabilizes erroneous K-MT attachments. Once a kinetochore lacks a MT attachment the SAC is activated [22–24]. SAC signals from unattached kinetochores prevent somatic cells from Anaphase onset until all proper K-MT attachments are established. Though the SAC is active in oocytes, it is less efficient compared to somatic cells. This inefficiency worsens with age, leading to oocytes advancing to Anaphase I onset despite the presence of unattached or mis-attached kinetochores. Consequently, this defect can result in a higher incidence of lagging chromosomes and increased risk of egg aneuploidy [25–28]. The continuous K-MT attachment changes occurring in oocytes is thought to contribute to a reduced strength of the SAC signaling (or reduced sensitivity to it) [29]. Lastly, cohesins localized at kinetochores and at centromeres of chromosomes protect sister chromatids from premature separation caused by poleward spindle forces that act on kinetochores during segregation [30, 31].

From spindle building to chromosomes alignment, all these processes work together to ensure that meiosis is successful. Uncovering more details about the regulation of meiosis will further our understanding of how egg viability is achieved.

Aurora Protein Kinases

The Aurora protein kinases (AURKs) are a family of three serine/threonine kinases involved in many aspects that control cellular division [32, 33]. AURKA and AURKB are both expressed in mitosis and meiosis, and AURKC is predominantly expressed in gametes, although it has also been detected in some cancer cell lines [1, 34]. Since the identification of the mammalian paralogs, the canonical roles of AURKA and AURKB in regulating spindle formation, chromosome alignment, and cytokinesis have been examined in mitosis and meiosis [35–37]. In somatic cells, both AURKA and AURKB are essential because of their non-overlapping roles in controlling spindle formation and regulating cell division, respectively [33, 38–40]. More specifically, AURKA, or the "polar" Aurora kinase, regulates spindle formation from its spindle pole localization [34, 41]. In mitotic cells, AURKB is involved in regulating chromosome alignment, destabilizing K-MT attachments, activating the SAC and triggering cohesin cleavage [39, 42–44]. Mitotic AURKB is the catalytic subunit of the CPC; however, in mouse oocytes, AURKC replaces AURKB [22, 34, 45]. Recent works demonstrate that each AURK has unique requirements in mouse oocytes [25, 34, 41, 46]. Their unique functions could arise from different affinities for binding partners and/or regulators, different levels of expression or from the unique features of oocyte meiosis, such as chromosome organization, unique spindle properties, and the absence of centrioles. In this chapter, we discuss the recent studies dissecting the roles of AURKA, AURKB, and AURKC in mouse oocytes. These studies employ genetic mouse models to demonstrate how female fertility is directly affected by AURK loss and broaden our understanding of AURK involvement in mechanisms driving oocyte meiosis.

AURKB and AURKC

The Aurora kinases participate in various pathways mediating chromosome alignment such as functioning in the CPC and activating the SAC. The CPC, primarily responsible for the coordination of chromosome alignment, is composed of three regulatory proteins (INCENP, Borealin, and Survivin), and one catalytic subunit (AURKB or AURKC). In oocytes, AURKC is the predominant AURK functioning in the CPC, whereas AURKB is the catalytic subunit in mitotic cells. The preferential binding of AURKC over AURKB to INCENP may occur because of differences in protein concentration in mammalian oocytes: AURKC is more abundant than AURKB [34]. The low expression level of AURKB posed a challenge when examining the localization of AURKB in mouse oocytes. The first localization study used overexpression of *Aurkb-Gfp* to demonstrate the capacity of AURKB to localize to kinetochores/centromeres [44]. In accordance with displacement of AURKB from the CPC, localization of endogenous AURKB via immunocytochemistry was demonstrated to be mostly cytoplasmic, with an enrichment at MI and MII spindles [22]. As expected, based on its association with the CPC, immunocytochemistry studies confirmed AURKC localization to centromeres and along chromosome arms at MI and to centromeres at MII [44]. Pharmacological methods to distinguish the individual roles of AURKB and AURKC were challenging due to high sequence homology shared between the two kinases; inhibitors were not AURKB or AURKC specific. Despite this obstacle, methods for identifying and dissecting the roles of these two kinases in oocytes have depended primarily on pharmacological inhibitors such as ZM447439 and AZD1152, and genetic methods involving knockout mouse models. Although pharmacological assays are limiting in terms of their specificity, genetic approaches revealed additional intricacies in the interplay between the AURKs, which are discussed in detail here [47–50]. Because of the aforementioned limitations, the possibility of a kinetochore/centromere-specific subpopulation of AURKB is not discarded, and AURKB and AURKC are discussed together unless clear distinctions of their functions are experimentally demonstrated.

To dissect the non-overlapping roles of AURKB and AURKC in oocyte meiosis and circumvent the limitations of pharmacological methods, knockout mouse models were established. In an AURKB conditional knockout (*Aurkb* cKO), *LoxP* Cre-recombinase recognition sites flank exons 2 and 6 of the *Aurkb* gene which generates mice with conditional KO alleles of *Aurkb* [51]. Gdf9-driven Cre recombinase was used to evaluate oocyte-specific functions of AURKB in the floxed genetic background [52]. The expression of CRE recombinase at postnatal day 3 leads to consequent excision of exons 2–6 of *Aurkb* specifically in oocytes of primordial follicles, a time after completion of meiotic recombination [53]. *Aurkc* is a germ cell-specific gene and a whole-body knockout (KO) does not affect viability and function of other cell types [54]. Therefore, mice harboring null alleles of *Aurkc* were generated by replacing exons 2–4 with a *LacZ/Neo* genes cassette [55]. To study the potential effects of losing both AURKB and AURKC, AURKB/C double knockouts (double KO; DKO) were generated by crossing *Aurkb* cKO and *Aurkc* KO mouse lines. These mouse models presented the unique opportunity to study how oocyte-specific loss of either AURKB, AURKC or both directly impacts female fertility. Whereas *Aurkc* KOs and DKO females were slightly subfertile, three out of five *Aurkb* cKO females developed more significant premature age-related infertility [34]. Specifically, *Aurkb* cKO female litter sizes were similar to WT at the beginning of the fertility trials but they declined rapidly as the mothers aged. These results indicate that requirements for AURKB and AURKC in female fertility might not be the same and that one cannot fully compensate for the other. Efforts to pin-point AURKB and AURKC mechanisms controlling fertility are discussed in the following section.

Chromosome Alignment and the Spindle Assembly Checkpoint (SAC)

In mitosis, AURKB is implicated in regulating the fidelity of chromosome segregation through controlling chromosome alignment [42, 56]. The prevailing model is that AURKB is a tension sensor that destabilizes microtubules to allow correction of K-MT attachment errors. This occurs because tension is low when there are improper K-MT attachments thereby bringing AURKB close to its substrates [57]. When tension is high, due to proper K-MT attachments, AURKB is not locally concentrated at kinetochores and cannot reach its substrates to destabilize MTs. However, whether AURKB plays a direct function in regulating the SAC or an indirect one via its role in microtubule attachment error correction is a topic of debate in the field (discussed in [58] and [59]). Similar attempts to dissect AURKB and AURKC involvement in error correction and SAC are ongoing in mouse oocytes. Previous works identified that both these AURKs are directly involved in destabilization of K-MT attachments in mouse oocytes as pharmacological inhibition starting at meiotic onset increased the frequency of incorrect K-MT attachments, and acute inhibition starting at early Metaphase I maintained only already-established attachments [22, 60, 61].

A chromosome tension-dependent pathway was identified in mouse oocytes that are regulated by AURKB/C: cells with decreased interkinetochore distance exhibited higher frequency of incorrect kinetochore-microtubule (K-MT) attachments, a sign of an active error-correction pathway [62]. The activation of this corrective pathway was confirmed by elevated phosphorylation levels of Highly Expressed in Cancer 1 (HEC1) at serine 55 in low-tension conditions. Furthermore, the role of AURKB/C in K-MT attachment correction was suggested when phosphorylated HEC1 levels decreased to levels comparable to the ones in high tension control cells upon AURKB/C inhibition with ZM447439 at 10 μM, a dose that also inhibits AURKA [47, 62]. Recent work in somatic cells demonstrated that both AURKA and AURKB are involved in regulating K-MT attachments by phosphorylating specific (AURKA at serine 69 and AURKB at serine 44) and shared sites (serine 55) of HEC1 [63]. As such, it is possible that use of ZM447439 phenocopied high tension conditions because of its pan-AURK effect. Therefore, further studies are necessary to determine if the overlapping roles of AURKs in K-MT correction are conserved in mammalian oocytes.

In mouse oocytes, the stabilization of K-MT attachments depends on the gradual localization of Protein Phosphatase 2A (PP2A) at the kinetochores to counteract AURKB/C activity. This was determined by measuring the phosphorylation levels of proteins HEC1 and Kinetochore Scaffold 1 (KNL1), two AURKB/C substrates [64]. The balanced localized activity between AURKB/C and PP2A observed in oocytes is distinct from what occurs in somatic cells. In somatic cells K-MT attachment induced tension leads to spatial separation of AURKB in the inner centromeres from the kinetochores, and consequently its substrates, leading to the stabilization of end-on attachments [65]. Therefore, the differences in AURKB/C--PP2A interactions could arise from the inherent geometry of chromosomes in gametes, which could alter the spatial arrangement and relative localizations of these proteins: centromeres of sister chromatids in meiosis I have cohesin proteins that add further compaction to the centromere-kinetochore region compared to centromeres in mitotic chromosomes [66, 67]. The difference in sister centromere structure allows for increased distance between kinetochores and centromeres of sister chromatids in somatic cells that does not occur in meiosis I sister chromatids that must remain fused as a unit to segregate together (Fig. 1). This difference likely explains the need for an alternative mechanisms in counteracting the K-MT attachment destabilization activity of AURKB/C observed in mouse oocytes.

Although involved in regulating K-MT attachments, AURKB/C likely do not play a direct role in SAC regulation: inhibition of AURKB/C activity in oocytes with an active SAC does not cause oocytes to progress through meiosis I[62]. Mouse oocytes with low kinetochore tension [induced by inhibiting kinesin EG5 with s-trityl-l-cysteine (STLC)] showed increased AURKB/C activity, as measured by increased phosphorylation of substrate HEC1 at serine 55, and delayed Anaphase I onset. However, inhibition of AURKB/C with 10 μM ZM447439 in oocytes with low or no tension did not satisfy the SAC and, consequently, arrested at Metaphase

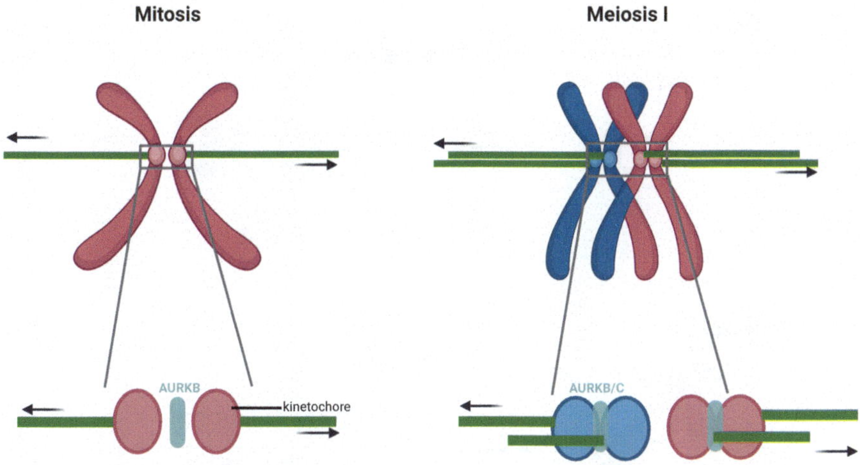

Fig. 1 Difference in kinetochore distance between mitosis and meiosis. In mitosis, there is a greater distance between kinetochores of sister chromatids compared to meiosis I. Figure made in biorender

I, suggesting that AURKB/C (and likely AURKA because of the dose of ZM447439 used [47]) are not directly involved in SAC regulation. Consistent with these observations, oocytes from reproductively young mice (4–6 weeks of age) lacking *Aurkb* or *Aurkc* do not show SAC defects and *Aurkb* cKO oocytes have comparable Anaphase I onset timing to oocytes from wild-type mice [25, 68]. Interestingly, oocytes lacking *Aurkc* have higher incidences of misaligned chromosomes, in addition to shorter spindles in Metaphase I and a delayed Anaphase onset, consistent with a direct role of AURKC in the K-MT error-correction pathway [25, 68]. Absence of AURKB in oocytes did not affect K-MT attachments, suggesting that AURKB has other functions in oocyte meiosis (discussed below).

In both mitosis and meiosis, the association of AURKB/C-CPC to chromatin depends on the phosphorylation of histone H3 at threonine 3 (H3pT3) by Haspin kinase (Haspin) [69, 70]. Surprisingly, inhibition of Haspin in mouse oocytes allowed the distinction of two populations of AURKB/C at chromosomes: one population along the interchromatid axis (ICA) dependent on H3pT3, and another population at the centromeres independent of this histone phosphorylation (Fig. 2) [1]. Furthermore, the effects of Haspin inhibition in meiotic progression were dependent on when during the cell-cycle oocytes were treated with the inhibitor. When Haspin was inhibited from the onset of meiotic resumption, Anaphase I onset, as measured by PBE, was delayed by about 4 h compared to controls. On the other hand, Haspin inhibition during pro-Metaphase I (after 5 h from meiotic resumption onset) accelerated Anaphase I onset by approximately 3 h. Haspin inhibition, and the consequent displacement of AURKB/C from the ICA, phenocopies the effects of AURKB/C inhibition in oocytes treated with ZM447439 (10 μM) at different times in meiotic maturation [71]. The kinetochore population of AURKB/C requires activity of both Haspin and BUB1 kinases because their dual inhibition prevents

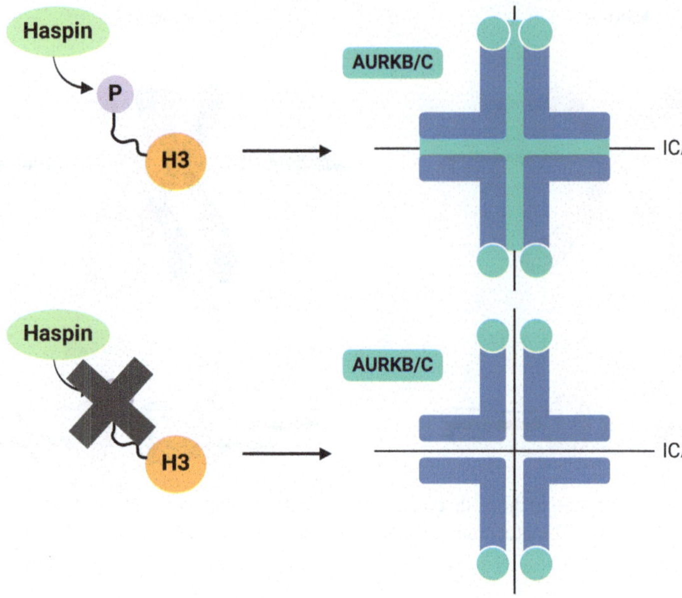

Fig. 2 AURKB/C localization is dependent on Haspin kinase. This figure was adapted from [1]. Haspin phosphorylation of histone H3 on threonine 3 directs AURKB/C localization at interchromosome axes. Figure made in Biorender

their localization and further abrogates the K-MT attachment pathway [72]. These results not only provide insight into temporally dependent functions of AURKB/C in regulating meiotic progression, but also suggest that AURKB/C subpopulations have distinct roles. Surprisingly, *Aurkc* KO oocytes had an intact SAC upon Haspin inhibition unless ZM447439 was also added, suggesting that AURKB is more efficient than AURKC at maintaining an active SAC or that the absence of AURKC leads to SAC activation [1]. This observation is consistent with the defects in recruitment of SAC components when AURKC is overexpressed in mitotically dividing cells [73]. The differences between AURKB and AURKC functions could arise from their structural differences upon INCENP binding, which in turn could alter affinities toward shared substrates [49, 74]. Structural comparison of crystallized forms of AURKB-INCENP and AURKC-INCENP showed that the substrate binding site and the activation loop of each AURK was distinct, suggesting that the binding sites accommodate distinct substrates [49]. However, the ability of AURKA to compensate by functioning in the CPC in absence of AURKC cannot be discarded as discussed later in this chapter [34].

Cohesin

Cohesin is a protein complex involved in chromosome condensation, recombination, and stabilization of homologous chromosomes and sister chromatids. The distinct functions of the predominant types of cohesin complexes in meiosis have been extensively reviewed by others [75], for which the details will not be covered here. Each cohesin complex is composed of at least four subunits: two structural maintenance of chromosomes (SMC) proteins, a kleisin protein, and an accessory protein [75]. The cohesin subunit composition differs between mitosis and meiosis, specifically by the identity of the kleisin subunit: RAD21 is expressed in both mitotic and meiotic cells, whereas REC8 and RAD21L are meiosis specific. While REC8-cohesin persists throughout meiosis, RAD21L-cohesin dissociates after completing its role in initial homologous chromosome pairing during Prophase I [76, 77].

In oocytes, immunofluorescence detected distinct subpopulations of REC8-cohesin in kinetochores, at centromeres, and on chromosome arms during Metaphase I, and between the centromeres of sister chromatids in Metaphase II[30]. Previous works demonstrated that phosphorylation of REC8 directs degradation of the cohesin complex [75]. To prevent premature separation of sister chromatids (PSSC), Shugoshin 2 (SGO2) complexes with PP2A and localizes to centromeres to prevent REC8 phosphorylation and cohesin cleavage [77]. Interestingly, oocytes from *Aurkb* cKO mice, which display premature oocyte aging phenotypes (~ 3 months of age), have increased incidences of aneuploidy from PSSC, which is hypothesized to derive from a decrease in the recruitment of SGO2 based on immunofluorescence analyzes [25, 34]. AURKC activity is increased in *Aurkb* cKO oocytes. Accordingly, these phenotypes are partially reverted in the absence of one allele of *Aurkc*, suggesting that increased activity of AURKC can lead to precocious REC8-cohesin cleavage, possibly by counterbalancing the phosphatase activity of PP2A [34]. A recent study proposes the role of AURKs in REC8-cohesin cleavage during oocyte meiosis I-II transition [78]. AURKB/C inhibition with either ZM447439 (20 μM) or AZD1152 (10 μM) led to a prevalence of bivalents, instead of univalents, in Metaphase II and a decrease in REC8phosphorylation levels at serine 482. However, because of the concentrations of inhibitors used in these experiments, we cannot rule out the possibility that AURKA could also regulate cohesin cleavage [47]. Consistent with this idea, oocytes lacking AURKB, AURKC, or both do not exhibit failure in bivalent separation [25, 34, 68]. Furthermore, inhibition of all three AURKs with ZM447439 (10 μM) reduces MPS1 phosphorylation and SGO2 localization, consistent with MPS1 activity being necessary for localization of SGO2 to prevent PSSC in mouse oocytes and suggesting an indirect role of AURKs in regulating cohesin cleavage via MPS1-dependent SGO2 recruitment [79].

A second population of SGO2 is recruited to pericentromeres by BUB1 kinase. In oocytes from transgenic mice encoding copies of *Bub1* gene lacking kinase activity (*Bub1KD*), SGO2 is reduced [79]. However, the role of pericentromeric SGO2 is currently undetermined because the absence of this population did not affect sister chromatid cohesion, and *Bub1KD* female mice are fertile [78, 80]. Nonetheless, the involvement of AURKB in recruiting BUB1 to centromeres is suggested because BUB1 persists at kinetochores when AURKC activity is specifically disrupted by ectopic expression of an ATP-gatekeeper mutated *Aurkc* allele [22]. Although oocytes from *Aurkb* cKO mice showed higher frequency of PSSC than oocytes from wild-type mice, MPS1 was recruited and at normal levels, consistent with the lack of a direct involvement of AURKB in SAC as previously described [25]. Therefore, further studies are necessary to uncover the roles of AURKB/C in both MPS1- and BUB1-dependent pathways of SGO2 recruitment in mouse oocytes. Other mechanisms by which AURKB/C regulate cohesion of sister chromatids should also be explored. For instance, the homolog of SGO2 in fission yeast is reported to recruit the CPC to the chromosomes, opening the possibility of a direct AURKB/C-SGO2 interaction and positive feedback to ensure chromosome segregation fidelity [81]. Alternatively, it is possible that AURKB regulates SGO2 recruitment and activity at centromeres more efficiently than AURKC, supporting the hypothesis that structural differences among AURKs affect their affinity for specific substrates, as discussed in the previous section. Nonetheless, the pleiotropic effect of AURKs makes it difficult to draw definite conclusions: for instance, overexpression of AURKB protein prevents Securin degradation and, consequently, Anaphase I onset; while overexpression of AURKC protein has the opposite effect [82, 83].

AURKA

In mouse oocytes, AURKA localizes to aMTOCs during Prophase I where it remains associated throughout Metaphase I and Metaphase II. Its active form, phosphorylated on T288 (pT288), is present before germinal vesicle (or nuclear envelope) breakdown (GVBD) and after [84]. AURKA pT288 is almost exclusively located at spindle poles at Metaphase I, suggesting an involvement of AURKA activity in establishing a bipolar spindle [84, 85]. *Aurka* is predominant in oocytes as its transcript expression is 12-fold higher than *Aurkb* and *Aurkc* when evaluated by Taqman-based qRT-PCR assays [86]. Evaluating the role of AURKA in mouse oocytes was possible with specific dosage of pharmacological inhibitor, MLN8237, and an AURKA oocyte-specific conditional KO (*Aurka* cKO) mouse line [41, 47]. Female mice with *Aurka* floxed alleles (*Aurka*$^{fl/fl}$), in which exon 3 is flanked by LoxP sequences, were crossed with *Aurka*$^{fl/fl}$ males harboring *Gdf9*-Cre [41]. Unlike the observations of AURKB/C single and double KO females which are subfertile, *Aurka* cKO females are sterile [41]. These results suggest AURKA is necessary and sufficient for female fertility in mouse. The following section will discuss meiotic changes occurring upon *Aurka* deletion explaining the sterility observed in *Aurka* cKO females.

aMTOC

AURKA is crucial for initiating aMTOC fragmentation, the first step to establishing spindle bipolarity. Overexpression of AURKA, by microinjection of GFP-tagged *Aurka* mRNA, induces aMTOC fragmentation in Prophase I arrested oocytes [84]. In contrast, when AURKA is conditionally removed in mouse oocytes by CRE-mediated excision, aMTOCs fail to fragment (Fig. 3). aMTOCs, marked by γ-tubulin, in *Aurka* cKO oocytes are fewer in number and larger in size, compared to aMTOCs in WT oocytes[41]. Based on these results, the current model is that AURKA drives aMTOC fragmentation in mouse oocytes. aMTOC decondensation, the first step in the fragmentation process, is triggered by Polo-like kinase 1 (PLK1) influencing the dissociation of Centrosomal NEK2-associated protein 1 (C-NAP1) from the aMTOCs [13, 85, 87]. Inhibition of PLK1 prevents C-NAP1 dissociation from aMTOCs which is required for aMTOC decondensation [13]. AURKA is involved in this process because it is required to activate PLK1 by phosphorylation at Threonine 210 [88]. In WT oocytes, C-NAP1 is associated with aMTOCs during Prophase I but is absent after nuclear envelope breakdown (NEBD) when activated

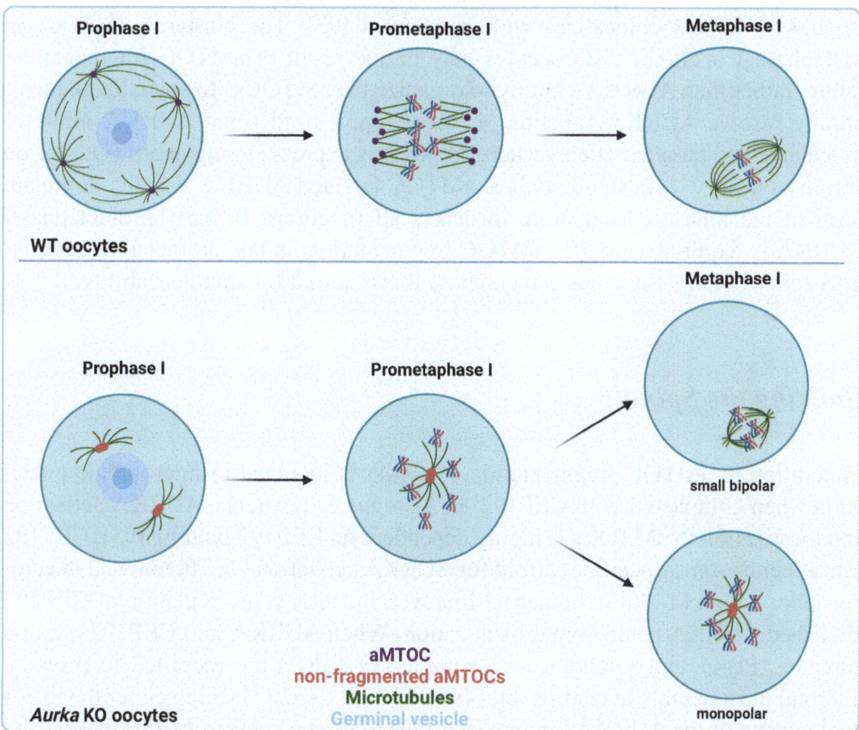

Fig. 3 Meiosis I in WT and *Aurka* KO oocytes. Upon meiotic resumption aMTOCs in *Aurka KO* oocytes fail to fragment. Oocytes will arrest at Metaphase I and have either monopolar or small bipolar spindles. Figure made in Biorender

PLK1 is present at aMTOCs. Depletion of AURKA in oocytes results in a significant decrease in PLK1 phosphorylation at aMTOCs causing an abnormal persistence of C-NAP1 after NEBD observed by immunofluorescence [41]. Together, lack of PLK1 phosphorylation and maintenance of C-NAP1 at the aMTOCs in *Aurka* cKO oocytes prevent aMTOC fragmentation.

Pericentrin localizes to aMTOCs during Prophase I and remains associated with them during Metaphases I and II[89]. The main function of pericentrin at aMTOCs is to anchor essential proteins, such as Neural precursor cell Expressed Developmentally Down-regulated protein 1 (NEDD1), γ-tubulin and centrosomal protein 215 (CEP215), to the aMTOCs to support meiotic spindle stability [89–91]. *Drosophila* CEP215 ortholog (Centrosomin) interacts with AURKA, which prompted exploration of this interaction in mammalian oocytes [92, 93]. At Metaphase II, CEP215 localizes to aMTOCs in a partial or full ring configuration. Inhibition of AURKA with MLN8237 does not displace CEP215 from aMTOCs- intensity measurements remain the same in comparison to untreated oocytes- but it obliterates the ring configuration and instead CEP215 is clustered at the spindle poles [93]. A similar effect is observed at Metaphase I in *Aurka* cKO oocytes, where CEP215 is clustered at aMTOCs [41]. Though these results suggest an additional function of AURKA may involve CEP215-dependent aMTOC organization, CEP215 aMTOC localization is dependent on and colocalizes with pericentrin [93]. The clustered aMTOCs at Metaphase I in *Aurka* cKO oocytes may be the result of aMTOC fragmentation failure rather than AURKA's ability to organize the aMTOCs. To spatially and temporally resolve AURKA function, a recent report used acute AURKA inhibitor treatments and targeted alleles where AURKA is expressed only at aMTOCs or on chromosomes [94]. In these studies, aMTOC-targeted AURKA was the dominant form of the kinase participating in nearly all functions. In early prometaphase, AURKA is required to trigger aMTOC fragmentation, in late prometaphase it controls spindle elongation, and in metaphase it is required for spindle stability.

Building the Spindle

In addition to aMTOC fragmentation, AURKA is involved in microtubule nucleation when complexed with CEP192. In *Xenopus* egg extracts, AURKA activation and localization to aMTOCs is highly dependent on CEP192 binding [95]. CEP192 acts as centrosome-specific scaffold for AURKA activation via Threonine 295 autophosphorylation (T-288 in human). Moreover, in HeLa cells, depletion of CEP192 abolished AURKA centrosomal localization. When AURKA and CEP192 are in a complex, PLK1 and γ-tubulin ring complex (γ-TuRC) are recruited to promote microtubule assembly at centrosomes [95]. Mutant *Cep192* lacking either the PLK1 docking site or the AURKA binding site, fails to rescue centrosomal formation and spindle assembly defects caused by depletion of CEP192. In mouse oocytes, *Cep192* siRNA-mediated knockdown results in impaired aMTOC formation, α-tubulin reduction in the spindle and higher incidence of small, multipolar, or nonpolar spindles compared to controls [96]. In addition to CEP192, CIP2A was identified as an integral part of AURKA and PLK1 recruitment to aMTOCs in mouse oocytes [97]. CIP2A is present throughout oocyte meiosis, but notably localizes to the spindle poles during MI and MII. Downregulation of CIP2A, with double-stranded RNA, does not affect meiotic resumption but significantly decreases polar body extrusion rates. Knockdown of CIP2A resulted in similar spindle defects observed in CEP192 knockdown oocytes. CEP192 levels decrease with CIP2A knockdown and vice versa, indicative of the mutual requirement for localization at aMTOCs. Inhibition of both AURKA and PLK1 decreases levels of CIP2A at spindle poles but only inhibition of PLK1 displaces CIP2A to the chromosomes [97]. Taken together, AURKA in a complex with PLK1, CIP2A and CEP192 promotes localization of these proteins to aMTOCs for spindle assembly during oocyte meiotic maturation.

AURKA and Polo-like kinase 4 (PLK4) work together to promote rapid growth of microtubules upon oocyte meiotic resumption [85]. Oocytes treated with centrione, a PLK4 inhibitor, or MLN8237, an AURKA inhibitor, had reduced spindle-building efficiency in comparison to control oocytes. Though delayed, treatment with either inhibitor did not prevent the formation of a bipolar spindle but resulted in a significantly reduced spindle area observed as oocytes reached Metaphase I. Inhibition of both AURKA and PLK4 with VX-680 treatment is even more detrimental to spindle-building kinetics. The decrease in spindle-building efficiency and area, observed when both kinases are inhibited, was rescued by exogenous expression of AURKA and PLK4 VX-680-resistant variants. Expression of drug-resistant AURKA was efficient in partially restoring spindle size and reducing the length of time taken to build the spindle, while drug-resistant PLK4 alone had limited rescue of spindle-kinetics. Co-expression of drug-resistant AURKA and PLK4 significantly improves spindle size and kinetics, more than either variant expressed alone, suggesting that PLK4 enhances the effects AURKA has on spindle-building efficiency [98, 99].

Liquid-Like Spindle Domain

A recent study described a dynamic structure that localizes or concentrates microtubule regulatory factors around the meiotic spindle. The structure, named the liquid-like meiotic spindle domain (LISD), includes a variety of centrosomal, centriolar, and microtubule-related proteins that dynamically redistribute around the spindle region as meiosis I progresses [100]. The LISD is described as a phase separated, liquid-like region based on experimental evaluations that use the aliphatic molecule 1, 6-hexanediol, which may have some non-specific inhibitory capacity on enzymes [101]. Therefore, further evaluation of a LISD is still needed.

LISD assembly requires AURKA activity and its regulation of TACC3 [41, 102]. TACC3, an AURKA substrate, is important for meiotic spindle stabilization [102]. When TACC3 is phosphorylated by AURKA, it localizes to the spindle in mitotic and meiotic cells [98, 103]. The LISD appears disassembled when TACC3 is depleted from mouse oocytes, which causes dispersion of multiple LISD-associated proteins and decreased length and stability of the spindles. Similar to TACC3 depletion, AURKA inhibition resulted in spindle defects and dispersion of microtubule regulatory factors normally enriched near the spindle [100]. In addition to reducing microtubule intensity and spindle volume, disassembly of the LISD causes a coalescence of aMTOCs, which are normally scattered at the spindle poles, into two separate foci. TACC3 localization at the Metaphase I spindle is lost in *Aurka* cKO oocytes, indicating that the LISD is also perturbed in oocytes lacking AURKA [41, 104]. AURKA is important for TACC3 localization to the spindle, directly driving the assembly of this newly described LISD domain that facilitates protein accessibility as the meiotic spindle is formed.

In oocytes, abnormal meiotic spindles and misaligned chromosomes are distinctive phenotypes of AURKA inhibition and depletion. Live imaging revealed that *Aurka* cKO oocytes fail to fragment aMTOCs, an important primary step in establishing a bipolar spindle. As a result, monopolar or short bipolar spindles form, and the oocytes are unable to proceed past Metaphase I (Fig. 3) [41]. Although AURKA involvement in aMTOC fragmentation can explain the inadequate spindle building in *Aurka* cKO oocytes, it is likely that AURKA is important for downstream processes needed to fully establish and stabilize the spindle. For example, because the LISD in *Aurka* cKO oocytes is perturbed, it is likely that AURKA plays a role in LISD establishment. Recent reports inhibiting AURKA later in meiotic maturation also suggests that AURKA is required to maintain LISD and spindle stability [94].

Beyond the Spindle and Chromosomes: Emerging Non-canonical Roles of the Aurora Kinases

As discussed in the previous sections, the Aurora kinases are important for mitotic and meiotic progression through their participation in aMTOC formation, spindle building, chromosome alignment and cohesion. However, recent findings link AURK activity to additional processes that also support the proper completion of mitosis and meiosis. These emerging non-canonical roles include mitochondrial regulation, translation regulation, and cytokinesis. The following section will discuss these new findings and reveal more about the compensatory and regulatory activity amongst the AURKs. Beyond their well-established functions in mitosis and meiosis, the non-canonical roles of the AURKs shed light on their multifaceted nature and their significance in various processes influencing cellular function.

Compensation and Regulation Amongst the AURKs

In many cases, the subcellular localization of proteins is predictive of their function. For instance, in mouse oocytes, AURKA is predominantly localized to spindle poles and AURKC localizes at the chromosomes, both localizations which correspond to their observed roles during mitosis and meiosis. These canonical localizations of AURKs are dependent on their binding partners, namely TPX2 for spindle and pole localization and INCENP for chromosome localization. Interestingly, mutation of a single amino acid in AURKA in HeLa cells (Glycine 198 to Asparagine) switched its binding affinity from TPX2 to INCENP and, consequently, AURKA localized to chromosomes [105, 106]. Surprisingly, localization of AURKC at aMTOCs is also observed in mouse oocytes, although whether AURKC localization at poles depends on similar binding partners as AURKA is currently unknown [107]. Haspin activity regulates AURKC localization at poles as pharmacological inhibition of Haspin causes reduction of the polar population of AURKC. Because aMTOC clustering was defective upon Haspin inhibition and AURKC reduction at the poles, it is possible that AURKC is involved in the process of aMTOC maturation during the Prophase I-to-Metaphase I transition in mouse oocytes. However, as previously mentioned, Haspin also disrupts the localization of AURKC at the chromosomes, so further dissection is needed to determine the role of AURKC at poles. A recent study found a significant difference in aMTOC fragmentation when WT and AURKB/C KO oocytes were treated with MLN[94]. aMTOC fragmentation during late prometaphase was similar to the untreated group, when WT oocytes were treated with MLN. Under this experimental condition, AURKC is the only Aurora kinase present at the poles (Fig. 4). Notably, upon treatment with MLN, over-clustering of aMOTCs occurred in AURKB/C KO oocytes. The difference in

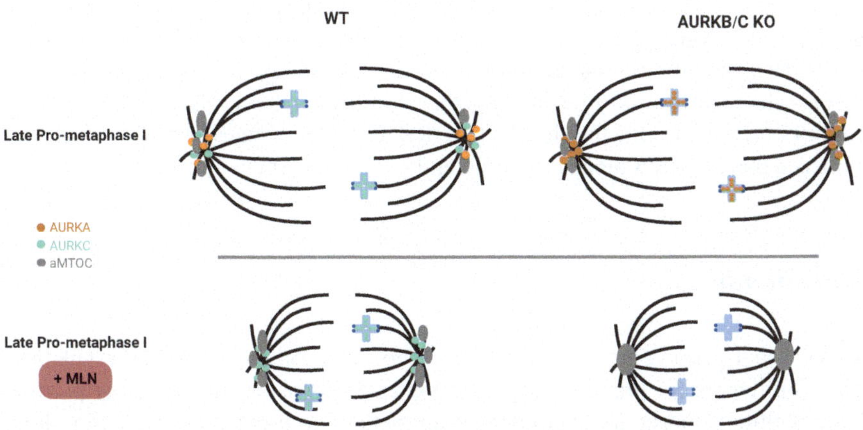

Fig. 4 AURKC is important for aMTOC separation during late prometaphase I. The difference in aMTOC phenotypes between WT and *Aurkb/c* knockout oocytes when treated with MLN in late prometaphase reveals a role for AURKC in aMTOC function. Figure made in Biorender

aMTOC phenotypes supports the model that AURKC regulates aMTOC separation [107]. Nonetheless, it is worth noting that these observations suggest that the function of AURKC is likely distinct from that of AURKA at spindle poles because AURKC cannot compensate for loss of AURKA [41].

The complexity of the overlapping roles and compensatory capacities of AURKs in oocytes is being unraveled by genetics-based studies. In AURKC KO and AURKB/C DKO oocytes, AURKA localizes to both chromosomes and aMTOCs, and compensates for the absence of AURKC [34]. Further evaluation of these oocytes that express only AURKA determined that, similar to AURKB/C, chromosomal localization of AURKA is dependent upon the CPC because INCENP depletion prevents localization of AURKA to chromosomes. Surprisingly, AURKB localization to chromosomes also depends on AURKC status: though normally localized to the spindle in WT oocytes, AURKB localizes to kinetochores when AURKC is depleted [22]. Consistent with this observation in mouse oocytes, exogenous expression of AURKC in somatic cells reduces the levels of centromeric AURKB [108]. Moreover, when either AURKB or AURKC are exogenously expressed in AURKB/C DKO oocytes, AURKA localization is maintained at aMTOCs but lost from chromosomes. These results suggest that AURKB/C outcompete AURKA for INCENP binding, providing insight into the binding affinity of each AURK to INCENP. Because AURKC is primarily associated with chromosomes, its higher affinity for INCENP could be indirectly regulating AURKA by keeping AURKA at poles and in the LISD. The changes in AURK localization upon introducing AURKC to the cellular landscape suggest that INCENP binding affinities dictate AURK subcellular function.

As described throughout this chapter, AURKs serve unique roles to support meiotic success in oocytes. Though some of their roles may be distinct from one another, some overlap (Fig. 5). These overlapping roles allow for some compensation when one or two AURKs are depleted. Inter-kinase regulation is observed in AURKB cKO oocytes, where levels of AURKA and AURKC increase compared to WT oocytes [46]. Although how these expression changes occur are not fully understood, these alterations pointed to studies toward assessing a role in regulating translation, which is discussed below. Together, fertility outcomes and in-depth oocyte studies performed with the different AURK KO mouse lines provide substantial evidence that AURKA alone can sufficiently maintain female fertility.

Mitochondria

Somatic and germ cell function is highly dependent on mitochondrial regulation. The requirement for mitochondrial function extends past adequate energy production, as mitochondria are also involved in metabolite production, Ca^{2+} flux, lipid synthesis, autophagy regulation and many other functions [109–111]. Regulation of mitochondrial dynamics, the process of fission and fusion, is particularly important

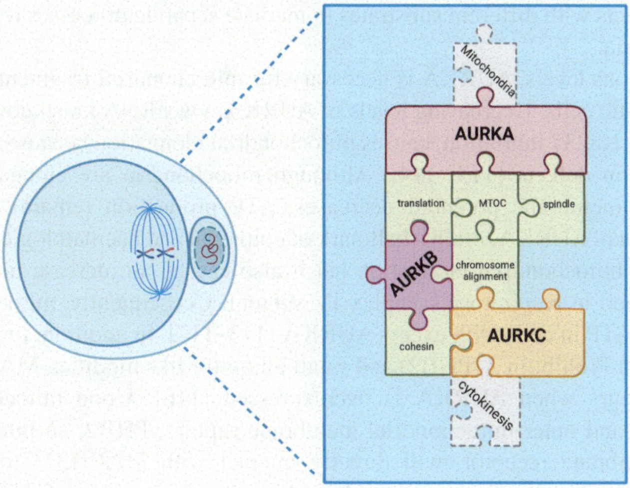

Fig. 5 Aurora kinases are involved in several meiotic processes. The AURKs ensure oocyte meiotic success by participating in chromosome alignment and spindle building processes. Some of these processes overlap as shown though connecting puzzle pieces. Though shown in somatic cells, their role in mitochondrial regulation and cytokinesis are unknown in oocytes. Figure made in Biorender

to alter the mitochondrial network with the goal to maintain cellular homeostasis [109, 112]. A role for AURKA in oocyte mitochondrial function is currently not known. Therefore, this section will review what is known in somatic cells and make connections to oocyte mitochondrial biology.

In somatic cells, AURKA is linked to the regulation of mitochondrial dynamics [113, 114]. Proteomic analysis suggests interaction of AURKA with Prohibitin (PHB), TOMM20 and TOMM70, mitochondrial proteins located in the inner and outer mitochondrial membranes, respectively [114]. AURKA association to mitochondria is further confirmed by co-localization with PMPCB, a mitochondrial matrix peptidase, and TOMM20, a mitochondrial import receptor subunit [113, 114]. Additionally, AURKA is detected in purified mitochondrial extracts from U2OS cells [114] and mitochondrial lysates of HEK293 cells [113]. Localization to mitochondria occurs via a mitochondrial targeting sequence (MTS) found in the amino-terminal region of AURKA. When the MTS is removed, AURKA no longer localizes to mitochondria [113, 114]. Transmission electron microscopy analysis refined the localization of AURKA to the mitochondrial matrix. More recently, dSTORM super-resolution microscopy not only confirmed that AURKA is present in the mitochondrial matrix but that there is also a pool of AURKA localizing at the outer mitochondrial membrane [115]. Some of AURKA is imported and processed in mitochondria, resulting in the detection of three different isoforms in mitochondrial lysates of HEK293 cells [113]. Based on localization in both the outer mitochondrial membrane and the inner mitochondrial matrix, it is possible that AURKA interacts with different substrates to mediate separate processes regulating the mitochondria.

At endogenous levels, AURKA is necessary for mitochondrial fragmentation or fission in somatic cells. Decreasing levels of AURKA, via siRNA knockdown or its activity via MLN8237 inhibition, causes mitochondrial elongation because of mitochondrial fission defects [113, 114]. Although mitochondria are elongated and mitochondrial membrane potential decreases, ATP production remains normal. Overexpression of AURKA, often a hallmark of epithelial and haematologic tumors, also results in mitochondrial elongation but it also caused an increase in oxygen consumption and in respiratory complex IV subunit. Consequently, mitochondria produce more ATP in cells with excess AURKA [113–117]. In addition, through its interaction with Prohibitin 2 (PHB2) and small ubiquitin-like modifier, MAP1LC3, mitophagy occurs when AURKA is overexpressed [116]. Upon mitochondrial depolarization and outer mitochondrial membrane rupture, PHB2, an inner mitochondrial membrane receptor, will directly interact with MAP1LC3 to trigger mitophagy of damaged mitochondria [116, 118]. Close proximity of AURKA to both PHB2 and MAP1LC3 was observed using FRET/FLIM. The interaction is absent in cells expressing an AURKA mutant lacking the amino terminus, the domain containing the MTS, indicating that AURKA mitochondrial import is necessary for its effects [116]. AURKA is required to phosphorylate PHB2 at Serine 39, which must occur for MAP1LC3 interaction [116]. Based on observations of cells overexpressing AURKA, it is possible that downregulation of AURKA could lead to accumulation of damaged mitochondria and therefore affect cellular function, but this speculation requires investigation.

A separate function of AURKA in mitochondrial regulation involves energy metabolism. For this function, AURKA interacts with ATP5F1A and ATP5F1B, mitochondrial Complex V subunits, to promote energy production in breast cancer cells [119]. Perturbing this interaction can result in two different outcomes, cellular arrest or death, depending on the primary metabolic process used to produce energy in the cell. Cells dependent on oxidative phosphorylation will arrest in G0/G1 when the AURKA/ATP5F1A/ATP5F1B interaction is blocked by inhibiting any of the three components. Along with that arrest, AURKA downregulation lowered mitochondrial membrane potential and cellular ATP levels [119]. In cells with glycolytic metabolism, targeting the AURKA/ATP5F1A/ATP5F1B interaction will result in cell death. Although mitochondria are critical for oocyte growth and maturation, there is still much to learn about how mitochondrial activity is regulated [111, 120, 121]. However, it is known that oocytes depend on oxidative phosphorylation to meet energy demands [122, 123]. How AURKA mitochondrial regulation in oocytes is similar or differs from somatic cells remains unknown. Uncovering the intricate mechanisms regulating mitochondrial activity is crucial to advance our understanding of mitochondrial dysfunction in aging oocytes.

Translation

Transcriptional quiescence occurs in fully grown mammalian oocytes, prior to resuming and successful completion of meiotic maturation. Chromatin condensation is one of the first processes to occur once the oocyte resumes meiosis and is an important pre-requisite to ensure chromosome segregation proceeds accurately [124]. Once transcription ceases and the chromatin is condensed, oocytes become dependent on temporally regulated translation to ensure sufficient availability of proteins required for successful meiotic completion, fertilization and early embryogenesis [125]. mRNAs stored in oocytes (commonly called maternal mRNAs) must undergo polyadenylation for their subsequent translation. Cytoplasmic polyadenylation binding element 1 (CPEB1) is an important regulator of mRNA polyadenylation in fully grown oocytes [125, 126]. In *Xenopus* oocytes, CPEB1 activation occurs through its phosphorylation by AURKA at Serine 174 [127]. Mitogen-activated protein kinase (MAPK) is also implicated CPEB1 phosphorylation [128]. Hyperphosphorylation of CPEB1 triggers its degradation, an event simple to monitor via western blotting. Therefore, current analyzes of CPEB1 associate its turnover with recently completed translational activity [126]. Oocytes lacking AURKB hyperphosphorylate and turnover CPEB1 with more rapid kinetics because CPEB1 is significantly decreased in *Aurkb* cKO eggs [46]. The rapid turnover and increase in Cyclin B1 (CCNB1) and AURKC proteins, both of which contain CPEB1-binding sites in their 3'UTRs, suggests translation is increased in *Aurkb* cKO oocytes. However, oocytes lacking AURKB also contain more activated AURKA. Therefore, elevated AURKA activity in *Aurkb* cKO oocytes may explain why CPEB1-dependent translation is increased. These results insinuate that an additional non-canonical role of AURKA involves translation regulation or regulating protein homeostasis in oocytes, but more evidence is required to rule out other protein kinases that also phosphorylate CPEB1 [125].

Cytokinesis

Following the alignment of chromosomes during metaphase and SAC silencing, cells proceed into cytokinesis to finalize cellular division. In mitosis, Centralspindlin, a tetramer composed of homodimers of RACGAP1 and MKLP1, concentrates and activates RHOA along the spindle equator during Anaphase via ECT2, a guanine nucleotide exchange factor [129–132]. The localized activity of RHOA is essential to recruit components of the actomyosin ring and microtubule crosslinkers, and disrupting its localization leads to cytokinesis failure [133, 134]. The compaction of the midzone spindle and furrowing of the membrane leads to the formation of a transient organelle called the midbody within the intercellular bridge between the two daughter cells (further details on the process can be found in [135], [136] and [137]). In most cells, cytokinesis finalizes in abscission of the intercellular bridge and cell membranes and results in two independent cells, but exceptions are found where the bridge is stabilized and cytokinesis is incomplete (reviewed in [138]).

Both AURKA and AURKB localize at the midzone region in somatic cells, and, through immunofluorescence, similar localization patterns of AURKB and AURKC have been described in mammalian oocytes [139–145]. In addition, phosphorylated INCENP, a marker of an active CPC, localizes at the midzone in mouse oocytes during cytokinesis [22]. In somatic cells, midzone-localized AURKB-CPC acts as a sensor to detect lagging chromosomes during cytokinesis and as an orchestrator of cytokinesis by delaying cytokinesis resolution (discussed in [146–148]). The delay, sometimes called an abscission checkpoint and imposed by AURKB-CPC during early stages of cytokinesis, is thought to be an essential step to recruit proteins required at later stages and ensures fidelity of cytokinesis completion [149, 150]. Upon Anaphase onset, the CPC translocates from kinetochores to the midzone spindle via MKLP2, and inhibiting this step prevents proper cytokinesis from taking place [151, 152]. The CPC also interacts with Centralspindlin by activating RACGAP1 and is required to maintain the fidelity of the process by delaying abscission of the membrane [153, 154]. In mitosis, AURKB phosphorylates one of the final components of the endosomal sorting complexes required for transport (ESCRT) pathway, CHMP4C, and delays its recruitment/polymerization to the abscission sites until phosphatases counteract the kinase activity [154–157]. In addition, the CPC subunit Borealin binds the homolog CHMP4B, which is essential for activation of the abscission pathway and is recruited immediately prior to abscission [156].

Further studies are needed to identify more precise localizations of each AURK and determine their roles in cytokinesis and abscission. Nonetheless, it is evident that the AURKs (particularly AURKA) can compensate for each other during cytokinesis because oocytes lacking either AURKB or AURKC, or both, form polar bodies and the mice are fertile [34, 68]. The absence of AURKA prevents meiotic progression past Metaphase I, and therefore will require additional approaches to distinguish the role of AURKA pre- and post-Anaphase I onset [41]. Pharmacological methods show that culturing of mouse oocytes in ZM447439 at concentrations that almost completely inhibit AURKC and that inhibit about 50% of AURKA still allows oocytes to complete cytokinesis [144]. The ability of oocytes to complete cytokinesis and abscission without active AURKB/C suggests their involvement in regulating the abscission timing, consistent with the accelerated abscission observed in somatic cells when AURKB was inhibited with Hesperadin or ZM447439 after midbody formation [158]. Interestingly, overexpression of AURKC in mouse oocytes, as opposed to overexpression of AURKB prevented successful cytokinesis [82]. From these studies, we speculate that AURKC is the principal player in abscission timing regulation, but more precise studies are necessary to extensively dissect the unique and overlapping roles of AURKs during meiotic cytokinesis.

Concluding Remarks

The canonical roles of the AURKs have become clearer through recent works in spindle assembly and chromosome dynamics by identification of novel structures, such as the LISD, or dissection of the specific phosphorylation sites of substrates like HEC1. In addition, the breadth of the processes in which each AURK is involved, and their context-specific roles demonstrate that there are further uncharted territories to explore. Identification of novel interacting partners and regulators might build onto the current understanding of AURK functions and unveil new roles and mechanisms by which these kinases are controlled. In the context of fertility, new studies will help us understand the evolutionary need for the diversification of the AURK family and provide insight into the molecular intricacies that unfold in meiosis.

References

1. Quartuccio SM, Dipali SS, Schindler K (2017) Haspin inhibition reveals functional differences of interchromatid axis-localized AURKB and AURKC. Mol Biol Cell 28:2233–2240. https://doi.org/10.1091/mbc.E16-12-0850
2. Krajnik K, Mietkiewska K, Skowronska A, Kordowitzki P, Skowronski MT (2023) Oogenesis in women: from molecular regulatory pathways and maternal age to stem cells. Int J Mol Sci 24:6837. https://doi.org/10.3390/ijms24076837
3. Sanchez F, Smitz J (2012) Molecular control of oogenesis. Biochim Biophys Acta 1822:1896–1912. https://doi.org/10.1016/j.bbadis.2012.05.013
4. MacLennan M, Crichton JH, Playfoot CJ, Adams IR (2015) Oocyte development, meiosis and aneuploidy. Semin Cell Dev Biol 45:68–76. https://doi.org/10.1016/j.semcdb.2015.10.005
5. Wang X, Pepling ME (2021) Regulation of meiotic prophase one in mammalian oocytes. Front Cell Dev Biol 9:667306. https://doi.org/10.3389/fcell.2021.667306
6. Mogessie B, Scheffler K, Schuh M (2018) Assembly and positioning of the oocyte meiotic spindle. Annu Rev Cell Dev Biol 34:381–403. https://doi.org/10.1146/annurev-cellbio-100616-060553
7. Charalambous C, Webster A, Schuh M (2023) Aneuploidy in mammalian oocytes and the impact of maternal ageing. Nat Rev Mol Cell Biol 24:27–44. https://doi.org/10.1038/s41580-022-00517-3
8. Rincon AM, Monje-Casas F (2020) A guiding torch at the poles: the multiple roles of spindle microtubule-organizing centers during cell division. Cell Cycle 19:1405–1421. https://doi.org/10.1080/15384101.2020.1754586
9. Blengini CS, Schindler K (2022) Acentriolar spindle assembly in mammalian female meiosis and the consequences of its perturbations on human reproductiondagger. Biol Reprod 106:253–263. https://doi.org/10.1093/biolre/ioab210
10. Sanchez AD, Feldman JL (2017) Microtubule-organizing centers: from the centrosome to non-centrosomal sites. Curr Opin Cell Biol 44:93–101. https://doi.org/10.1016/j.ceb.2016.09.003
11. Avidor-Reiss T, Fishman EL (2019) It takes two (centrioles) to tango. Reproduction 157:R33–R51. https://doi.org/10.1530/REP-18-0350
12. Gruhn JR et al (2019) Chromosome errors in human eggs shape natural fertility over reproductive life span. Science 365:1466–1469. https://doi.org/10.1126/science.aav7321

13. Clift D, Schuh M (2015) A three-step MTOC fragmentation mechanism facilitates bipolar spindle assembly in mouse oocytes. Nat Commun 6:7217. https://doi.org/10.1038/ncomms8217
14. Schuh M, Ellenberg J (2007) Self-organization of MTOCs replaces centrosome function during acentrosomal spindle assembly in live mouse oocytes. Cell 130:484–498. https://doi.org/10.1016/j.cell.2007.06.025
15. Namgoong S, Kim NH (2018) Meiotic spindle formation in mammalian oocytes: implications for human infertility. Biol Reprod 98:153–161. https://doi.org/10.1093/biolre/iox145
16. Severance AL, Latham KE (2018) Meeting the meiotic challenge: specializations in mammalian oocyte spindle formation. Mol Reprod Dev 85:178–187. https://doi.org/10.1002/mrd.22967
17. Bunning AR, Gupta ML Jr (2023) The importance of microtubule-dependent tension in accurate chromosome segregation. Front Cell Dev Biol 11:1096333. https://doi.org/10.3389/fcell.2023.1096333
18. Lara-Gonzalez P, Westhorpe FG, Taylor SS (2012) The spindle assembly checkpoint. Curr Biol 22:R966–R980. https://doi.org/10.1016/j.cub.2012.10.006
19. Cimini D, Moree B, Canman JC, Salmon ED (2003) Merotelic kinetochore orientation occurs frequently during early mitosis in mammalian tissue cells and error correction is achieved by two different mechanisms. J Cell Sci 116:4213–4225. https://doi.org/10.1242/jcs.00716
20. Brunet S, Pahlavan G, Taylor S, Maro B (2003) Functionality of the spindle checkpoint during the first meiotic division of mammalian oocytes. Reproduction 126:443–450. https://doi.org/10.1530/rep.0.1260443
21. Kitajima TS, Ohsugi M, Ellenberg J (2011) Complete kinetochore tracking reveals error-prone homologous chromosome biorientation in mammalian oocytes. Cell 146:568–581. https://doi.org/10.1016/j.cell.2011.07.031
22. Balboula AZ, Schindler K (2014) Selective disruption of aurora C kinase reveals distinct functions from aurora B kinase during meiosis in mouse oocytes. PLoS Genet 10:e1004194. https://doi.org/10.1371/journal.pgen.1004194
23. Carmena M, Wheelock M, Funabiki H, Earnshaw WC (2012) The chromosomal passenger complex (CPC): from easy rider to the godfather of mitosis. Nat Rev Mol Cell Biol 13:789–803. https://doi.org/10.1038/nrm3474
24. Kitagawa M, Lee SH (2015) The chromosomal passenger complex (CPC) as a key orchestrator of orderly mitotic exit and cytokinesis. Front Cell Dev Biol 3:14. https://doi.org/10.3389/fcell.2015.00014
25. Blengini CS, Nguyen AL, Aboelenain M, Schindler K (2021) Age-dependent integrity of the meiotic spindle assembly checkpoint in females requires Aurora kinase B. Aging Cell 20:e13489. https://doi.org/10.1111/acel.13489
26. Chiang T, Duncan FE, Schindler K, Schultz RM, Lampson MA (2010) Evidence that weakened centromere cohesion is a leading cause of age-related aneuploidy in oocytes. Curr Biol 20:1522–1528. https://doi.org/10.1016/j.cub.2010.06.069
27. Mihajlovic AI, Haverfield J, FitzHarris G (2021) Distinct classes of lagging chromosome underpin age-related oocyte aneuploidy in mouse. Dev Cell 56:2273–2283.e2273. https://doi.org/10.1016/j.devcel.2021.07.022
28. Nakagawa S, FitzHarris G (2017) Intrinsically defective microtubule dynamics contribute to age-related chromosome segregation errors in mouse oocyte meiosis-I. Curr Biol 27:1040–1047. https://doi.org/10.1016/j.cub.2017.02.025
29. Marston AL, Wassmann K (2017) Multiple duties for spindle assembly checkpoint kinases in meiosis. Front Cell Dev Biol 5:109. https://doi.org/10.3389/fcell.2017.00109
30. Llano E et al (2008) Shugoshin-2 is essential for the completion of meiosis but not for mitotic cell division in mice. Genes Dev 22:2400–2413. https://doi.org/10.1101/gad.475308
31. Cairo G, Lacefield S (2020) Establishing correct kinetochore-microtubule attachments in mitosis and meiosis. Essays Biochem 64:277–287. https://doi.org/10.1042/EBC20190072

32. Bolanos-Garcia VM (2005) Aurora kinases. Int J Biochem Cell Biol 37:1572–1577. https://doi.org/10.1016/j.biocel.2005.02.021
33. Carmena M, Earnshaw WC (2003) The cellular geography of aurora kinases. Nat Rev Mol Cell Biol 4:842–854. https://doi.org/10.1038/nrm1245
34. Nguyen AL et al (2018) Genetic interactions between the Aurora kinases reveal new requirements for AURKB and AURKC during oocyte meiosis. Curr Biol 28:3458–3468 e3455. https://doi.org/10.1016/j.cub.2018.08.052
35. Zhou H et al (1998) Tumour amplified kinase STK15/BTAK induces centrosome amplification, aneuploidy and transformation. Nat Genet 20:189–193. https://doi.org/10.1038/2496
36. Bischoff JR et al (1998) A homologue of Drosophila aurora kinase is oncogenic and amplified in human colorectal cancers. EMBO J 17:3052–3065. https://doi.org/10.1093/emboj/17.11.3052
37. Terada Y et al (1998) AIM-1: a mammalian midbody-associated protein required for cytokinesis. EMBO J 17:667–676. https://doi.org/10.1093/emboj/17.3.667
38. Nigg EA (2001) Mitotic kinases as regulators of cell division and its checkpoints. Nat Rev Mol Cell Biol 2:21–32. https://doi.org/10.1038/35048096
39. Nguyen AL, Schindler K (2017) Specialize and divide (twice): functions of three aurora kinase homologs in mammalian oocyte meiotic maturation. Trends Genet 33:349–363. https://doi.org/10.1016/j.tig.2017.03.005
40. Willems E et al (2018) The functional diversity of Aurora kinases: a comprehensive review. Cell Div 13:7. https://doi.org/10.1186/s13008-018-0040-6
41. Blengini CS et al (2021) Aurora kinase A is essential for meiosis in mouse oocytes. PLoS Genet 17:e1009327. https://doi.org/10.1371/journal.pgen.1009327
42. Santaguida S, Vernieri C, Villa F, Ciliberto A, Musacchio A (2011) Evidence that Aurora B is implicated in spindle checkpoint signalling independently of error correction. EMBO J 30:1508–1519. https://doi.org/10.1038/emboj.2011.70
43. Nezi L, Musacchio A (2009) Sister chromatid tension and the spindle assembly checkpoint. Curr Opin Cell Biol 21:785–795. https://doi.org/10.1016/j.ceb.2009.09.007
44. Shuda K, Schindler K, Ma J, Schultz RM, Donovan PJ (2009) Aurora kinase B modulates chromosome alignment in mouse oocytes. Mol Reprod Dev 76:1094–1105. https://doi.org/10.1002/mrd.21075
45. Balboula AZ, Blengini CS, Gentilello AS, Takahashi M, Schindler K (2017) Maternal RNA regulates Aurora C kinase during mouse oocyte maturation in a translation-independent fashion. Biol Reprod 96:1197–1209. https://doi.org/10.1093/biolre/iox047
46. Aboelenain M, Schindler K (2021) Aurora kinase B inhibits aurora kinase A to control maternal mRNA translation in mouse oocytes. Development 148:dev199560. https://doi.org/10.1242/dev.199560
47. Blengini CS, Ik Jung G, Aboelenain M, Schindler K (2022) A field guide to Aurora kinase inhibitors: an oocyte perspective. Reproduction 164:V5–V7. https://doi.org/10.1530/REP-22-0292
48. de Groot CO et al (2015) A cell biologist's field guide to Aurora kinase inhibitors. Front Oncol 5:285. https://doi.org/10.3389/fonc.2015.00285
49. Abdul Azeez KR et al (2019) Structural mechanism of synergistic activation of Aurora kinase B/C by phosphorylated INCENP. Nat Commun 10:3166. https://doi.org/10.1038/s41467-019-11085-0
50. Kratka C, Drutovic D, Blengini CS, Schindler K (2022) Using ZINC08918027 inhibitor to determine Aurora kinase-chromosomal passenger complex isoforms in mouse oocytes. BMC Res Notes 15:96. https://doi.org/10.1186/s13104-022-05987-4
51. Fernandez-Miranda G et al (2011) Genetic disruption of aurora B uncovers an essential role for aurora C during early mammalian development. Development 138:2661–2672. https://doi.org/10.1242/dev.066381

52. Wellard SR, Schindler K, Jordan PW (2020) Aurora B and C kinases regulate chromosome desynapsis and segregation during mouse and human spermatogenesis. J Cell Sci 133:jcs248831. https://doi.org/10.1242/jcs.248831
53. Lan ZJ, Xu X, Cooney AJ (2004) Differential oocyte-specific expression of Cre recombinase activity in GDF-9-iCre, Zp3cre, and Msx2Cre transgenic mice. Biol Reprod 71:1469–1474. https://doi.org/10.1095/biolreprod.104.031757
54. Tseng TC, Chen SH, Hsu YP, Tang TK (1998) Protein kinase profile of sperm and eggs: cloning and characterization of two novel testis-specific protein kinases (AIE1, AIE2) related to yeast and fly chromosome segregation regulators. DNA Cell Biol 17:823–833. https://doi.org/10.1089/dna.1998.17.823
55. Kimmins S et al (2007) Differential functions of the Aurora-B and Aurora-C kinases in mammalian spermatogenesis. Mol Endocrinol 21:726–739. https://doi.org/10.1210/me.2006-0332
56. DeLuca JG et al (2006) Kinetochore microtubule dynamics and attachment stability are regulated by Hec1. Cell 127:969–982. https://doi.org/10.1016/j.cell.2006.09.047
57. Lampson MA, Cheeseman IM (2011) Sensing centromere tension: Aurora B and the regulation of kinetochore function. Trends Cell Biol 21:133–140. https://doi.org/10.1016/j.tcb.2010.10.007
58. Broad AJ, DeLuca JG (2020) The right place at the right time: Aurora B kinase localization to centromeres and kinetochores. Essays Biochem 64:299–311. https://doi.org/10.1042/EBC20190081
59. Musacchio A (2015) The molecular biology of spindle assembly checkpoint Signaling dynamics. Curr Biol 25:R1002–R1018. https://doi.org/10.1016/j.cub.2015.08.051
60. Davydenko O, Schultz RM, Lampson MA (2013) Increased CDK1 activity determines the timing of kinetochore-microtubule attachments in meiosis I. J Cell Biol 202:221–229. https://doi.org/10.1083/jcb.201303019
61. Rattani A et al (2013) Sgol2 provides a regulatory platform that coordinates essential cell cycle processes during meiosis I in oocytes. elife 2:e01133. https://doi.org/10.7554/eLife.01133
62. Vallot A et al (2018) Tension-induced error correction and not kinetochore attachment status activates the SAC in an Aurora-B/C-dependent manner in oocytes. Curr Biol 28:130–139 e133. https://doi.org/10.1016/j.cub.2017.11.049
63. DeLuca KF et al (2018) Aurora A kinase phosphorylates Hec1 to regulate metaphase kinetochore-microtubule dynamics. J Cell Biol 217:163–177. https://doi.org/10.1083/jcb.201707160
64. Yoshida S, Kaido M, Kitajima TS (2015) Inherent instability of correct kinetochore-microtubule attachments during meiosis I in oocytes. Dev Cell 33:589–602. https://doi.org/10.1016/j.devcel.2015.04.020
65. Hauf S, Watanabe Y (2004) Kinetochore orientation in mitosis and meiosis. Cell 119:317–327. https://doi.org/10.1016/j.cell.2004.10.014
66. Sakuno T, Tada K, Watanabe Y (2009) Kinetochore geometry defined by cohesion within the centromere. Nature 458:852–858. https://doi.org/10.1038/nature07876
67. Watanabe Y (2012) Geometry and force behind kinetochore orientation: lessons from meiosis. Nat Rev Mol Cell Biol 13:370–382. https://doi.org/10.1038/nrm3349
68. Schindler K, Davydenko O, Fram B, Lampson MA, Schultz RM (2012) Maternally recruited Aurora C kinase is more stable than Aurora B to support mouse oocyte maturation and early development. Proc Natl Acad Sci U S A 109:E2215–E2222. https://doi.org/10.1073/pnas.1120517109
69. Wang F et al (2010) Histone H3 Thr-3 phosphorylation by Haspin positions Aurora B at centromeres in mitosis. Science 330:231–235. https://doi.org/10.1126/science.1189435
70. Nguyen AL et al (2014) Phosphorylation of threonine 3 on histone H3 by haspin kinase is required for meiosis I in mouse oocytes. J Cell Sci 127:5066–5078. https://doi.org/10.1242/jcs.158840

71. Lane SI, Chang HY, Jennings PC, Jones KT (2010) The Aurora kinase inhibitor ZM447439 accelerates first meiosis in mouse oocytes by overriding the spindle assembly checkpoint. Reproduction 140:521–530. https://doi.org/10.1530/REP-10-0223
72. Cairo G et al (2023) Distinct Aurora B pools at the inner centromere and kinetochore have different contributions to meiotic and mitotic chromosome segregation. Mol Biol Cell 34:ar43. https://doi.org/10.1091/mbc.E23-01-0014
73. Lin BW et al (2014) Overexpression of Aurora-C interferes with the spindle checkpoint by promoting the degradation of Aurora-B. Cell Death Dis 5:e1106. https://doi.org/10.1038/cddis.2014.37
74. Elkins JM, Santaguida S, Musacchio A, Knapp S (2012) Crystal structure of human aurora B in complex with INCENP and VX-680. J Med Chem 55:7841–7848. https://doi.org/10.1021/jm3008954
75. Ishiguro KI (2019) The cohesin complex in mammalian meiosis. Genes Cells 24:6–30. https://doi.org/10.1111/gtc.12652
76. Lee J, Hirano T (2011) RAD21L, a novel cohesin subunit implicated in linking homologous chromosomes in mammalian meiosis. J Cell Biol 192:263–276. https://doi.org/10.1083/jcb.201008005
77. Lee J et al (2008) Unified mode of centromeric protection by shugoshin in mammalian oocytes and somatic cells. Nat Cell Biol 10:42–52. https://doi.org/10.1038/ncb1667
78. Nikalayevich E et al (2022) Aurora B/C-dependent phosphorylation promotes Rec8 cleavage in mammalian oocytes. Curr Biol 32:2281–2290.e2284. https://doi.org/10.1016/j.cub.2022.03.041
79. El Yakoubi W et al (2017) Mps1 kinase-dependent Sgo2 centromere localisation mediates cohesin protection in mouse oocyte meiosis I. Nat Commun 8:694. https://doi.org/10.1038/s41467-017-00774-3
80. Ricke RM, Jeganathan KB, Malureanu L, Harrison AM, van Deursen JM (2012) Bub1 kinase activity drives error correction and mitotic checkpoint control but not tumor suppression. J Cell Biol 199:931–949. https://doi.org/10.1083/jcb.201205115
81. Vanoosthuyse V, Prykhozhij S, Hardwick KG (2007) Shugoshin 2 regulates localization of the chromosomal passenger proteins in fission yeast mitosis. Mol Biol Cell 18:1657–1669. https://doi.org/10.1091/mbc.e06-10-0890
82. Sharif B et al (2010) The chromosome passenger complex is required for fidelity of chromosome transmission and cytokinesis in meiosis of mouse oocytes. J Cell Sci 123:4292–4300. https://doi.org/10.1242/jcs.067447
83. Li W et al (2017) Differential regulation of H3S10 phosphorylation, mitosis progression and cell fate by Aurora kinase B and C in mouse preimplantation embryos. Protein Cell 8:662–674. https://doi.org/10.1007/s13238-017-0407-5
84. Saskova A et al (2008) Aurora kinase A controls meiosis I progression in mouse oocytes. Cell Cycle 7:2368–2376. https://doi.org/10.4161/cc.6361
85. Bury L et al (2017) Plk4 and Aurora A cooperate in the initiation of acentriolar spindle assembly in mammalian oocytes. J Cell Biol 216:3571–3590. https://doi.org/10.1083/jcb.201606077
86. Swain JE, Ding J, Wu J, Smith GD (2008) Regulation of spindle and chromatin dynamics during early and late stages of oocyte maturation by aurora kinases. Mol Hum Reprod 14:291–299. https://doi.org/10.1093/molehr/gan015
87. Kalous J, Aleshkina D (2023) Multiple roles of PLK1 in mitosis and meiosis. Cells 12:187. https://doi.org/10.3390/cells12010187
88. Macurek L et al (2008) Polo-like kinase-1 is activated by aurora A to promote checkpoint recovery. Nature 455:119–123. https://doi.org/10.1038/nature07185
89. Ma W, Viveiros MM (2014) Depletion of pericentrin in mouse oocytes disrupts microtubule organizing center function and meiotic spindle organization. Mol Reprod Dev 81:1019–1029. https://doi.org/10.1002/mrd.22422

90. Ma W, Baumann C, Viveiros MM (2010) NEDD1 is crucial for meiotic spindle stability and accurate chromosome segregation in mammalian oocytes. Dev Biol 339:439–450. https://doi.org/10.1016/j.ydbio.2010.01.009
91. Baumann C, Wang X, Yang L, Viveiros MM (2017) Error-prone meiotic division and subfertility in mice with oocyte-conditional knockdown of pericentrin. J Cell Sci 130:1251–1262. https://doi.org/10.1242/jcs.196188
92. Terada Y, Uetake Y, Kuriyama R (2003) Interaction of Aurora-A and centrosomin at the microtubule-nucleating site in drosophila and mammalian cells. J Cell Biol 162:757–763. https://doi.org/10.1083/jcb.200305048
93. Wang X, Baumann C, De La Fuente R, Viveiros MM (2020) CEP215 and AURKA regulate spindle pole focusing and aMTOC organization in mouse oocytes. Reproduction 159:261–274. https://doi.org/10.1530/REP-19-0263
94. Blengini CS, Vaskovicova M, Schier J, Drutovic D, Schindler K (2024) Spatio-temporal requirements of Aurora kinase A in mouse oocyte meiotic spindle building. iScience 27:110451. https://doi.org/10.1016/j.isci.2024.110451
95. Joukov V, Walter JC, De Nicolo A (2014) The Cep192-organized aurora A-Plk1 cascade is essential for centrosome cycle and bipolar spindle assembly. Mol Cell 55:578–591. https://doi.org/10.1016/j.molcel.2014.06.016
96. Lee IW et al (2018) Distinct roles of Cep192 and Cep152 in acentriolar MTOCs and spindle formation during mouse oocyte maturation. FASEB J 32:625–638. https://doi.org/10.1096/fj.201700559RR
97. Wang H et al (2017) CIP2A acts as a scaffold for CEP192-mediated microtubule organizing center assembly by recruiting Plk1 and aurora A during meiotic maturation. Development 144:3829–3839. https://doi.org/10.1242/dev.158584
98. Burgess SG et al (2018) Mitotic spindle association of TACC3 requires Aurora-A-dependent stabilization of a cryptic alpha-helix. EMBO J 37:e97902. https://doi.org/10.15252/embj.201797902
99. LeRoy PJ et al (2007) Localization of human TACC3 to mitotic spindles is mediated by phosphorylation on Ser558 by Aurora A: a novel pharmacodynamic method for measuring Aurora A activity. Cancer Res 67:5362–5370. https://doi.org/10.1158/0008-5472.CAN-07-0122
100. So C et al (2019) A liquid-like spindle domain promotes acentrosomal spindle assembly in mammalian oocytes. Science 364:eaat9557. https://doi.org/10.1126/science.aat9557
101. Duster R, Kaltheuner IH, Schmitz M, Geyer M (2021) 1,6-Hexanediol, commonly used to dissolve liquid-liquid phase separated condensates, directly impairs kinase and phosphatase activities. J Biol Chem 296:100260. https://doi.org/10.1016/j.jbc.2021.100260
102. Mahdipour M et al (2015) TACC3 is important for correct progression of meiosis in bovine oocytes. PLoS One 10:e0132591. https://doi.org/10.1371/journal.pone.0132591
103. Burgess SG et al (2015) Aurora-A-dependent control of TACC3 influences the rate of mitotic spindle assembly. PLoS Genet 11:e1005345. https://doi.org/10.1371/journal.pgen.1005345
104. Wellard SR et al (2021) Overlapping roles for PLK1 and Aurora A during meiotic centrosome biogenesis in mouse spermatocytes. EMBO Rep 22:e51023. https://doi.org/10.15252/embr.202051023
105. Hans F, Skoufias DA, Dimitrov S, Margolis RL (2009) Molecular distinctions between Aurora A and B: a single residue change transforms Aurora A into correctly localized and functional Aurora B. Mol Biol Cell 20:3491–3502. https://doi.org/10.1091/mbc.e09-05-0370
106. Fu J, Bian M, Liu J, Jiang Q, Zhang C (2009) A single amino acid change converts Aurora-A into Aurora-B-like kinase in terms of partner specificity and cellular function. Proc Natl Acad Sci U S A 106:6939–6944. https://doi.org/10.1073/pnas.0900833106
107. Balboula AZ et al (2016) Haspin kinase regulates microtubule-organizing center clustering and stability through Aurora kinase C in mouse oocytes. J Cell Sci 129:3648–3660. https://doi.org/10.1242/jcs.189340

108. Sasai K et al (2004) Aurora-C kinase is a novel chromosomal passenger protein that can complement Aurora-B kinase function in mitotic cells. Cell Motil Cytoskeleton 59:249–263. https://doi.org/10.1002/cm.20039
109. Monzel AS, Enriquez JA, Picard M (2023) Multifaceted mitochondria: moving mitochondrial science beyond function and dysfunction. Nat Metab 5:546–562. https://doi.org/10.1038/s42255-023-00783-1
110. Nunnari J, Suomalainen A (2012) Mitochondria: in sickness and in health. Cell 148:1145–1159. https://doi.org/10.1016/j.cell.2012.02.035
111. Adhikari D, Lee IW, Yuen WS, Carroll J (2022) Oocyte mitochondria-key regulators of oocyte function and potential therapeutic targets for improving fertility. Biol Reprod 106:366–377. https://doi.org/10.1093/biolre/ioac024
112. Giacomello M, Pyakurel A, Glytsou C, Scorrano L (2020) The cell biology of mitochondrial membrane dynamics. Nat Rev Mol Cell Biol 21:204–224. https://doi.org/10.1038/s41580-020-0210-7
113. Bertolin G et al (2018) Aurora kinase a localises to mitochondria to control organelle dynamics and energy production. elife 7:e38111. https://doi.org/10.7554/eLife.38111
114. Grant R et al (2018) Constitutive regulation of mitochondrial morphology by Aurora A kinase depends on a predicted cryptic targeting sequence at the N-terminus. Open Biol 8:170272. https://doi.org/10.1098/rsob.170272
115. Durel B, Kervrann C, Bertolin G (2021) Quantitative dSTORM super-resolution microscopy localizes Aurora kinase A/AURKA in the mitochondrial matrix. Biol Cell 113:458–473. https://doi.org/10.1111/boc.202100021
116. Bertolin G et al (2021) Mitochondrial Aurora kinase A induces mitophagy by interacting with MAP1LC3 and Prohibitin 2. Life Sci Alliance 4:e202000806. https://doi.org/10.26508/lsa.202000806
117. Bertolin G, Tramier M (2020) Insights into the non-mitotic functions of Aurora kinase A: more than just cell division. Cell Mol Life Sci 77:1031–1047. https://doi.org/10.1007/s00018-019-03310-2
118. Wei Y, Chiang WC, Sumpter R Jr, Mishra P, Levine B (2017) Prohibitin 2 is an inner mitochondrial membrane mitophagy receptor. Cell 168:224–238.e210. https://doi.org/10.1016/j.cell.2016.11.042
119. Sharma RK, Chafik A, Bertolin G (2023) Aurora kinase a/AURKA functionally interacts with the mitochondrial ATP synthase to regulate energy metabolism and cell death. Cell Death Discov 9:203. https://doi.org/10.1038/s41420-023-01501-2
120. Kirillova A, Smitz JEJ, Sukhikh GT, Mazunin I (2021) The role of mitochondria in oocyte maturation. Cells 10:2484. https://doi.org/10.3390/cells10092484
121. van der Reest J, Nardini Cecchino G, Haigis MC, Kordowitzki P (2021) Mitochondria: their relevance during oocyte ageing. Ageing Res Rev 70:101378. https://doi.org/10.1016/j.arr.2021.101378
122. Sugiura K et al (2007) Oocyte-derived BMP15 and FGFs cooperate to promote glycolysis in cumulus cells. Development 134:2593–2603. https://doi.org/10.1242/dev.006882
123. Richani D, Dunning KR, Thompson JG, Gilchrist RB (2021) Metabolic co-dependence of the oocyte and cumulus cells: essential role in determining oocyte developmental competence. Hum Reprod Update 27:27–47. https://doi.org/10.1093/humupd/dmaa043
124. Swain JE, Ding J, Brautigan DL, Villa-Moruzzi E, Smith GD (2007) Proper chromatin condensation and maintenance of histone H3 phosphorylation during mouse oocyte meiosis requires protein phosphatase activity. Biol Reprod 76:628–638. https://doi.org/10.1095/biolreprod.106.055798
125. Conti M, Kunitomi C (2024) A genome-wide perspective of the maternal mRNA translation program during oocyte development. Semin Cell Dev Biol 154(Pt B):88–98. https://doi.org/10.1016/j.semcdb.2023.03.003

126. Han SJ et al (2017) The translation of cyclin B1 and B2 is differentially regulated during mouse oocyte reentry into the meiotic cell cycle. Sci Rep 7:14077. https://doi.org/10.1038/s41598-017-13688-3
127. Mendez R et al (2000) Phosphorylation of CPE binding factor by Eg2 regulates translation of c-mos mRNA. Nature 404:302–307. https://doi.org/10.1038/35005126
128. Keady BT, Kuo P, Martinez SE, Yuan L, Hake LE (2007) MAPK interacts with XGef and is required for CPEB activation during meiosis in Xenopus oocytes. J Cell Sci 120:1093–1103. https://doi.org/10.1242/jcs.03416
129. Bement WM, Benink HA, von Dassow G (2005) A microtubule-dependent zone of active RhoA during cleavage plane specification. J Cell Biol 170:91–101. https://doi.org/10.1083/jcb.200501131
130. Yonemura S, Hirao-Minakuchi K, Nishimura Y (2004) Rho localization in cells and tissues. Exp Cell Res 295:300–314. https://doi.org/10.1016/j.yexcr.2004.01.005
131. Tatsumoto T, Xie X, Blumenthal R, Okamoto I, Miki T (1999) Human ECT2 is an exchange factor for Rho GTPases, phosphorylated in G2/M phases, and involved in cytokinesis. J Cell Biol 147:921–928. https://doi.org/10.1083/jcb.147.5.921
132. Nishimura Y, Yonemura S (2006) Centralspindlin regulates ECT2 and RhoA accumulation at the equatorial cortex during cytokinesis. J Cell Sci 119:104–114. https://doi.org/10.1242/jcs.02737
133. Matsumura F (2005) Regulation of myosin II during cytokinesis in higher eukaryotes. Trends Cell Biol 15:371–377. https://doi.org/10.1016/j.tcb.2005.05.004
134. Watanabe S et al (2008) mDia2 induces the actin scaffold for the contractile ring and stabilizes its position during cytokinesis in NIH 3T3 cells. Mol Biol Cell 19:2328–2338. https://doi.org/10.1091/mbc.e07-10-1086
135. Green RA, Paluch E, Oegema K (2012) Cytokinesis in animal cells. Annu Rev Cell Dev Biol 28:29–58. https://doi.org/10.1146/annurev-cellbio-101011-155718
136. Fededa JP, Gerlich DW (2012) Molecular control of animal cell cytokinesis. Nat Cell Biol 14:440–447. https://doi.org/10.1038/ncb2482
137. Andrade V, Echard A (2022) Mechanics and regulation of cytokinetic abscission. Front Cell Dev Biol 10:1046617. https://doi.org/10.3389/fcell.2022.1046617
138. Chaigne A, Brunet T (2022) Incomplete abscission and cytoplasmic bridges in the evolution of eukaryotic multicellularity. Curr Biol 32:R385–R397. https://doi.org/10.1016/j.cub.2022.03.021
139. Reboutier D et al (2013) Aurora A is involved in central spindle assembly through phosphorylation of Ser 19 in P150Glued. J Cell Biol 201:65–79. https://doi.org/10.1083/jcb.201210060
140. Lioutas A, Vernos I (2013) Aurora A kinase and its substrate TACC3 are required for central spindle assembly. EMBO Rep 14:829–836. https://doi.org/10.1038/embor.2013.109
141. Ramkumar N, Patel JV, Anstatt J, Baum B (2021) Aurora B-dependent polarization of the cortical actomyosin network during mitotic exit. EMBO Rep 22:e52387. https://doi.org/10.15252/embr.202152387
142. Fuller BG et al (2008) Midzone activation of aurora B in anaphase produces an intracellular phosphorylation gradient. Nature 453:1132–1136. https://doi.org/10.1038/nature06923
143. Yang KT et al (2010) Aurora-C kinase deficiency causes cytokinesis failure in meiosis I and production of large polyploid oocytes in mice. Mol Biol Cell 21:2371–2383. https://doi.org/10.1091/mbc.e10-02-0170
144. Vogt E, Kipp A, Eichenlaub-Ritter U (2009) Aurora kinase B, epigenetic state of centromeric heterochromatin and chiasma resolution in oocytes. Reprod Biomed Online 19:352–368. https://doi.org/10.1016/s1472-6483(10)60169-1
145. Uzbekova S et al (2008) Spatio-temporal expression patterns of aurora kinases a, B, and C and cytoplasmic polyadenylation-element-binding protein in bovine oocytes during meiotic maturation. Biol Reprod 78:218–233. https://doi.org/10.1095/biolreprod.107.061036
146. Petsalaki E, Zachos G (2021) The abscission checkpoint: a guardian of chromosomal stability. Cells 10:3350. https://doi.org/10.3390/cells10123350

147. Nahse V, Christ L, Stenmark H, Campsteijn C (2017) The abscission checkpoint: making it to the final cut. Trends Cell Biol 27:1–11. https://doi.org/10.1016/j.tcb.2016.10.001
148. Hadders MA, Lens SMA (2022) Changing places: chromosomal passenger complex relocation in early anaphase. Trends Cell Biol 32:165–176. https://doi.org/10.1016/j.tcb.2021.09.008
149. D'Avino PP, Capalbo L (2016) Regulation of midbody formation and function by mitotic kinases. Semin Cell Dev Biol 53:57–63. https://doi.org/10.1016/j.semcdb.2016.01.018
150. Christ L et al (2016) ALIX and ESCRT-I/II function as parallel ESCRT-III recruiters in cytokinetic abscission. J Cell Biol 212:499–513. https://doi.org/10.1083/jcb.201507009
151. van der Horst A et al (2015) Inter-domain cooperation in INCENP promotes Aurora B relocation from centromeres to microtubules. Cell Rep 12:380–387. https://doi.org/10.1016/j.celrep.2015.06.038
152. Gruneberg U, Neef R, Honda R, Nigg EA, Barr FA (2004) Relocation of Aurora B from centromeres to the central spindle at the metaphase to anaphase transition requires MKlp2. J Cell Biol 166:167–172. https://doi.org/10.1083/jcb.200403084
153. Douglas ME, Davies T, Joseph N, Mishima M (2010) Aurora B and 14-3-3 coordinately regulate clustering of centralspindlin during cytokinesis. Curr Biol 20:927–933. https://doi.org/10.1016/j.cub.2010.03.055
154. Capalbo L et al (2016) Coordinated regulation of the ESCRT-III component CHMP4C by the chromosomal passenger complex and centralspindlin during cytokinesis. Open Biol 6:160248. https://doi.org/10.1098/rsob.160248
155. Capalbo L et al (2019) The midbody interactome reveals unexpected roles for PP1 phosphatases in cytokinesis. Nat Commun 10:4513. https://doi.org/10.1038/s41467-019-12507-9
156. Carlton JG, Caballe A, Agromayor M, Kloc M, Martin-Serrano J (2012) ESCRT-III governs the Aurora B-mediated abscission checkpoint through CHMP4C. Science 336:220–225. https://doi.org/10.1126/science.1217180
157. Capalbo L et al (2012) The chromosomal passenger complex controls the function of endosomal sorting complex required for transport-III Snf7 proteins during cytokinesis. Open Biol 2:120070. https://doi.org/10.1098/rsob.120070
158. Steigemann P et al (2009) Aurora B-mediated abscission checkpoint protects against tetraploidization. Cell 136:473–484. https://doi.org/10.1016/j.cell.2008.12.020

Meiotic Spindle Organization and Function in *Drosophila* Female Oocytes

Siwen Wu and Kim S. McKim

Abstract *Drosophila* has long served as a model system for studying spindle assembly and chromosome segregation in female oocyte meiosis. *Drosophila* has the benefits of powerful genetics and cytology, along with striking similarities to mammals, including an extended prophase, acentrosomal spindle assembly, and programmed arrest points. This chapter reviews advances in understanding the mechanisms that regulate meiotic spindle assembly and organization in the oocyte meiotic cell divisions, which is critical for accurate segregation of homologous chromosomes. We review several topics that are fundamental to oocyte meiosis, including the regulation of acentrosomal spindle assembly, kinetochore assembly, the meiotic central spindle, the involvement of microtubule-associated motor proteins and spindle pole proteins, the mechanisms of kinetochore attachments to microtubules, force generation and chromosome segregation, and the role of cytoskeletal elements. Finally, we speculate on significant unsolved questions, future prospects, and emerging technologies in the study of oocyte meiotic spindles.

Keywords Drosophila · Oocyte · Meiosis · Chromosome segregation · Spindle assembly

Introduction

Oocytes undergo meiosis, a specialized form of cell division distinct from mitosis. While mitosis involves a single round of chromosome replication and division to form identical daughter cells [1], meiosis differs by initiating with chromosome replication followed by two rounds of divisions: meiosis I and meiosis II [2, 3]. In the first meiotic division, homologous chromosome pairs separate [4]. Meiosis II

S. Wu · K. S. McKim (✉)
Waksman Institute and Department of Genetics, Rutgers, the State University of New Jersey, Piscataway, NJ, USA
e-mail: mckim@waksman.rutgers.edu

B. Mogessie (ed.), *Cellular Architecture and Dynamics in Female Meiosis*,
https://doi.org/10.1007/978-3-031-97173-0_9

results in daughter cells with single chromatids. Prophase I is the longest phase in meiosis I. During this stage, the diploid cell's homologous chromosomes first pair, recombine (genetic material exchange through crossing over) and form a compact cluster known as karyosome [5] (Fig. 1). Subsequently, the nuclear envelope breaks down (NEB) and microtubule filaments nucleated in the cytoplasm during prophase are recruited by the chromosomes to form bipolar spindles in prometaphase I (Fig. 1) [6]. These spindles attach to kinetochores—multi-subunit protein complexes located at the centromeres—connecting the microtubules to the chromosomes [5, 7–9]. During metaphase I, homologous chromosome pairs align at the spindle equator to prepare for segregation. In anaphase I, homologous chromosomes separate and move to opposite poles, facilitated by spindle fibers. Telophase I result in two daughter cells, each with a set of unpaired homologous chromosomes [4]. These diploid cells then go through meiosis metaphase II to form spindles again and further divide during anaphase II to form haploid cells which only with one set of chromatids at telophase II (Fig. 1).

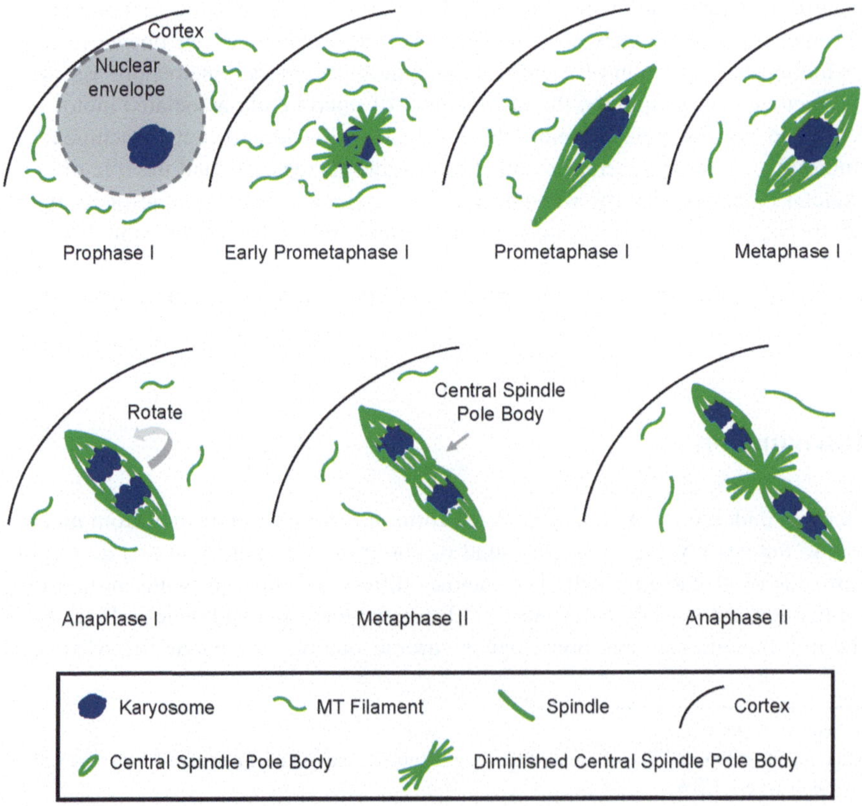

Fig. 1 Spindle assembly and dynamics during different stages of meiosis I and II in *Drosophila* oocytes

In meiosis, accurate chromosome segregation requires precise spatial and temporal regulation of spindle assembly [5, 8]. This key process is essential for maintaining genome integrity and ensuring the proper development of organisms. Defects in this process can lead to aneuploidy, causing infertility, miscarriages, and catastrophic birth defects [10]. A unique feature of meiosis in oocytes of females is acentrosomal spindle assembly, where organisms form spindles in the absence of centrosomes [8, 11–13]. The centrosomes, composed of centrioles, are often acting as the dominant microtubule nucleators in mitosis but are eliminated prior to meiosis I. As a result, there are no centrosome-based cytoplasmic microtubule-organizing centers (MTOCs) at spindle poles during female meiosis [5, 14]. Instead, chromosomes-driven meiotic spindle assembly occurs [5, 8, 14–16], particularly notable in the female germlines of a diverse array of species. For instance, in organisms like *Homo sapiens, Nematodes, Drosophila melanogaster,* and *Caenorhabditis elegans*, functional centrioles are absent in oocytes yet retained in sperm cells throughout meiosis [5, 8, 13, 14, 16–19]. Interestingly, *M. musculus* lose centrioles in both oocytes and sperm [20, 21].

This review emphasizes intracellular processes of spindle organization and function in the fruit fly *Drosophila melanogaster* female oocytes during meiosis, with the aim of contributing deeper thought and experimentation. We also refer readers to a review of all stages of meiosis in *Drosophila* females, which includes a comprehensive list of genes known to be involved in oocyte spindle assembly and function [22].

Part I: Overview of *Drosophila* Female Oogenesis

In *Drosophila melanogaster*, the ovary is the single largest organ in the female fly. Each ovary typically consists of 16–20 ovarioles. Each ovariole forms a tubular-like egg production line with a string of sequentially developing egg chambers. The germarium (housing germline stem cells and initiating oogenesis) is at the anterior end and consists of three regions. In the anterior Region 1, germline stem cells divide mitotically to produce one cystoblast, which then divides mitotically four times to develop into a 16-cell cyst. In Region 2, oocytes within these cysts enter early pachytene. In Region 3, each 16 cell cyst contains one oocyte maturing into late pachytene [23].

Region 3 is also stage 1 of the vitellarium. The vitellarium contains a series of developing egg chambers, with an oocyte and 15 nurse cells within a follicle (with early-stage oocytes toward the anterior and later stage oocytes posterior). Ovarioles are staged based on the size and position of the nurse cells and oocyte. The nurse cells provide essential nutrition and molecules for the embryo. The development of the oocyte within the *Drosophila* ovariole progresses through 14 morphological stages, starting with stage 1 budding from the germarium and ending with stage 14, representing the mature meiosis I oocyte. In the early stages 2–6, the bivalent chromosomes of the oocyte form a condensed karyosome and the synaptonemal complex (SC) disappears. In stages 7–10, the egg chambers start to elongate, and oocytes accumulate yolk. The latter stages 11–13 involve the dumping of nurse cell contents into the oocyte, oocyte growth, and elimination of the oocyte's centrosome before meiotic spindle assembly [14]. Nuclear envelope breakdown (NEB) and initiation of meiotic spindle formation occurs at stage 13, transitioning into prometaphase I and metaphase I [14, 24]. As stage 14, mature oocytes all undergo a metaphase I arrest [2, 5, 24]. The metaphase I arrest can last from 2 h to several days [7, 25]. All the ovarioles connect to a common oviduct at the posterior end. Following stage 14, the mature oocyte passes through the oviduct, which activates the oocytes so that it undergoes meiosis II in preparation for fertilization and exit via the uterus [26–28].

Oocytes are the single largest cell in fruit fly (around 400μm × 100μm) [29] and are a powerful model system for studying meiosis mechanisms through genetic, cell biology and biochemistry tools. In the following sections, we will focus on spindle assembly and regulation during *Drosophila* female meiosis.

Part II: Regulation of Acentrosomal Spindle Assembly in *Drosophila* Female Oocytes

The Formation of Meiotic Spindles in Drosophila *Oocytes*

Drosophila oocytes utilize a chromosome-dependent mechanism for spindle assembly, initiating at stage 13 [8]. During prophase I, oocyte chromosomes coalesce to form a structure called karyosome [8], which is crucial for initiating spindle assembly [30]. Lacking centrosomes, spindle formation during meiosis I begins with microtubules assembling around the condensed chromosomes after NEB [5, 7] (Fig. 1). Microtubules in the cytoplasm have been observed to form acentrosomal asters, which then migrate toward the condensed chromosomes and then bundled and integrated into the developing spindle [31].

During spindle formation, microtubule organization and growth depend on the chromosomal passenger complex (CPC)-mediated pathway in *Drosophila* female oocytes [9, 24, 32–34]. While depletion of the RanGTP-dependent pathway [35] and the Augmin pathway can result in chromosome misalignment and female infertility, they are dispensable for acentrosomal spindle assembly in meiosis [35, 36], similar to mouse oocytes [37]. Instead, the RanGTP pathway [35] promotes spindle pole organization and achiasmate chromosome segregation in homologous chromosomes lacking crossovers [35], while the Augmin pathway facilitates microtubule assembly toward poles by recruiting γ-tubulin in *Drosophila* female oocytes [38, 39]. This evidence suggests there are three distinct pathways for spindle assembly.

The CPC pathway employs multiple mechanisms to interact with both chromosomes and microtubules during spindle assembly [32]. The CPC is composed of four components: the inner centromere protein (INCENP), Aurora B Kinase,

Borealin and Survivin (known as Deterin in *Drosophila*) [19, 32, 40]. It has been proposed that CPC recruitment involves interaction with heterochromatin protein 1 (HP1, also known as Su(var)205) [41], a highly conserved non-histone chromosomal protein associated with heterochromatin [41, 42]. HP1 is enriched in adult ovaries [43, 44] and it localizes on the karyosome of oocytes during meiotic prophase [43]. It was previously reported that HP1 can directly bind to the CPC through its interactions with INCENP and Borealin [45–48]. These physical interactions are achieved by the binding of the HP1 chromo shadow domain to PXVXL/I motifs within INCENP and the C-terminus of Borealin [47, 49–51]. During early prometaphase I, a trimeric CPC complex containing INCENP, Borealin, and Survivin is recruited to the chromosomes through HP1 binding [41, 52]. However, further investigation is needed to confirm the mechanism of CPC-HP1 interaction and other mechanisms in recruiting the CPC to the chromosomes.

The Formation of Meiotic K-Fibers and Kinetochore

Drosophila meiotic spindles consist of two microtubule types, kinetochore fibers (k-fibers) and interpolar or central spindle fibers. K-fibers make end attachments to the kinetochores and attach the chromosomes to the spindle poles. Central spindle fibers are microtubules connected to one pole that make antiparallel overlaps with microtubules from the other pole (Fig. 1). The catalytic subunit Aurora B kinase is recruited by the chromatin-localized trimeric complex INCENP/Borealin/Deterin. Aurora B activation then initiates a phosphorylation cascade, targeting INCENP [53], kinetochore proteins, and other spindle assembly factors that accumulate microtubules around karyosome [8] (Fig. 1). Aurora B phosphorylation also activates kinetochore assembly [9, 41, 54], a protein complex forms on top of the centromeres and attaches to spindle fibers [55, 56]. The observation of star-like microtubule clusters lacking a defined pole on the karyosome marks a crucial stage of spindle assembly at early prometaphase I [8, 31] (Fig. 1). As the plus-ends of microtubules are stabilized on the chromosomes, enrichment of Augmin and other factors drives early spindle assembly by recruiting γ-tubulin for microtubule nucleation and pole assembly [36, 38, 39, 57]. The spindle axis is established by the central spindle (see below). Short discontinuous spindle fibers elongate rapidly, are bundled by the kinesin-14 Nonclaret disjuctional (NCD), and form a bipolar spindle with tapered poles around the karyosome [8, 31, 57–60]. Spindle formation away from the chromosomes is inhibited by the phospho-docking protein 14-3-3 [29, 61]. This process is regulated by a phosphoregulatory pathway in which Aurora B phosphorylation releases the inhibitory effects of 14-3-3 on spindle proteins like NCD and Borealin to initiate spindle assembly on chromosomes [61, 62]. As a result, the condensation of the CPC bundles parallel microtubules to form kinetochore fibers [63]. Meanwhile, Aurora B phosphorylates H3S10, leading to the inhibition of H3K9me3 [64, 65], which disrupts the interaction between HP1 and chromosomes. Additionally, Aurora B can directly phosphorylate HP1 [66]. Both result in the exit of HP1 and CPC from chromosomes, relocating them to microtubules and kinetochores, thereby facilitating bipolar spindle and kinetochore formation. However, the mechanism behind the CPC and HP1 movement from chromosomes to microtubules is unclear.

The Formation of Meiotic Inner and Outer Kinetochore

Drosophila centromeres are clustered and paired prior to NEB [67]. During prometaphase, the kinetochores of homologous centromere pairs orient and stabilize in opposite directions, capturing microtubules to form a k-fibers [24]. Bipolar spindle assembly does not depend on kinetochores but is driven by chromatin and the central spindle [8, 9]. Centromeres assemble kinetochores, and depends on the H3 variant CENP-A (known as CID in *Drosophila*), CENP-C and CAL1 [68]. CENP-C and MIS12 are also loaded during prophase, which is required for kinetochore assembly [68]. During kinetochore assembly, SPC105R, the Mis12 complex (Mis12c), and the Ndc80 complex (Ndc80c) collectively constitute the KMN network [56, 69, 70], regulating both the kinetochore assembly and kinetochore-microtubule attachment. Upon NEB, CENP-C recruits the Mis12 complex (Mis12, Nnf1, and Nsl1 in *Drosophila*) [69–71], providing a foundation for outer kinetochore assembly. The *Drosophila* genome lacks a Dsn1 orthologue [72]. Later, outer kinetochore component SPC105R (human homolog KNL1) recruitment and outer kinetochore assembly depend on Aurora B [72]. SPC105R is sufficient for attachment of kinetochores along the sides of microtubules (lateral attachments), which precedes the more stable end-on attachments and initiates homologous chromosome biorientation [9]. SPC105R provides the docking site for the Ndc80 complex. The Spc24-Spc25 heterodimer links the Ndc80 complex to SPC105R, while the Ndc80 complex's other components, Ndc80 and Nuf2, make end-on attachments to microtubules. Regulating the transition from lateral attachment to end-on attachments is an important step in ensuring accurate chromosome segregation [73, 74].

The Formation of Meiotic Central Spindles at Prometaphase I

The interpolar or central spindle fibers form from the overlap of antiparallel microtubules in the center of the spindle, or midzone. During early prometaphase I, the CPC and HP1 are transported by the kinesin-6 motor protein Subito (orthologue of MKLP2 in mammals) from the microtubules to central spindles [25, 41], forming a concentrated ring structure essential for bipolar spindle assembly and accurate chromosome segregation [32]. Subito is also essential for recruiting the γ-tubulin complex to promote microtubule assembly within the central spindle [75]. After the CPC and HP1 localize to the central spindle, HP1 maintains homolog chromosome biorientation via interactions with INCENP and Borealin [41]. Within the CPC,

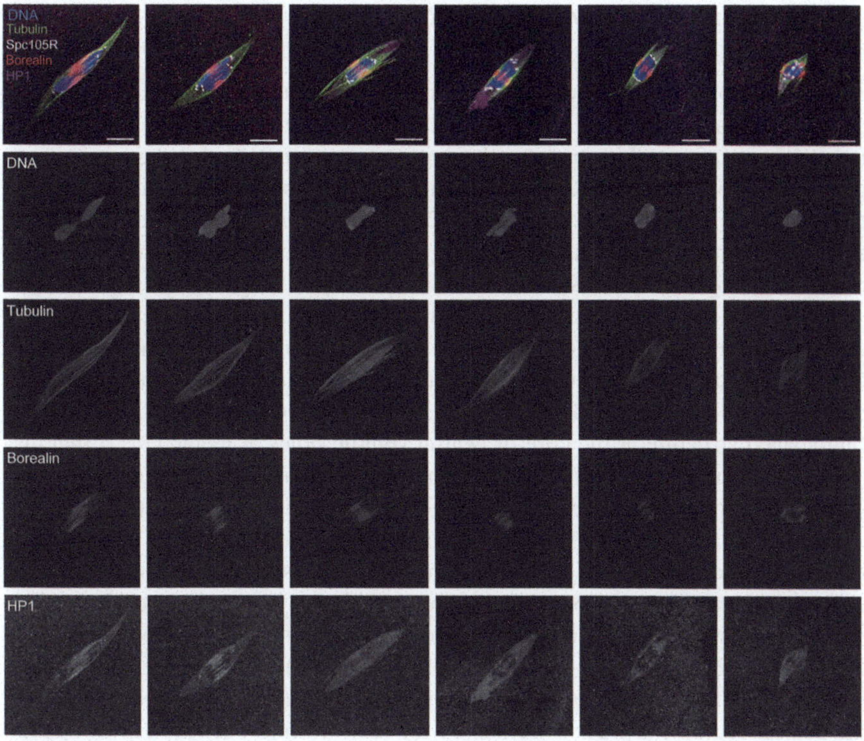

Fig. 2 Six immunostained *Drosophil*a female oocytes arranged in order based on spindle length. Because the spindle first elongates, then shortens and thickens between prometaphase and metaphase I, these are also probably in temporal order. Tubulin is in green, Spc105R is in white, the CPC is in red, HP1 is in purple, and DNA is in blue. Scale bars, 5μm

Borealin builds the central spindle through its interaction with HP1, while Deterin stabilizes the central spindle during metaphase I arrest [11, 42, 57, 76]. All these processes happen dramatically within the initial 2-h period from NEB to metaphase I [57].

Between prometaphase I and metaphase I, the meiotic spindle shortens [57, 77] (Fig. 2). This spindle morphology change is regulated by several motor proteins, including kinesin-6 MKlp1/Pav at central spindle, kinesin-13 KLP10A at the spindle poles, and other microtubule regulators [57, 78]. Metaphase I arrest follows, lasting approximately 6 h or more at room temperature in *Drosophila* oocytes [57]. During metaphase I arrest, the spindle aligns parallel to the cortex with all chromosomes aligned on the metaphase plate [5, 7, 79]. This stage is at a pivotal point in regulating the meiotic cell cycle as the spindle transitions from promoting chromosome alignment to facilitating segregation. Also of note, during meiosis the CPC relocates to central spindles at prometaphase I, whereas in mitosis, this movement occurs later, at anaphase.

Meiotic Spindle Positioning and Remodeling in Meiosis II

Accurate chromosome segregation not only requires functional spindle assembly and remodeling, but also a correct spindle positioning (Fig. 1). Upon ovulation-induced activation, all regulatory checkpoints are satisfied, and the oocyte progresses into anaphase I [26–28]. The spindle rotates, allowing one spindle pole to be positioned closer to the oocyte surface [2, 80]. The spindle further elongates and extends the midzone central spindles in anaphase, generating force to separate paired homologous chromosomes toward the poles [80] (Fig. 1).

In meiosis II, spindles inherited from meiosis I undergo reorganization without disassembly. The meiosis II oocyte has two spindles arrayed in tandem, with a "central spindle pole body" formed between that arises from the midzone of the anaphase I spindle [81, 82] (Fig. 1). The central spindle pole body is a disk- or ring-shaped structure at the core of central spindle poles at metaphase II that disperses by anaphase II [82] (Fig. 1). The central spindle pole body includes γ-tubulin [82], CP60 and CP190 [80, 83]. Kinesin-14 motor protein NCD recruits γ-tubulin at anaphase I to assemble the γ-TuRC microtubule nucleation complex and form the central spindle pole body [80]. Although the central spindle pole body contains centrosomal components [39, 84], like meiosis I, oocytes lack centrioles during meiosis II. Meiosis II spindle assembly occurs after central spindle pole body formation [80, 82, 85]. During this process, the plus-end microtubule filaments detach from the initial pole and are replaced by minus-ends, altering spindle polarity within the central region [82]. The mechanism for this "sorting out" system of spindle reorganization is unclear. Remarkably, meiosis II spindle formation is rapid, typically completing within approximately 2 min. Once the array of microtubules is established between twin spindles, NCD stabilizes mature spindles for chromosome segregation in metaphase II [82]. There is no direct evidence showing that the central spindle pole body is essential for meiosis II spindle assembly. Even when the central spindle pole body fails to form during anaphase I, spindle formation proceeds in meiosis II [85]. Thus, as in meiosis I, chromatin plays a primary role in driving spindle assembly in meiosis II. As the spindle is matured, the twin spindles elongate and taper during anaphase II, resulting in the thinning and shortening of the spindle shape (Fig. 1). This process leads to the segregation of sister chromatids by the end of meiosis II [80].

The Spindle Mechanics That Drive Chromosome Segregation

The forces in meiotic spindles that drive chromosome segregation are poorly understood. This question can be divided into the source of the force, when and where the forces are generated and how it's transduced. Interestingly, when chiasma are absent, resulting in a lack of tension on the kinetochores, oocytes proceed directly into a precocious anaphase I, bypassing the metaphase I arrest. Thus, the metaphase

I arrest may be maintained only by chiasma. In contrast, the absence of tension does not cause a metaphase I arrest [15, 86].

Drosophila oocytes undergo a mixture of anaphase A and B at anaphase as the distance between the poles and kinetochores decreases (anaphase A) during chromosome segregation while spindle elongates, pushing the chromosomes apart (anaphase B). However, it remains unclear which motion dominates. Identifying the sources of forces on the anaphase spindle poses a challenge. Potential origins, such as poles, kinetochores microtubules, or central spindles, are still under investigation. Given that there are two primary types of forces involved -pulling and pushing, we propose a model wherein spindle poles and k-fibers can generate forces that pull chromosomes apart while the central spindle generates pushing forces along kinetochore microtubules, guiding chromosomes toward the opposite poles.

The Pulling Forces on the Kinetochores

Given the absence of centrioles in fly oocytes, the role of poles in force generation remains poorly understood. In anaphase A, spindle length shortens while central spindle length is unchanged, indicating shortening of k-fibers that pulls chromosomes toward the poles [87]. This force originates from either the poles or the kinetochores. Spindles generate forces through the dynamic processes of microtubule polymerization and depolymerization [88, 89], promoting chromosome movement [90–92]. Depletion of kinesin-13 KLP10A, a microtubule-depolymerizing protein, results in spindle elongation, disorganization, and chromosome segregation defects in oocytes. This defect is attributed to the loss of microtubule depolymerization at the poles [78, 93]. Other pole-associated proteins like D-TACC, Augmin, and ASP [38, 57, 75, 84, 94] functioning in pole stabilization may also play critical roles in force contribution.

Anaphase A could result from increased activity of depolymerization at the poles combined with decreased incorporation of microtubule subunits at the kinetochores [95]. Indeed, the metaphase I spindle could be maintained by a balance of depolymerization at the poles by KLP10A and incorporation at the kinetochores by the microtubule stabilizer CLASP [95, 96]. Anaphase A could involve the loss of CLASP activity at the kinetochores, resulting in depolymerization of microtubules and pulling chromosomes toward the poles. However, studies regarding the forces sustained by kinetochore-microtubule junctions in *Drosophila* oocytes and how do kinetochores maintain connections with depolymerizing microtubules remain limited. In addition, minus-end directed motor proteins like Dynein may bind to kinetochores, sliding them toward the minus-end poles to generate a pulling force that separates chromosomes. Knocking down the outer kinetochore protein SPC105R eliminates kinetochore-microtubule attachments and chromosomes don't move apart at anaphase [9, 97]. However, in the absence of Ndc80, anaphase can still occur, showing that lateral attachments to central spindle microtubules are sufficient for poleward chromosome movement [97].

The Pushing Force Generated by Central Spindle

As in *C. elegans,* anaphase I in *Drosophila* oocytes could be combination of pulling and pushing forces [98–100]. Central spindle microtubules likely generate pushing forces on chromosomes at anaphase B. These forces can be achieved through sliding of the antiparallel microtubules and motor proteins within the central spindles [87, 92] while motor proteins slide apart overlapping non-kinetochore fibers toward the poles to push apart chromosomes toward the opposite poles [87]. Together, the pulling and pushing forces generated by spindle poles, kinetochores microtubules, and central spindles provide the drive for chromosome segregation. However, this model proposed several questions. First, the structure of meiotic spindles, characterized by short and discontinuous filaments, pose challenges in understanding the force transduction mechanisms in meiosis. Second, measuring tension within spindles is particularly difficult. Addressing this, experiments using photobleaching fluorescence-tagged microtubules to compare microtubule recovery rates with chromosome movement rates in live cells are crucial [38]. Additionally, manipulating spindle tension through laser ablation experiments will also help elucidating spindle tension dynamics.

Part III: The Role of Cytoskeletal Elements in Spindle Architecture

While extensive research has focused on microtubule dynamics in spindle formation during cell division, few studies addressed the role of other cytoskeletal components in spindle assembly and chromosome segregation in female meiosis. Recent studies across several species, mostly mammals and starfish, discovered the involvement of the actin cytoskeleton in meiotic spindle assembly and accurate chromosome segregation [101–105]. Indeed, the actin cytoskeleton appears to regulate many aspects of spindle assembly, including nuclear positioning, vesicle transport, chromosome congression, actin-microtubule crosstalk for k-fiber formation, and restraining chromatid segregation and realignment associated with aging-related aneuploidy [102, 103]. During prophase I arrest of mouse oocytes, a dense actin meshwork formed by actin nucleators spire and Formin 2 regulates vesicle transportation in a myosin-dependent manner [106]. Actin also regulates oocyte rotation and positioning during meiotic anaphase I [107–109]. Actin-based pushing forces driven by Formin 2 and the Arp2/3 complex further regulate chromosome migration [109, 110].

Research on whether actin regulates microtubule spindle assembly process in *Drosophila* female oocytes is limited. In *Drosophila* oocytes, nuclear positioning occurs from stages 5 to 7. During *Drosophila* oogenesis, actin filaments, nucleated by Cappuccino (the *Drosophila* homologue of Formin-2), chickadee (a *Drosophila* Profilin that binds to monomeric actin), and spire (an actin nucleation factor) regulate microtubule organization in the developing oocyte [111]. The actin meshwork indirectly regulates microtubules by inhibiting kinesin-dependent cytoplasmic flows, thus acting as a brake on streaming [111]. Interestingly, mutations in actin-associated proteins do not impair nuclear positioning [112], suggesting that actin may not be essential for regulating this process. Recently it has been shown that spindle-associated actin has pivotal role in regulating microtubule spindle formation and chromosome segregation during the first meiotic metaphase-anaphase transition in *Drosophila* oocytes [113]. Calcium may also play an important role in maintaining microtubule and spindle actin [113].

Additional evidence for a role of the actin cytoskeleton comes from studies of the central spindle in *Drosophila.* The Centralspindlin component RacGAP50C localizes to the central spindle in *Drosophila* oocytes [57, 76] and some of the proteins it regulates, such as Rho1, Sticky kinase, the kinesin KLP38B and RhoGEF2 may interact with spindle actin and are required for homologous chromosome biorientation in metaphase I oocytes [114]. These results suggest a novel function for proteins normally involved in mitotic cell division and cytokinesis in the regulation of microtubule-kinetochore interactions. Interactions between the microtubule and actin networks may not only regulate spindle position, but also the accuracy of chromosome segregation.

Part IV: Microtubule-Associated Motor Proteins in Spindle Assembly

Microtubule-dependent transportation relies heavily on both kinesin and dynein motor proteins. In addition, spindle bipolarity relies on two key processes: the formation of bivalent centromeres and the bundling of microtubules into spindle filaments. Kinesin motors cooperate together to regulate spindle assembly, chromosome alignment, and segregation in oocytes [71]. In *Drosophila* oocytes, the central spindle is required for chromosome biorientation [9, 24]. The plus-end kinesin-6 Subito (*Drosophila* ortholog of MKLP2) has role to establish and stabilize antiparallel microtubules and spindle bipolarity required for chromosome segregation during meiotic metaphase I [24, 71, 75, 76]. In oocytes, Subito is required for organizing a bipolar spindle, central spindle formation, and accurate chromosome segregation [71, 76, 118]. Subito consists of an acidic N-terminal region, a conserved central motor domain, and a C-terminal coiled-coil domain connected to the motor region by a "neck-linker" [115, 116] (Fig. 3). The central motor domain may coordinate with the C-terminal domain to bundle antiparallel microtubules [117] (Fig. 3). One of the most significant roles of Subito is its ability to bundle microtubules. By interacting with the CPC component INCENP [33], Subito transports the CPC to central spindles microtubules, where they cooperatively form and stabilize the central spindle [33, 114]. In addition, along with Augmin-mediated microtubule nucleation at the spindle poles [38], Subito promotes microtubules nucleation in the central

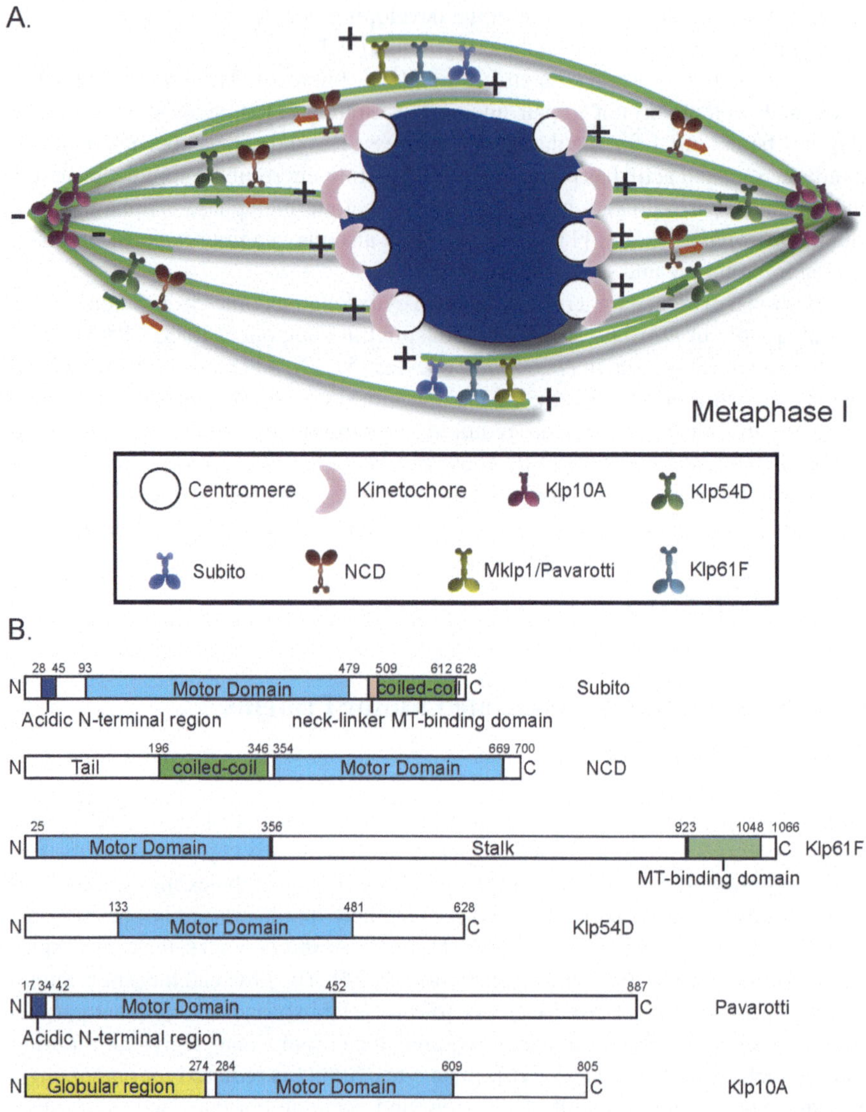

Fig. 3 Motor protein in *Drosophila* oocytes. (**a**) Schematic showing the localization of motor proteins at meiotic metaphase I. (**b**) Structure of different kinesins with conserved domains. We used FlyBase and UniProt to obtain the data

spindle through recruiting the γ-tubulin complex [75, 76]. Interestingly, the N-terminal of Subito may suppress microtubule nucleation by regulating localization to microtubules and central spindle assembly [75, 115, 117]. The N-terminal domain inhibits the interaction of Subito with the γ-tubulin complex through Grip71/NEDD1, thereby preventing ectopic microtubule nucleation [75]. However, the underlying mechanism of this negative regulation remains unclear, and may be depend on nuclear- or chromosome-based signal from a distance [117]. With both microtubule bundling and microtubule nucleation ability, Subito is a powerful regulator of spindle assembly downstream of the CPC.

Drosophila NCD is a minus-end kinesin-14 motor protein also involved in bundling k-fiber and interpolar microtubules [2, 5, 38, 71, 116]. In early meiotic spindle assembly, NCD colocalizes with short microtubules on the karyosome. Later, during bipolar spindle formation, NCD is observed on k-fibers and, possibly in lower amounts, on the central spindle [5] (Fig. 3). NCD has three domains: a core composed of a central β-sheet and six α-helices, and a small lobe with three antiparallel β-strands [118]; Its motor domain is located at the C-terminus [118] and a proline-rich globular domain at the N-terminus [119] (Fig. 3). The NCD tail has specialized elements for ATP-independent microtubule binding, including a lysine-rich so-called E-Hook-binding sequence that binds negatively charged residues in the tubulin C-terminal tails (known as the E-Hook) [5, 120]. NCD can bind to microtubules at both ends, allowing it to cross-link parallel or antiparallel arrays of microtubules and slide them relative to each other [120]. The NCD stalk is proposed to coordinate the action of multiple NCD proteins on microtubules [119, 120], contributing to its function in establishing and stabilizing spindle poles for bipolar spindle assembly in the absence of MTOCs during meiosis I [5, 121]. Loss of function of NCD indicates its role in chromosome congression, spindle elongation, and stability of bipolar spindle fibers during meiotic division [2, 5, 121]. As a result, it impairs microtubule–microtubule sliding and disrupts force production for chromosome segregation, resulting in dispersed or unstable chromosomes, short multipolar spindles, and prolonged spindle formation in meiosis [5, 121]. Additionally, NCD may influence polar body formation after the meiotic divisions, which are needed to prevent continued spindle-associated divisions of the maternal chromosomes after the initial two divisions [121].

Spatial and temporal regulation of motor protein localization between early and late metaphase may play distinct roles in spindle formation in female oocytes. Indeed, Klp61F (kinesin-5) is a plus-end motor protein essential for spindle and centromere symmetry in oocytes [57, 71]. It initially localizes at spindle poles during stage-13 but shifts significantly to central spindles by stage14 [57] (Fig. 3). Loss of Klp61F results in asymmetric spindles and associated chromosome disorganization [57, 71], indicating a role in maintaining even distribution of kinetochores and microtubules during metaphase I. The asymmetry observed in the absence of KLP61F depends on NCD, the kinesin-12 KLP54D, and the spindle pole protein ASP [57]. ASP is associated with the minus-end acentriolar spindle poles at metaphase I. Collectively, NCD, KLP54D, and ASP contribute to generating asymmetries within the acentrosomal spindle, while KLP61F collaborates with Subito to balance these forces, ensuring accurate chromosome segregation [71].

Another kinesin that exhibits dynamic localization is the kinesin-6 motor protein MKlp1/Pavarotti. During stage 13, MKlp1/Pavarotti shows no or weak spindle localization, but by stage 14, it becomes prominently enriched on the central spindles [57] (Fig. 3). This localization pattern is slightly different from Subito, as Subito begins expression in stage13 and accumulates on central spindles shortly after NEB [57]. This delayed enrichment of MKlp1/Pavarotti suggests that it is not required for central spindle assembly, but its microtubule cross-linking activity may contribute to remodeling and shortening the spindle during the prometaphase-metaphase I transition [57] (Fig. 1).

KLP10A, a *Drosophila* plus-end kinesin-13 motor protein, also dynamically regulates bipolar spindle stability in oocytes during metaphase arrest [78]. Previous studies have shown that KLP10A concentrated toward the spindle poles at early and late stage of metaphase [57, 78] (Fig. 3) and controls spindle length through microtubule depolymerization independently of the CPC pathway in spindle assembly [24, 78, 122, 123]. Aurora B negatively regulates this process, thereby maintaining a balance between microtubule depolymerization and polymerization to control spindle length [93]. Based on the observation that the other Kinesin 13 motors (KLP59C and KLP59D) are expressed only in males and are not required for female meiosis (unpublished), it is likely that Klp10A is the only Kinesin-13 functioning in oocyte meiosis.

Together, the coordinated microtubule bundling activities of Subito, KLP54D, NCD, Klp61F, and MKlp1/Pavarotti, along with the microtubule depolymerization ability of MCAK/KLP10A, contribute to the function of the oocyte spindle [71, 117]. However, several unresolved questions persist regarding spindle assembly and the roles of motor proteins. Firstly, the functions of other kinesin motor proteins in this process, such as Kinesin-8 KLP67A, remain incompletely understood. Secondly, the mechanisms responsible for the localization and transport cargos of motor proteins require further investigation. Thirdly, the activities of kinesin and dynein may be interdependent [124, 125]. Loss of kinesin-5 in Xenopus extract meiotic spindles can be compensated for by the absence of dynein at the poles [126, 127], suggesting that dynein plays a role in transporting to opposite microtubule ends. Therefore, it's reasonable to hypothesize that minus-end directed dynein may also plays a role in spindle dynamics and potentially interact with kinesins to regulate spindle assembly. Additionally, the involvement of the actin motor protein myosin in spindle assembly is yet to be determined.

Part V: Chromosome-Associated Spindle Proteins

Spindle Assembly Checkpoint (SAC) Is Essential for Microtubule-Kinetochore Attachment

During meiotic cell division, oocytes undergo a series of key events, including nuclear envelope breakdown, spindle assembly, chromosome-spindle attachment, metaphase I arrest, chromosome segregation, and cell cycle exit. In mitotic cells, these events are typically regulated by the spindle assembly checkpoint (SAC) machinery. One of the primary components is the Anaphase Promoting Complex or Cyclosome (APC/C), which controls many meiotic processes through regulation of

CDK1, the sister chromatid cohesion regulator, Separase, and other targets [128, 129]. In *Drosophila*, APC/C activity is regulated by two co-activators, Cdc20 and Fzr1/Cdh1 [129, 130]. Mutations in fzr1 cause both meiosis I and meiosis II arrests in *Drosophila* oocytes, leading to infertility [131]. Until stabilized microtubule-kinetochore attachments established, APC/C is activated by Cdc20 and Fzr1/Cdh1, both the members of the Cdc20 protein family, to progress chromosome segregation [128–130]. Cyclin-dependent kinase 1 (CDK1) is another dominant regulator of the cell cycle with three cyclin partners, cyclins A, B, and B3 in *Drosophila* oocytes [132–134]. Cyclins A, B, and B3 play overlapping roles in NEB, with cyclin B functioning independently from cyclins A and B3 for most of meiosis. Cyclin A is also required for biorientation of homologs in meiosis I, while cyclin B3 is required for anaphase progression in both meiosis I and II. Cyclin A promotes localization of Mps1 and the RZZ complex (see below), which may contribute to its requirement in biorientation [134, 135]. Cyclin B is required for maintaining metaphase I arrest in mature oocytes, organizing the spindle, and ensuring timely progression through the second meiotic division, ultimately contributing to polar body formation upon completion of meiosis [134]. In mouse oocytes, slow increase in CDK1 activity in meiosis I allow stable K-MT attachments only after bipolar spindle formation, thus preventing attachment errors [132].

The *Drosophila* genome encodes homologs of Monopolar Spindles 1 (Mps1), Mitotic arrest-deficient (Mad) proteins (Mad1, Mad2, and Mad3) and the Budding uninhibited by benzimidazole (Bub) proteins (Bub1, BubR1, and Bub3) [128, 136, 137]. Interestingly, these proteins may not regulate APC/C activity in *Drosophila* female meiosis, suggesting the metaphase I arrest is not under SAC control [138]. In addition, many of these proteins are not required for accurate chromosomes segregation in meiosis I [135]. BubR1 is required for maintaining sister chromatid cohesion and for normal maintenance of SC during meiotic prophase in *Drosophila* female oocytes [93, 139]. Thus, while the SAC in mitosis has been extensively discussed [140], its molecular regulation remains incompletely understood in *Drosophila* female meiosis. Two of these proteins, however, Mps1 and the Rod-Zw10-Zeste (RZZ) complex, are required for accurate biorientation of homologous chromosomes during meiosis I [135]. Mps1 is a conserved serine/threonine kinase [25, 140–142]. Mps1 is localized to the meiotic kinetochores and numerous cytoplasmic filaments alongside Polo [25, 135].

The Chromosome Passenger Complex as the Main Regulator of Spindle Assembly

The CPC begins meiosis on the chromosomes, where it promotes centromeres separation, kinetochore, and spindle assembly [9]. The CPC then moves to the microtubules where it is crucial for spindle bipolarity and regulation of attachments for chromosome biorientation [24, 32]. This activity may be related to the observation that the CPC can bundle microtubules to form parallel bundles in vitro [63], However, the localization of the CPC is dynamic, which reflects its interactions with chromosome and microtubules substrates [24, 32, 40]. Critical for meiosis I is that Subito is required for assembly of the central spindle, and movement of the CPC to antiparallel microtubules. Thus, the CPC functions in spindle assembly while Subito functions in spindle bipolarity. Depletion of any CPC components causes a complete failure of meiotic spindle assembly and kinetochore assembly in oocytes [9, 24, 32, 33, 41]. More research is needed to understand the mechanisms for how the three non-catalytic subunits (INCENP, Borealin, and Deterin) and Aurora B kinase, recruit spindle assembly factors to promotes spindle assembly and establish the spindle axis and kinetochores.

Drosophila INCENP is the scaffolding protein providing the platform for binding other components. The C-terminal contains the IN-box domain which binds and activates Aurora B while the N-terminal harbors the BS or CEN domain which interact with Borealin and Survivin for centromere targeting [32, 40]. Borealin, Deterin, and the N-terminal coiled-coil domain (CEN domain, first 54 amino acids) of INCENP are the three centromere-targeting regions [63]. They work together to target the CPC to chromosomes [32, 40]. INCENP also contains a PXVXL/I region required for binding to HP1 [40, 47, 49]. The middle region has a short coiled-coil domain with unknown function [32, 52]. In addition, the single α-helix (SAH) domain interacts with microtubules while the STD binding domain may recruit Subito to the microtubules to establish and stabilize antiparallel spindles on central spindles [33, 143] (Fig. 4a). Both Borealin and Deterin are conserved centromere-targeting proteins. Borealin is recruited to the outer centromere/inner kinetochore region by SAC protein Bub1 kinase-mediated phosphorylation of H2AT120 [32]. Its N-terminus contains a microtubule (MT)-binding domain which may regulate central spindle formation [32, 144]. It also contains a N-terminal domain that interacts with INCENP and Deterin [32, 52, 144]. The C-terminal region harbors the Borealin dimerization domain, which also binds Sgo1 and HP1 and likely contributes to centromere targeting [32, 41, 47, 144, 145] (Fig. 4a).

In *Drosophila*, Deterin is the smallest component in the CPC [146] and is necessary for recruiting the CPC to the chromosomes, and yet not sufficient for translocation of the CPC to the microtubules and spindle assembly [32, 41]. Its middle region, baculoviral IAP repeat (BIR) domain, is important for protein interaction and dimerization [147, 148] (Fig. 4a). Deterin is recruited to the inner centromere by Hapsin kinase phosphorylation of H3T3 [32, 149]. In mouse mitosis and meiosis, both Haspin and BUB1 kinases phosphorylate histones to provide the docking site for the CPC [150–153]. Deterin's BIR domain binds the H3T3ph site created by Haspin [149] to recruit the CPC [32, 154]. Meanwhile, Bub1 phosphorylates H2AT120, which is then recognized by Sgo1 which can recruit Borealin [155]. However, in *Drosophila* oocytes, depletion of Haspin, BUB1 or MEI-S332/SGO doesn't disrupt CPC chromosomal localization or spindle assembly, indicating that these kinases are not required for this process in *Drosophila* female meiosis [32, 41]. Deterin's role appears to be in recruiting INCENP to the chromosomes, and

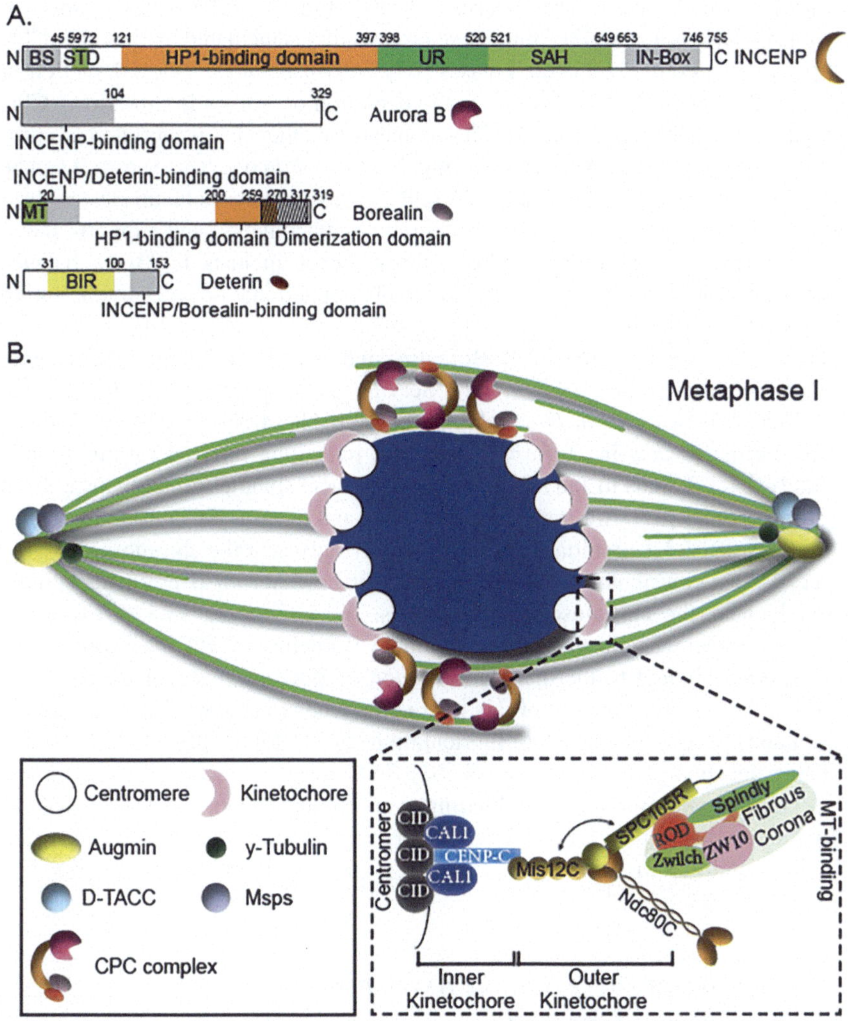

Fig. 4 (**a**) Structure of *Drosophila* CPC Components. The structure of *Drosophila* CPC components is depicted, with predicted domains based on literature references. (**b**) Spidle-associated proteins' localization on *Drosophila* female oocytes. In general, outer kinetochore components depend on inner kinetochore components. Notably, Mis12C and SPC105R exhibit mutual interdependency

potentially recruiting Borealin to the CPC to facilitate movement to the microtubules [41].

Aurora B is a highly conserved Ser/Thr kinase that plays a critical role in regulating the phosphorylation and localization of INCENP during meiosis. It consists of N-lobe and C-lobe domains, with its active site positioned between them [63]. Aurora B interacts with INCENP through its N-lobe region and shows no direct interaction with Borealin and Deterin [32, 40] (Fig. 4a). The full activation of Aurora B kinase is a complex, multistage process that is mediated by the other CPC components and other cell cycle kinases [63]. In *Drosophila* oocytes, depletion of INCENP disrupts CPC localization, while depletion of Aurora B still allows CPC localization on chromosomes [41]. This suggests that the CPC is recruited to chromosomes independent of Aurora B binding. In *Drosophila* oocytes, Aurora B can be antagonized by Protein Phosphatase 2A (PP2A) [93], which maintains sister chromatid cohesion, establishes end-on microtubule attachments, and supports metaphase I arrest by regulating spindle stability during meiosis I [93]. A balance between Aurora B and PP2A activity maintains a functional oocyte spindle during meiosis I [93].

Three CPC subunits INCENP, Borealin, and Aurora B kinase interact with phospho-docking protein 14-3-3 [62]. Serving as a phosphor-docking protein, 14-3-3 has two isoforms in *Drosophila*, ε and ξ [61]. 14-3-3 knockdown showed severe spindle defects in *Drosophila* oocytes, indicating its vital role in spindle assembly. 14-3-3 binds to and inhibits NCD to prevent spindle assembly away from the chromosomes [61]. Aurora B phosphorylates and releases NCD from the inhibitory effects of 14-3-3 to initiate spindle assembly only near the chromosomes [61]. Hence, 14-3-3 contributes to the mechanism for how Aurora B promotes spindle assembly near the chromosomes in oocytes. Indeed, during spindle assembly, 14-3-3 also interact with the CPC component Borealin. 14-3-3 can suppress the CPC through binding to the disordered region of Borealin, particularly the S163 phosphorylation site. Deletion of this interaction leads to abnormal Borealin localization and defects in centromere biorientation [62]. Additionally, the S161 site is probably phosphorylated by Aurora B [62]. The interaction between Borealin and 14-3-3 may prevent Borealin from binding to microtubules in the oocytes cytoplasm until Aurora B phosphorylates CPC subunits on chromosomes, thereby initiating spindle assembly [62].

Multiple Kinases Drive Spindle Assembly and Chromosome Segregation

As mentioned above, CDK1 is an important regulator of meiotic progression. Indeed, in the absence of CDK1, nuclear envelope breakdown does not occur (unpublished, [134]). Polo kinase is required for multiple processes, including spindle assembly, chromosome segregation, cohesin removal, and kinetochore orientation [156, 157]. In *Drosophila* female oocytes, Polo localizes to kinetochores and central spindles during meiotic prometaphase and metaphase I to both stabilize and destabilize kinetochore-microtubule end-on attachment [156, 158]. Its activity can be inhibited by the female meiosis-specific protein Matrimony (Mtrm) [156]. Following nuclear envelope breakdown, Mtrm accumulates along the meiotic spindle to restrict Polo away from centromeres and chromosomes by localizing it to the meiotic spindle, preventing the premature release of sister chromatid cohesion [156]. Aurora A is also required for biorientation of homologous chromosomes at metaphase I but not spindle assembly (unpublished) and is also critical for meiosis in mouse oocytes [40]. Thus, Aurora A, Aurora B, CDK1, and PLK work together to regulate the overall progression of meiosis.

Microtubule-Kinetochore Attachment Regulators

Functional spindle assembly requires proper spindle nucleation and clustering, bipolar formation, and stable microtubule-chromosome attachment. Dysregulation of this process can result in chromosome mis-segregation, leading to aneuploidy [159]. Kinetochores serve as a dock to establish microtubule attachments in meiosis, first lateral and then end-on MT-kinetochore attachments [9, 31]. Kinetochore assembly and function requires various regulators located at different layers, from the chromosome side to the microtubule side: (i) CENP-C and CID at inner kinetochores and centromeres; (ii) the KMN network at the outer kinetochore; (iii) as well as the kinesin Centromere-associated Protein E (CENP-E), dynein, and the ROD-ZW10-Zwilch (RZZ) complex [32, 160–162] (Fig. 4).

Unlike other species, which possess a constitutive centromere-associated network (CCAN), *Drosophila* relies solely on the single CCAN component, CENP-C, for assembly of the kinetochore [163]. CENP-C binds to CENP-A/CID and CAL1 to stabilize them on the centromeres (Fig. 4b), maintaining sister centromere cohesion and centromere clustering, and serves as the central platform within the inner kinetochore to recruit KMN network to the outer kinetochore [68–71]. The KMN network in *Drosophila* oocytes comprises three distinct sub-complexes: SPC105R (the *Drosophila* KNL1 homolog), MIS12c, and NDC80c, each serving unique functions [161, 164]. The Mis12 complex, composed of Nsl1, Mis12, Nnf1a and Nnf1b in *Drosophila*, binds to CENP-C during early prophase I, and recruits the rest of the KMN network to the kinetochores [162] (Fig. 4b). Although lacking a Dsn1 subunit, the Mis12c complex recruits SPC105R [72]. The situation is complex, however, because SPC105R is also required for recruiting the Mis12 complex [9, 164, 165].

SPC105R has three functions in meiosis [166]. First, it is essential for outer kinetochore assembly by recruiting the NDC80 complex [9]. Second, SPC105R is required for co-orientation through fusion of sister centromeres by regulating microtubule attachments and protecting centromeric cohesion [167]. This ensures that sister kinetochores attach to the same spindle poles at prometaphase I. Third, SPC105R is sufficient for lateral KT-MT attachments during prometaphase I, an early step in chromosome segregation [68, 164, 168]. Polo kinase (Polo/PLK) is also required for localization of SPC105R [9, 164, 165]. SPC105R is composed of several domains. The C-terminal domain contains coiled-coil motifs which recruits the NDC80 complex for outer kinetochore assembly. This domain also recruits BUBR1, which in turn recruits the cohesion protection proteins MEI-S332/

Shugoshin and PP2A [169, 170]. PP2A can also be directly recruited by SPC105R through its central domain, although it is not known if these are redundant [93]. The central domain also recruits SAC proteins, including those important for biorientation like Mps1 and RZZ. Interestingly, a region within the central domain is sufficient for lateral interactions with microtubules. Thus, the KMN network binds to microtubules through both SPC105R and NDC80 [9, 93].

The Ndc80c complex, consisting of Ndc80, Nuf2, SPC24, and SPC25 [56], plays a crucial role in kinetochore function because it is required for end-on attachments [9]. It is likely that Aurora B kinase destabilizes NDC80-microtubule attachment, resulting in lateral attachments. Subsequently, PP2A antagonizes Aurora B phosphorylation at kinetochores, allowing for establishment of stable end-on attachments [9, 93]. While end-on attachment is dispensable for prometaphase chromosome movement which facilitates chromosome alignment, it is essential for stable biorientation of homologs [9]. Therefore, both NDC80 and PP2A are crucial for forming end-on attachments, ensuring the accuracy of chromosome segregation [93].

The RZZ complex, composed of Rough Deal (Rod), Zeste-white 10 (ZW10) and Zwilch, are in the outer kinetochore, building what is sometimes referred to as the fibrous corona (Fig. 4b). The RZZ complex has been shown to have roles in MT-KT attachment, k-fiber nucleation, and the SAC, all ensuring the fidelity of chromosome biorientation and accurate segregation [160]. The RZZ complex accomplishes these roles by interacting with protein like the motors dynein and CENP-E and SAC components Mad1-Mad2 [171, 172]. Spindly is a Dynein recruitment factor that interacts with RZZ [173] (Fig. 4b). Little is known, however, about RZZ function in meiosis. Recently it has been shown that ROD is recruited by the central domain of SPC105R in *Drosophila* oocytes and is required for biorientation of homologous chromosomes [135]. Indeed, both CENP-E and Spindly are also required for biorientation, which may reflect an interaction of these two motors with RZZ [9, 135].

Part VI: Spindle Pole Proteins

Even though oocyte spindles are acentrosomal, there is an enrichment of pole-associated proteins that is important for spindle assembly and accurate chromosome segregation. For example, γ-tubulin, which normally contributes to the microtubule nucleation, is required during meiosis I spindle assembly and proper kinetochore attachmens in *Drosophila* oocytes [39, 174–176]. The Augmin complex recruits γ-tubulin to the pre-existing microtubules to generate new microtubules [177, 178] (Fig. 4b). During spindle assembly, the CPC probably activates Augmin [36, 174], which then recruits the γ-tubulin complex via Grip71 for microtubule nucleation at the spindle poles [75] (Fig. 4b). Depletion of Augmin component (Wac) in *Drosophila* oocytes leads to untapered poles, reduced microtubule density, elongated spindles, and chromosome misalignment [36, 38], indicating that spindle poles depend on Augmin for accurate chromosome segregation.

Along with Augmin, transforming acidic coiled-coil-containing protein (D-TACC) and Mini-spindles (Msps) co-localize at meiotic spindle poles in *Drosophila* oocytes (Fig. 4b). In HeLa cells, TACC3 is found along K-fibers and centrosomes and is essential for acentrosomal microtubule nucleation, proper K-fiber formation, and chromosome alignment during mitosis [179, 180]. In contrast to its role in mitosis, TACC exhibits different localization in meiosis, suggesting distinct functions during this process. In *Drosophila* female meiosis, RanGTP appears to be important for tapering microtubules at the spindle poles [180]. It is proposed that TACC interacts with RanGTP to help organize the meiosis I spindle poles, although TACC is not be essential for initiating spindle assembly [35]. However, it is still unclear how TACC is being recruited to the spindle poles in meiosis. Msps, of the dis1/TOG family, is a crucial microtubule-associated protein concentrated at the spindle poles of meiotic *Drosophila* oocytes [84]. The polar localization of Msps is mediated by NCD, which transports it to the spindle poles. D-TACC further restricts Msps to these poles, stabilizing microtubules in *Drosophila* oocytes [84] (Fig. 4b). Another pole-localized protein, Abnormal spindle (ASP), aggregates microtubules to form spindle poles during meiosis [181].

Microtubules are dynamic structures that rapidly polymerize and depolymerize at the plus-end and depolymerize at the minus-end, allowing for quick adjustments in shape and length while maintaining a balanced state. While several spindle proteins have been discovered and/or characterized in *Drosophila*, not all have been characterized in meiosis. EB1-a conserved protein found at the growing plus-end of microtubules, plays a crucial role in regulating their dynamics. EB1 positively regulates Aurora B kinase activity by inhibiting PP2A, which prevents the dephosphorylation and inactivation of Aurora B [182, 183]. In *Drosophila* mitotic cells, depletion of EB1 results in less dynamic microtubules, reduced astral microtubules, spindle defects, defocused spindle pole, impaired spindle elongation, mispositioned spindles, and chromosome mis-segregation [184]. These observations highlight EB1's essential function in promoting microtubule growth and interactions within the central spindle and at the cell cortex [184]. While no studies have directly examined EB1 in *Drosophila* oocytes, Sentin, a microtubule catastrophe promoter, interacts with EB1 to regulate microtubule plus-end dynamics and delay stable kinetochore-microtubule attachment in oocytes. By destabilizing kinetochore-microtubules attachments, Sentin prevents premature and incorrect KT-MT attachments [185, 186].

Several other spindle pole proteins have been characterized in *Drosophila*, although mostly in mitotic cells [57]. In addition to those described above, Aurora kinase A (unpublished) and MEI-38/TPX2 [187–189] have been shown to be important for meiosis I spindle function. However, the mechanisms underlying how these proteins are localized at poles, how they regulate spindle stability, and why they are important for accurate chromosomes segregation in meiosis I are unclear. As with Sentin, they could delay end-on attachments through an unknown mechanism. Spindle pole proteins could also regulate microtubule plus-end dynamics, such as shown for Chb/CLASP in *Drosophila* [95]. Another possibility is that spindle pole proteins are required for the kinase activity of Aurora A to regulates microtubule attachments similar to Aurora B [190]. A recent study has shown that if chromosomes get close to the spindle poles, Aurora A is activated by kinetochore bound MPS1 [191]. This is an important area of study because instability of spindle poles has been associated with high aneuploidy in human oocytes [192]. Spindle pole stability in human oocytes has been shown to depend on proteins like NUMA, which is known as MUD in *Drosophila*. MUD has been shown to be important for meiosis II in *Drosophila*, but functions in meiosis I have not been thoroughly studied [193].

Part VII: Unresolved Questions, Future Prospects, and Emerging Technologies for Studying Meiotic Spindles

There are still questions regarding the chromosome-based signals that recruit microtubules and initiate acentrosomal spindle assembly in oocytes lacking centrosomes. Additionally, in oocytes lacking MTOCs like *Drosophila* and humans, how spindle poles are formed, how pole-associated proteins are recruited, and how spindle pole structures are regulated and maintained are not known. It is also not known whether the differences between oocytes and mitotic cells are due to the lack of MTOCs or other features of oocytes and female meiosis, such as the generation of polar bodies and only one functional product of meiosis.

During meiotic metaphase I, the arrangement of the central spindle around the karyosome is robust. Microscopy studies show that rather than forming complete ring-like structures around the entire karyosome, the central spindle typically organizes into multiple filament bundles distributed across its surfaces in *Drosophila* oocytes. There is often an enrichment of spindle filament bundles on one surface of the karyosome, accompanied by enriched spindle bundles on the opposite surface. This specific arrangement raises questions about its functional significance. It's possible that this arrangement physically provides potential benefits for scaffolding motor proteins and central spindle-associated proteins. A robust central spindle during prometaphase I may provide a region or environment where the forces are conducive to establishing biorientation as the precondition for anaphase I chromosome segregation. However, further investigations are needed to understand the reasons behind this organization. Furthermore, the mechanisms governing the tandem formation of spindles and the maintenance of bipolarity in meiosis II are still unknown.

Numerous genetic tools, such as the UAS-GAL4 system, have been extremely valuable to manipulate gene expression in *Drosophila* oocytes to study their roles in meiosis. Additionally, methods to create fusion genes and mutants with targeted deletions have allowed observing the effects of specific changes to deviate the function of a gene of interest. It would be valuable, however, to be able to manipulate protein level and activity more rapidly. While there are some drugs for this, more tools like degron systems are needed [194]. Live imaging of *Drosophila* female oocytes is a valuable tool [2, 4, 5, 195]. However, isolating and imaging oocytes at the appropriate stages is time consuming. There is a need for better culture and synchronization protocols.

The metaphase I oocyte is arrested at a very convenient time, where one can visualize the most important goal of meiosis I, the biorientation of homologous chromosomes centromeres. This can now be combined with recent advances in super-resolution microscopy to observe protein localization and spindle dynamics at specific stages of oocytes, offering insights into the roles of spindle dynamics, protein loading timing, expression level changes, and the localization of various proteins. Previously, studies have relied on yeast two-hybrid assays and Co-IP to investigate protein-protein interactions. Additionally, the *Drosophila* ovary is suitable for Mass Spectrometry approaches, such as IP-MS [196] and Pull-down-MS [61, 62]. Future work will include developing TurboID to enhance the detection of weak protein interactions and improve our mapping of these interactions. With emerging technologies, understanding spindle assembly and chromosome segregation regulation will offer crucial insights into oocyte infertility mechanisms and identify potential therapeutic targets for reproductive disorders associated with spindle assembly.

Acknowledgements Image acquisition and analysis was made possible by the Waksman Institute Shared Imaging Core Facility and The Human Genetics Institute Imaging Core Facility at Rutgers. This work was supported by NIH grant GM101955 to K.S.M and a Waksman Institute Busch Predoctoral fellowship to Siwen Wu.

References

1. Heald R, Khodjakov A (2015) Thirty years of search and capture: the complex simplicity of mitotic spindle assembly. J Cell Biol 211:1103–1111. https://doi.org/10.1083/jcb.201510015
2. Endow SA, Komma DJ (1997) Spindle dynamics during meiosis in Drosophila oocytes. J Cell Biol 137:1321–1336
3. Puro J, Nokkala S (1977) Meiotic segregation of chromosomes in Drosophila melanogaster oocytes. Chromosoma 63:273–286
4. Hughes SE, Hawley RS (2017) Live imaging of meiosis I in late-stage Drosophila melanogaster oocytes. Methods Mol Biol 1471:255–264. https://doi.org/10.1007/978-1-4939-6340-9_14
5. Matthies HJ, McDonald HB, Goldstein LS, Theurkauf WE (1996) Anastral meiotic spindle morphogenesis: role of the non-claret disjunctional kinesin-like protein. J Cell Biol 134:455–464
6. Theurkauf WE, Alberts BM, Jan YN, Jongens TA (1993) A central role for microtubules in the differentiation of Drosophila oocytes. Development 118:1169–1180
7. Theurkauf WE, Hawley RS (1992) Meiotic spindle assembly in Drosophila females: behavior of nonexchange chromosomes and the effects of mutations in the nod kinesin-like protein. J Cell Biol 116:1167–1180
8. McKim KS, Hawley RS (1995) Chromosomal control of meiotic cell division. Science 270:1595–1601
9. Radford SJ, Hoang TL, Głuszek AA, Ohkura H, McKim KS (2015) Lateral and end-on kinetochore attachments are coordinated to achieve bi-orientation in Drosophila oocytes. PLoS Genet 11:e1005605. https://doi.org/10.1371/journal.pgen.1005605

10. Compton DA (2011) Mechanisms of aneuploidy. Curr Opin Cell Biol 23:109–113. https://doi.org/10.1016/j.ceb.2010.08.007
11. Radford SJ, Nguyen AL, Schindler K, McKim KS (2017) The chromosomal basis of meiotic acentrosomal spindle assembly and function in oocytes. Chromosoma 126:351–364. https://doi.org/10.1007/s00412-016-0618-1
12. Bennabi I, Terret ME, Verlhac MH (2016) Meiotic spindle assembly and chromosome segregation in oocytes. J Cell Biol 215:611–619. https://doi.org/10.1083/jcb.201607062
13. Ohkura H (2015) Meiosis: an overview of key differences from mitosis. Cold Spring Harb Perspect Biol 7(5):a015859. https://doi.org/10.1101/cshperspect.a015859
14. Pimenta-Marques A, Bento I, Lopes CA, Duarte P, Jana SC, Bettencourt-Dias M (2016) A mechanism for the elimination of the female gamete centrosome in Drosophila melanogaster. Science 353:aaf4866. https://doi.org/10.1126/science.aaf4866
15. McKim KS, Jang JK, Theurkauf WE, Hawley RS (1993) Mechanical basis of meiotic metaphase arrest. Nature 362:364–366
16. Namgoong S, Kim NH (2018) Meiotic spindle formation in mammalian oocytes: implications for human infertility. Biol Reprod 98:153–161. https://doi.org/10.1093/biolre/iox145
17. Severson AF, von Dassow G, Bowerman B (2016) Oocyte meiotic spindle assembly and function. Curr Top Dev Biol 116:65–98. https://doi.org/10.1016/bs.ctdb.2015.11.031
18. Dumont J, Desai A (2012) Acentrosomal spindle assembly and chromosome segregation during oocyte meiosis. Trends Cell Biol 22:241–249. https://doi.org/10.1016/j.tcb.2012.02.007
19. Ruchaud S, Carmena M, Earnshaw WC (2007) Chromosomal passengers: conducting cell division. Nat Rev Mol Cell Biol 8:798–812
20. Manandhar G, Schatten H, Sutovsky P (2005) Centrosome reduction during gametogenesis and its significance. Biol Reprod 72:2–13. https://doi.org/10.1095/biolreprod.104.031245
21. Szollosi D, Calarco P, Donahue RP (1972) Absence of centrioles in the first and second meiotic spindles of mouse oocytes. J Cell Sci 11:521–541
22. Hughes SE, Miller DE, Miller AL, Hawley RS (2018) Female meiosis: synapsis, recombination, and segregation in Drosophila melanogaster. Genetics 208:875–908. https://doi.org/10.1534/genetics.117.300081
23. Waghmare I, Page-McCaw A (2018) Wnt signaling in stem cell maintenance and differentiation in the Drosophila germarium. Genes (Basel) 9(3):127. https://doi.org/10.3390/genes9030127
24. Radford SJ, Jang JK, McKim KS (2012) The chromosomal passenger complex is required for meiotic acentrosomal spindle assembly and chromosome bi-orientation. Genetics 192:417–429
25. Gilliland WD, Hughes SE, Cotitta JL, Takeo S, Xiang Y, Hawley RS (2007) The multiple roles of mps1 in Drosophila female meiosis. PLoS Genet 3:e113. https://doi.org/10.1371/journal.pgen.0030113
26. Mahowald AP, Goralski TJ, Caulton JH (1983) In vitro activation of Drosophila eggs. Dev Biol 98:437–445
27. Page AW, Orr-Weaver TL (1997) Activation of the meiotic divisions in Drosophila oocytes. Dev Biol 183:195–207. https://doi.org/10.1006/dbio.1997.8506
28. Heifetz Y, Yu J, Wolfner MF (2001) Ovulation triggers activation of Drosophila oocytes. Dev Biol 234:416–424
29. Fellmeth JE, McKim KS (2022) A brief history of Drosophila (female) meiosis. Genes (Basel) 13:775. https://doi.org/10.3390/genes13050775
30. Cullen CF, Brittle AL, Ito T, Ohkura H (2005) The conserved kinase NHK-1 is essential for mitotic progression and unifying acentrosomal meiotic spindles in Drosophila melanogaster. J Cell Biol 171:593–602
31. Skold HN, Komma DJ, Endow SA (2005) Assembly pathway of the anastral Drosophila oocyte meiosis I spindle. J Cell Sci 118:1745–1755
32. McKim KS (2022) Highway to hell-thy meiotic divisions: chromosome passenger complex functions driven by microtubules: CPC interactions with both the chromosomes and

microtubules are important for spindle assembly and function. BioEssays 44:e2100202. https://doi.org/10.1002/bies.202100202
33. Colombié N, Cullen CF, Brittle AL, Jang JK, Earnshaw WC, Carmena M, McKim K, Ohkura H (2008) Dual roles of Incenp crucial to the assembly of the acentrosomal metaphase spindle in female meiosis. Development 135:3239–3246
34. Sampath SC, Ohi R, Leismann O, Salic A, Pozniakovski A, Funabiki H (2004) The chromosomal passenger complex is required for chromatin-induced microtubule stabilization and spindle assembly. Cell 118:187–202
35. Cesario J, McKim KS (2011) RanGTP is required for meiotic spindle organization and the initiation of embryonic development in Drosophila. J Cell Sci 124:3797–3810
36. Meireles AM, Fisher KH, Colombie N, Wakefield JG, Ohkura H (2009) Wac: a new Augmin subunit required for chromosome alignment but not for acentrosomal microtubule assembly in female meiosis. J Cell Biol 184:777–784
37. Dumont J, Petri S, Pellegrin F, Terret ME, Bohnsack MT, Rassinier P, Georget V, Kalab P, Gruss OJ, Verlhac MH (2007) A centriole- and RanGTP-independent spindle assembly pathway in meiosis I of vertebrate oocytes. J Cell Biol 176:295–305
38. Colombié N, Gluszek AA, Meireles AM, Ohkura H (2013) Meiosis-specific stable binding of augmin to acentrosomal spindle poles promotes biased microtubule assembly in oocytes. PLoS Genet 9(6):e1003562. https://doi.org/10.1371/journal.pgen.1003562
39. Hughes SE, Beeler JS, Seat A, Slaughter BD, Unruh JR, Bauerly E, Matthies JG, Hawley RS (2011) Gamma-tubulin is required foir bipolar spindle assembly and for proper kinetochore microtubule attachments during prometaphase in Drosophila oocytes. PLoS Genet 7:e1002209
40. Carmena M, Wheelock M, Funabiki H, Earnshaw WC (2012) The chromosomal passenger complex (CPC): from easy rider to the godfather of mitosis. Nat Rev Mol Cell Biol 13:789–803
41. Wang LI, DeFosse T, Jang JK, Battaglia RA, Wagner VF, McKim KS (2021) Borealin directs recruitment of the CPC to oocyte chromosomes and movement to the microtubules. J Cell Biol 220:e202006018. https://doi.org/10.1083/jcb.202006018
42. Eissenberg JC, Elgin SC (2000) The HP1 protein family: getting a grip on chromatin. Curr Opin Genet Dev 10:204–210
43. Casale AM, Cappucci U, Fanti L, Piacentini L (2019) Heterochromatin protein 1 (HP1) is intrinsically required for post-transcriptional regulation of Drosophila germline stem cell (GSC) maintenance. Sci Rep 9:4372. https://doi.org/10.1038/s41598-019-40152-1
44. Giauque CC, Bickel SE (2016) Heterochromatin-associated proteins HP1a and Piwi collaborate to maintain the association of achiasmate homologs in Drosophila oocytes. Genetics 203:173–189. https://doi.org/10.1534/genetics.115.186460
45. Samejima K, Platani M, Wolny M, Ogawa H, Vargiu G, Knight PJ, Peckham M, Earnshaw WC (2015) The inner centromere protein (INCENP) coil is a single alpha-helix (SAH) domain that binds directly to microtubules and is important for chromosome passenger complex (CPC) localization and function in mitosis. J Biol Chem 290:21460–21472. https://doi.org/10.1074/jbc.M115.645317
46. Ruppert JG, Samejima K, Platani M, Molina O, Kimura H, Jeyaprakash AA, Ohta S, Earnshaw WC (2018) HP1alpha targets the chromosomal passenger complex for activation at heterochromatin before mitotic entry. EMBO J 37(6):e97677. https://doi.org/10.15252/embj.201797677
47. Liu X, Song Z, Huo Y, Zhang J, Zhu T, Wang J, Zhao X, Aikhionbare F, Zhang J, Duan H et al (2014) Chromatin protein HP1 interacts with the mitotic regulator borealin protein and specifies the centromere localization of the chromosomal passenger complex. J Biol Chem 289:20638–20649. https://doi.org/10.1074/jbc.M114.572842
48. Alekseyenko AA, Gorchakov AA, Zee BM, Fuchs SM, Kharchenko PV, Kuroda MI (2014) Heterochromatin-associated interactions of Drosophila HP1a with dADD1, HIPP1, and repetitive RNAs. Genes Dev 28:1445–1460. https://doi.org/10.1101/gad.241950.114

49. Kang J, Chaudhary J, Dong H, Kim S, Brautigam CA, Yu H (2011) Mitotic centromeric targeting of HP1 and its binding to Sgo1 are dispensable for sister-chromatid cohesion in human cells. Mol Biol Cell 22:1181–1190. https://doi.org/10.1091/mbc.E11-01-0009
50. Nozawa RS, Nagao K, Masuda HT, Iwasaki O, Hirota T, Nozaki N, Kimura H, Obuse C (2010) Human POGZ modulates dissociation of HP1alpha from mitotic chromosome arms through Aurora B activation. Nat Cell Biol 12:719–727. https://doi.org/10.1038/ncb2075
51. Ainsztein AM, Kandels-Lewis SE, Mackay AM, Earnshaw WC (1998) INCENP centromere and spindle targeting: identification of essential conserved motifs and involvement of heterochromatin protein HP1. J Cell Biol 143:1763–1774
52. Jeyaprakash AA, Klein UR, Lindner D, Ebert J, Nigg EA, Conti E (2007) Structure of a Survivin-Borealin-INCENP core complex reveals how chromosomal passengers travel together. Cell 131:271–285
53. Salimian KJ, Ballister ER, Smoak EM, Wood S, Panchenko T, Lampson MA, Black BE (2011) Feedback control in sensing chromosome biorientation by the Aurora B kinase. Curr Biol 21:1158–1165. https://doi.org/10.1016/j.cub.2011.06.015
54. Emanuele MJ, Lan W, Jwa M, Miller SA, Chan CS, Stukenberg PT (2008) Aurora B kinase and protein phosphatase 1 have opposing roles in modulating kinetochore assembly. J Cell Biol 181:241–254
55. McAinsh AD, Marston AL (2022) The four causes: the functional architecture of centromeres and kinetochores. Annu Rev Genet 56:279–314. https://doi.org/10.1146/annurev-genet-072820-034559
56. Musacchio A, Desai A (2017) A molecular view of kinetochore assembly and function. Biology (Basel) 6:5. https://doi.org/10.3390/biology6010005
57. Costa MFA, Ohkura H (2019) The molecular architecture of the meiotic spindle is remodeled during metaphase arrest in oocytes. J Cell Biol 218:2854–2864. https://doi.org/10.1083/jcb.201902110
58. Doubilet S, McKim KS (2007) Spindle assembly in the oocytes of mouse and Drosophila—similar solutions to a problem. Chromosom Res 15:681–696
59. Burbank KS, Groen AC, Perlman ZE, Fisher DS, Mitchison TJ (2006) A new method reveals microtubule minus ends throughout the meiotic spindle. J Cell Biol 175:369–375
60. Liang ZY, Hallen MA, Endow SA (2009) Mature Drosophila meiosis I spindles comprise microtubules of mixed polarity. Curr Biol 19:163–168
61. Beaven R, Bastos RN, Spanos C, Rome P, Cullen CF, Rappsilber J, Giet R, Goshima G, Ohkura H (2017) 14-3-3 regulation of Ncd reveals a new mechanism for targeting proteins to the spindle in oocytes. J Cell Biol 216:3029–3039. https://doi.org/10.1083/jcb.201704120
62. Repton C, Cullen CF, Costa MFA, Spanos C, Rappsilber J, Ohkura H (2022) The phospho-docking protein 14-3-3 regulates microtubule-associated proteins in oocytes including the chromosomal passenger Borealin. PLoS Genet 18:e1009995. https://doi.org/10.1371/journal.pgen.1009995
63. Niedzialkowska E, Truong TM, Eldredge LA, Ali A, Redemann S, Stukenberg PT (2024) Chromosomal passenger complex condensates generate parallel microtubule bundles in vitro. J Biol Chem 300:105669. https://doi.org/10.1016/j.jbc.2024.105669
64. Duan Q, Chen H, Costa M, Dai W (2008) Phosphorylation of H3S10 blocks the access of H3K9 by specific antibodies and histone methyltransferase. Implication in regulating chromatin dynamics and epigenetic inheritance during mitosis. J Biol Chem 283:33585–33590. https://doi.org/10.1074/jbc.M803312200
65. Fischle W, Tseng BS, Dormann HL, Ueberheide BM, Garcia BA, Shabanowitz J, Hunt DF, Funabiki H, Allis CD (2005) Regulation of HP1-chromatin binding by histone H3 methylation and phosphorylation. Nature 438:1116–1122. https://doi.org/10.1038/nature04219
66. Williams MM, Mathison AJ, Christensen T, Greipp PT, Knutson DL, Klee EW, Zimmermann MT, Iovanna J, Lomberk GA, Urrutia RA (2019) Aurora kinase B-phosphorylated HP1α functions in chromosomal instability. Cell Cycle 18:1407–1421. https://doi.org/10.1080/15384101.2019.1618126

67. Dernburg AF, Sedat JW, Hawley RS (1996) Direct evidence of a role for heterochromatin in meiotic chromosome segregation. Cell 86:135–146
68. Fellmeth JE, Jang JK, Persaud M, Sturm H, Changela N, Parikh A, McKim KS (2023) A dynamic population of prophase CENP-C is required for meiotic chromosome segregation. PLoS Genet 19:e1011066. https://doi.org/10.1371/journal.pgen.1011066
69. Liu Y, Petrovic A, Rombaut P, Mosalaganti S, Keller J, Raunser S, Herzog F, Musacchio A (2016) Insights from the reconstitution of the divergent outer kinetochore of Drosophila melanogaster. Open Biol 6:150236. https://doi.org/10.1098/rsob.150236
70. Singleton MR (2016) Getting to the heart of an unusual kinetochore. Open Biol 6:160040. https://doi.org/10.1098/rsob.160040
71. Radford SJ, Go AM, McKim KS (2017) Cooperation between kinesin motors promotes spindle symmetry and chromosome organization in oocytes. Genetics 205:517–527. https://doi.org/10.1534/genetics.116.194647
72. Przewloka MR, Venkei Z, Glover DM (2009) Searching for Drosophila Dsn1 kinetochore protein. Cell Cycle 8:1292–1293. https://doi.org/10.4161/cc.8.8.8159
73. Wei RR, Sorger PK, Harrison SC (2005) Molecular organization of the Ndc80 complex, an essential kinetochore component. Proc Natl Acad Sci U S A 102:5363–5367. https://doi.org/10.1073/pnas.0501168102
74. Wei RR, Al-Bassam J, Harrison SC (2007) The Ndc80/HEC1 complex is a contact point for kinetochore-microtubule attachment. Nat Struct Mol Biol 14:54–59. https://doi.org/10.1038/nsmb1186
75. Rome P, Ohkura H (2018) A novel microtubule nucleation pathway for meiotic spindle assembly in oocytes. J Cell Biol 217:3431–3445. https://doi.org/10.1083/jcb.201803172
76. Jang JK, Rahman T, McKim KS (2005) The kinesinlike protein Subito contributes to central spindle assembly and organization of the meiotic spindle in Drosophila oocytes. Mol Biol Cell 16:4684–4694. https://doi.org/10.1091/mbc.E04-11-0964
77. Gilliland WD, Hughes SF, Vietti DR, Hawley RS (2009) Congression of achiasmate chromosomes to the metaphase plate in Drosophila melanogaster oocytes. Dev Biol 325:122–128. https://doi.org/10.1016/j.ydbio.2008.10.003
78. Radford SJ, Harrison AM, McKim KS (2012) Microtubule-depolymerizing kinesin KLP10A restricts the length of the acentrosomal meiotic spindle in Drosophila females. Genetics 192:431–440
79. White-Cooper H, Alphey L, Glover DM (1993) The cdc25 homologue twine is required for only some aspects of the entry into meiosis in Drosophila. J Cell Sci 106(Pt 4):1035–1044
80. Riparbelli MG, Callaini G (1996) Meiotic spindle organization in fertilized Drosophila oocyte: presence of centrosomal components in the meiotic apparatus. J Cell Sci 109:911–918
81. Huettner AF (1924) Maturation and fertilization in Drosophila melanogaster. J Morphol Physiol 39:249–265. https://doi.org/10.1002/jmor.1050390108
82. Endow SA, Komma DJ (1998) Assembly and dynamics of an anastral:astral spindle: the meiosis II spindle of Drosophila oocytes. J Cell Sci 111:2487–2495
83. Brent AE, MacQueen A, Hazelrigg T (2000) The Drosophila wispy gene is required for RNA localization and other microtubule-based events of meiosis and early embryogenesis. Genetics 154:1649–1662
84. Cullen CF, Ohkura H (2001) Msps protein is localized to acentrosomal poles to ensure bipolarity of Drosophila meiotic spindles. Nat Cell Biol 3:637–642. https://doi.org/10.1038/35083025
85. Riparbelli MG, Callaini G (2005) The meiotic spindle of the Drosophila oocyte: the role of centrosomin and the central aster. J Cell Sci 118:2827–2836
86. Jang JK, Messina L, Erdman MB, Arbel T, Hawley RS (1995) Induction of metaphase arrest in Drosophila oocytes by chiasma-based kinetochore tension. Science 268:1917–1919
87. Burbank KS, Mitchison TJ, Fisher DS (2007) Slide-and-cluster models for spindle assembly. Curr Biol 17:1373–1383. https://doi.org/10.1016/j.cub.2007.07.058
88. Dogterom M, Yurke B (1997) Measurement of the force-velocity relation for growing microtubules. Science 278:856–860. https://doi.org/10.1126/science.278.5339.856

89. Lombillo VA, Stewart RJ, Mcintosh JR (1995) Minus-end-directed motion of kinesin-coated microspheres driven by microtubule depolymerization. Nature 373:161–164. https://doi.org/10.1038/373161a0
90. Mullen TJ, Davis-Roca AC, Wignall SM (2019) Spindle assembly and chromosome dynamics during oocyte meiosis. Curr Opin Cell Biol 60:53–59. https://doi.org/10.1016/j.ceb.2019.03.014
91. Anjur-Dietrich MI, Kelleher CP, Needleman DJ (2021) Mechanical mechanisms of chromosome segregation. Cells 10(2):465. https://doi.org/10.3390/cells10020465
92. Zhao TT, Graham OS, Raposo A, St Johnston D (2012) Growing microtubules push the oocyte nucleus to polarize the dorsal-ventral axis. Science 336:999–1003. https://doi.org/10.1126/science.1219147
93. Jang JK, Gladstein AC, Das A, Shapiro JG, Sisco ZL, McKim KS (2021) Multiple pools of PP2A regulate spindle assembly, kinetochore attachments, and cohesion in Drosophila oocytes. J Cell Sci 134(14):jcs254037. https://doi.org/10.1242/jcs.254037
94. Riparbelli MG, Callaini G, Glover DM, Avides MC (2002) A requirement for the abnormal spindle protein to organise microtubules of the central spindle for cytokinesis in Drosophila. J Cell Sci 115:913–922
95. Maiato H, Khodjakov A, Rieder CL (2005) Drosophila CLASP is required for the incorporation of microtubule subunits into fluxing kinetochore fibres. Nat Cell Biol 7:42–47. https://doi.org/10.1038/ncb1207
96. Laycock JE, Savoian MS, Glover DM (2006) Antagonistic activities of Klp10A and Orbit regulate spindle length, bipolarity and function in vivo. J Cell Sci 119:2354–2361
97. Joshi JN, Changela N, Mahal L, Jang J, Defosse T, Wang LI, Das A, Shapiro JG, McKim K (2024) Meiosis-specific functions of kinetochore protein SPC105R required for chromosome segregation in Drosophila oocytes. Mol Biol Cell 35(8):ar105. https://doi.org/10.1091/mbc.E24-02-0067
98. Muscat CC, Torre-Santiago KM, Tran MV, Powers JA, Wignall SM (2015) Kinetochore-independent chromosome segregation driven by lateral microtubule bundles. elife 4:e06462. https://doi.org/10.7554/eLife.06462
99. Laband K, Le Borgne R, Edwards F, Stefanutti M, Canman JC, Verbavatz JM, Dumont J (2017) Chromosome segregation occurs by microtubule pushing in oocytes. Nat Commun 8:1499. https://doi.org/10.1038/s41467-017-01539-8
100. Danlasky BM, Panzica MT, McNally KP, Vargas E, Bailey C, Li W, Gong T, Fishman ES, Jiang X, McNally FJ (2020) Evidence for anaphase pulling forces during C. elegans meiosis. J Cell Biol 219:e202005179. https://doi.org/10.1083/jcb.202005179
101. Lenart P, Bacher CP, Daigle N, Hand AR, Eils R, Terasaki M, Ellenberg J (2005) A contractile nuclear actin network drives chromosome congression in oocytes. Nature 436:812–818
102. Mogessie B (2020) Advances and surprises in a decade of oocyte meiosis research. Essays Biochem 64:263–275. https://doi.org/10.1042/EBC20190068
103. Mogessie B, Scheffler K, Schuh M (2018) Assembly and positioning of the oocyte meiotic spindle. Annu Rev Cell Dev Biol 34:381–403. https://doi.org/10.1146/annurev-cellbio-100616-060553
104. Dunkley S, Mogessie B (2023) Actin limits egg aneuploidies associated with female reproductive aging. Sci Adv 9:eadc9161. https://doi.org/10.1126/sciadv.adc9161
105. Mogessie B, Schuh M (2017) Actin protects mammalian eggs against chromosome segregation errors. Science 357:eaal1647. https://doi.org/10.1126/science.aal1647
106. Schuh M (2011) An actin-dependent mechanism for long-range vesicle transport. Nat Cell Biol 13:1431–1436. https://doi.org/10.1038/ncb2353
107. Field CM, Lenart P (2011) Bulk cytoplasmic actin and its functions in meiosis and mitosis. Curr Biol 21:R825–R830. https://doi.org/10.1016/j.cub.2011.07.043
108. Schuh M, Ellenberg J (2008) A new model for asymmetric spindle positioning in mouse oocytes. Curr Biol 18:1986–1992

109. Yi K, Rubinstein B, Unruh JR, Guo F, Slaughter BD, Li R (2013) Sequential actin-based pushing forces drive meiosis I chromosome migration and symmetry breaking in oocytes. J Cell Biol 200:567–576. https://doi.org/10.1083/jcb.201211068
110. Azoury J, Lee KW, Georget V, Rassinier P, Leader B, Verlhac MH (2008) Spindle positioning in mouse oocytes relies on a dynamic meshwork of actin filaments. Curr Biol 18:1514–1519. https://doi.org/10.1016/j.cub.2008.08.044
111. Dahlgaard K, Raposo AA, Niccoli T, St Johnston D (2007) Capu and Spire assemble a cytoplasmic actin mesh that maintains microtubule organization in the Drosophila oocyte. Dev Cell 13:539–553. https://doi.org/10.1016/j.devcel.2007.09.003
112. Manseau L, Calley J, Phan H (1996) Profilin is required for posterior patterning of the Drosophila oocyte. Development 122:2109–2116. https://doi.org/10.1242/dev.122.7.2109
113. Wood BW, Shi X, Weil TT (2024) F-actin coordinates spindle morphology and function in Drosophila meiosis. PLoS Genet 20:e1011111. https://doi.org/10.1371/journal.pgen.1011111
114. Das A, Shah SJ, Fan B, Paik D, DiSanto DJ, Hinman AM, Cesario JM, Battaglia RA, Demos N, McKim KS (2016) Spindle assembly and chromosome segregation requires central spindle proteins in Drosophila oocytes. Genetics 202:61–75. https://doi.org/10.1534/genetics.115.181081
115. Das A, Cesario J, Hinman AM, Jang JK, McKim KS (2018) Kinesin 6 regulation in Drosophila female meiosis by the non-conserved N- and C- terminal domains. G3 (Bethesda, MD) 8:1555–1569. https://doi.org/10.1534/g3.117.300571
116. Giunta KL, Jang JK, Manheim EA, Subramanian G, McKim KS (2002) Subito encodes a kinesin-like protein required for meiotic spindle pole formation in Drosophila melanogaster. Genetics 160:1489–1501
117. Jang JK, Rahman T, Kober VS, Cesario J, McKim KS (2007) Misregulation of the kinesin-like protein subito induces meiotic spindle formation in the absence of chromosomes and centrosomes. Genetics 177:267–280
118. Sablin EP, Kull FJ, Cooke R, Vale RD, Fletterick RJ (1996) Crystal structure of the motor domain of the kinesin-related motor NCD. Nature 380:555–559
119. McDonald HB, Stewart RJ, Goldstein LS (1990) The kinesin-like NCD protein of Drosophila is a minus end-directed microtubule motor. Cell 63:1159–1165
120. Fink G, Hajdo L, Skowronek KJ, Reuther C, Kasprzak AA, Diez S (2009) The mitotic kinesin-14 Ncd drives directional microtubule-microtubule sliding. Nat Cell Biol 11:717–723. https://doi.org/10.1038/ncb1877
121. Hatsumi M, Endow SA (1992) Mutants of the microtubule motor protein, nonclaret disjunctional, affect spindle structure and chromosome movement in meiosis and mitosis. J Cell Sci 101:547–559
122. Mennella V, Rogers GC, Rogers SL, Buster DW, Vale RD, Sharp DJ (2005) Functionally distinct kinesin-13 family members cooperate to regulate microtubule dynamics during interphase. Nat Cell Biol 7:235–245. https://doi.org/10.1038/ncb1222
123. Rogers GC, Rogers SL, Schwimmer TA, Ems-McClung SC, Walczak CE, Vale RD, Scholey JM, Sharp DJ (2004) Two mitotic kinesins cooperate to drive sister chromatid separation during anaphase. Nature 427:364–370. https://doi.org/10.1038/nature02256
124. Duncan JE, Warrior R (2002) The cytoplasmic dynein and kinesin motors have interdependent roles in patterning the Drosophila oocyte. Curr Biol 12:1982–1991. https://doi.org/10.1016/s0960-9822(02)01303-9
125. Januschke J, Gervais L, Dass S, Kaltschmidt JA, Lopez-Schier H, St Johnston D, Brand AH, Roth S, Guichet A (2002) Polar transport in the Drosophila oocyte requires dynein and kinesin I cooperation. Curr Biol 12:1971–1981. https://doi.org/10.1016/s0960-9822(02)01302-7
126. Miyamoto DT, Perlman ZE, Burbank KS, Groen AC, Mitchison TJ (2004) The kinesin Eg5 drives poleward microtubule flux in Xenopus laevis egg extract spindles. J Cell Biol 167:813–818. https://doi.org/10.1083/jcb.200407126
127. Mitchison TJ, Maddox P, Gaetz J, Groen A, Shirasu M, Desai A, Salmon ED, Kapoor TM (2005) Roles of polymerization dynamics, opposed motors, and a tensile element in

governing the length of Xenopus extract meiotic spindles. Mol Biol Cell 16:3064–3076. https://doi.org/10.1091/mbc.E05-02-0174
128. Gorbsky GJ (2015) The spindle checkpoint and chromosome segregation in meiosis. FEBS J 282:2471–2487. https://doi.org/10.1111/febs.13166
129. Pesin JA, Orr-Weaver TL (2008) Regulation of APC/C activators in mitosis and meiosis. Annu Rev Cell Dev Biol 24:475–499. https://doi.org/10.1146/annurev.cellbio.041408.115949
130. Zur A, Brandeis M (2002) Timing of APC/C substrate degradation is determined by FZY/FZR specificity of destruction boxes. EMBO J 21:4500–4510. https://doi.org/10.1093/emboj/cdf452
131. Swan A, Schupbach T (2007) The Cdc20 (Fzy)/Cdh1-related protein, Cort, cooperates with Fzy in cyclin destruction and anaphase progression in meiosis I and II in Drosophila. Development 134:891–899
132. Davydenko O, Schultz RM, Lampson MA (2013) Increased CDK1 activity determines the timing of kinetochore-microtubule attachments in meiosis I. J Cell Biol 202:221–229. https://doi.org/10.1083/jcb.201303019
133. Kishimoto T (2003) Cell-cycle control during meiotic maturation. Curr Opin Cell Biol 15:654–663. https://doi.org/10.1016/j.ceb.2003.10.010
134. Bourouh M, Dhaliwal R, Rana K, Sinha S, Guo Z, Swan A (2016) Distinct and overlapping requirements for cyclins A, B, and B3 in Drosophila female meiosis. G3 (Bethesda, MD) 6:3711–3724. https://doi.org/10.1534/g3.116.033050
135. Shapiro JG, Changela N, Jang JK, Joshi JN, McKim KS (2024) Distinct checkpoint and homolog biorientation pathways regulate meiosis I in Drosophila oocytes. PLoS Genet 21, e1011400. https://doi.org/10.1371/journal.pgen.1011400.
136. Logarinho E, Bousbaa H, Dias JM, Lopes C, Amorim I, Antunes-Martins A, Sunkel CE (2004) Different spindle checkpoint proteins monitor microtubule attachment and tension at kinetochores in Drosophila cells. J Cell Sci 117:1757–1771. https://doi.org/10.1242/jcs.01033
137. Sun SC, Kim NH (2012) Spindle assembly checkpoint and its regulators in meiosis. Hum Reprod Update 18:60–72. https://doi.org/10.1093/humupd/dmr044
138. Batiha O, Swan A (2012) Evidence that the spindle assembly checkpoint does not regulate APC(Fzy) activity in Drosophila female meiosis. Genome 55:63–67. https://doi.org/10.1139/g11-079
139. Malmanche N, Owen S, Gegick S, Steffensen S, Tomkiel JE, Sunkel CE (2007) Drosophila BubR1 is essential for meiotic sister-chromatid cohesion and maintenance of synaptonemal complex. Curr Biol 17:1489–1497
140. Guo Z, Batiha O, Bourouh M, Fifield E, Swan A (2016) Role of securin, separase and cohesins in female meiosis and polar body formation in Drosophila. J Cell Sci 129:531–542. https://doi.org/10.1242/jcs.179358
141. Abrieu A, Magnaghi-Jaulin L, Kahana JA, Peter M, Castro A, Vigneron S, Lorca T, Cleveland DW, Labbe JC (2001) Mps1 is a kinetochore-associated kinase essential for the vertebrate mitotic checkpoint. Cell 106:83–93. https://doi.org/10.1016/s0092-8674(01)00410-x
142. Marston AL, Wassmann K (2017) Multiple duties for spindle assembly checkpoint kinases in meiosis. Front Cell Dev Biol 5:109. https://doi.org/10.3389/fcell.2017.00109
143. van der Horst A, Vromans MJ, Bouwman K, van der Waal MS, Hadders MA, Lens SM (2015) Inter-domain cooperation in INCENP promotes Aurora B relocation from centromeres to microtubules. Cell Rep 12:380–387. https://doi.org/10.1016/j.celrep.2015.06.038
144. Trivedi P, Zaytsev AV, Godzi M, Ataullakhanov FI, Grishchuk EL, Stukenberg PT (2019) The binding of Borealin to microtubules underlies a tension independent kinetochore-microtubule error correction pathway. Nat Commun 10:682. https://doi.org/10.1038/s41467-019-08418-4
145. Bonner MK, Haase J, Saunders H, Gupta H, Li BI, Kelly AE (2020) The Borealin dimerization domain interacts with Sgo1 to drive Aurora B-mediated spindle assembly. Mol Biol Cell 31:2207–2218. https://doi.org/10.1091/mbc.E20-05-0341

146. Bryan NW, Ali A, Niedzialkowska E, Mayne L, Stukenberg PT, Black BE (2024) Structural basis for the phase separation of the chromosome passenger complex. elife 13:e92709. https://doi.org/10.7554/eLife.92709
147. Chantalat L, Skoufias DA, Kleman JP, Jung B, Dideberg O, Margolis RL (2000) Crystal structure of human survivin reveals a bow tie-shaped dimer with two unusual alpha-helical extensions. Mol Cell 6:183–189
148. Verdecia MA, Huang H, Dutil E, Kaiser DA, Hunter T, Noel JP (2000) Structure of the human anti-apoptotic protein survivin reveals a dimeric arrangement. Nat Struct Biol 7:602–608. https://doi.org/10.1038/76838
149. Kelly AE, Ghenoiu C, Xue JZ, Zierhut C, Kimura H, Funabiki H (2010) Survivin reads phosphorylated histone H3 threonine 3 to activate the mitotic kinase Aurora B. Science 330:235–239
150. Soupsana K, Karanika E, Kiosse F, Christogianni A, Sfikas Y, Topalis P, Batistatou A, Kanaki Z, Klinakis A, Politou AS, Georgatos S (2021) Distinct roles of haspin in stem cell division and male gametogenesis. Sci Rep 11(1):19901. https://doi.org/10.1038/s41598-021-99307-8
151. Hadders MA, Hindriksen S, Truong MA, Mhaskar AN, Wopken JP, Vromans MJM, Leas SMA (2020) Untangling the contribution of Haspin and Bub1 to Aurora B function during mitosis. J Cell Biol 219(3):e201907087. https://doi.org/10.1083/jcb.201907087
152. Nguyen AL, Gentilello AS, Balboula AZ, Shrivastava V, Ohring J, Schindler K (2014) Phosphorylation of threonine 3 on histone H3 by haspin kinase is required for meiosis I in mouse oocytes. J Cell Sci 127:5066–5078. https://doi.org/10.1242/jcs.158840
153. Cairo G, Lacefield S (2020) Establishing correct kinetochore-microtubule attachments in mitosis and meiosis. Essays Biochem 64:277–287. https://doi.org/10.1042/EBC20190072
154. Zhou LL, Tian XY, Zhu CL, Wang FW, Higgins JMG (2014) Polo-like kinase-1 triggers histone phosphorylation by Haspin in mitosis. EMBO Rep 15:273–281. https://doi.org/10.1002/embr.201338080
155. Schmitz ML, Higgins JMG, Seibert M (2020) Priming chromatin for segregation: functional roles of mitotic histone modifications. Cell Cycle 19:625–641. https://doi.org/10.1080/15384101.2020.1719585
156. Bonner AM, Hughes SE, Hawley RS (2020) Regulation of polo kinase by matrimony is required for Cohesin maintenance during Drosophila melanogaster female meiosis. Curr Biol 30:715–722.e713. https://doi.org/10.1016/j.cub.2019.12.027
157. Clyne RK, Katis VL, Jessop L, Benjamin KR, Herskowitz I, Lichten M, Nasmyth K (2003) Polo-like kinase Cdc5 promotes chiasmata formation and cosegregation of sister centromeres at meiosis I. Nat Cell Biol 5:480–485. https://doi.org/10.1038/ncb977
158. Barbosa J, Sunkel CE, Conde C (2022) The role of mitotic kinases and the RZZ complex in kinetochore-microtubule attachments: doing the right link. Front Cell Dev Biol 10:787294. https://doi.org/10.3389/fcell.2022.787294
159. Malmanche N, Maia A, Sunkel CE (2006) The spindle assembly checkpoint: preventing chromosome mis-segregation during mitosis and meiosis. FEBS Lett 580:2888–2895
160. Karess R (2005) Rod-Zw10-Zwilch: a key player in the spindle checkpoint. Trends Cell Biol 15:386–392. https://doi.org/10.1016/j.tcb.2005.05.003
161. Przewloka MR, Glover DM (2009) The kinetochore and the centromere: a working long distance relationship. Annu Rev Genet 43:439–465
162. Cheeseman IM, Chappie JS, Wilson-Kubalek EM, Desai A (2006) The conserved KMN network constitutes the core microtubule-binding site of the kinetochore. Cell 127:983–997
163. Barth TK, Schade GO, Schmidt A, Vetter I, Wirth M, Heun P, Thomae AW, Imhof A (2014) Identification of novel Drosophila centromere-associated proteins. Proteomics 14:2167–2178. https://doi.org/10.1002/pmic.201400052
164. Przewloka MR, Zhang W, Costa P, Archambault V, D'Avino PP, Lilley KS, Laue ED, McAinsh AD, Glover DM (2007) Molecular analysis of core kinetochore composition and assembly in Drosophila melanogaster. PLoS One 2:e478. https://doi.org/10.1371/journal.pone.0000478

165. Schittenhelm RB, Chaleckis R, Lehner CF (2009) Intrakinetochore localization and essential functional domains of Drosophila Spc105. EMBO J 28:2374–2386. https://doi.org/10.1038/emboj.2009.188
166. Miyazaki S, Kim J, Yamagishi Y, Ishiguro T, Okada Y, Tanno Y, Sakuno T, Watanabe Y (2017) Meikin-associated polo-like kinase specifies Bub1 distribution in meiosis I. Genes Cells 22:552–567. https://doi.org/10.1111/gtc.12496
167. Wang LI, Das A, McKim KS (2019) Sister centromere fusion during meiosis I depends on maintaining cohesins and destabilizing microtubule attachments. PLoS Genet 15:e1008072. https://doi.org/10.1371/journal.pgen.1008072
168. Richter MM, Poznanski J, Zdziarska A, Czarnocki-Cieciura M, Lipinszki Z, Dadlez M, Glover DM, Przewloka MR (2016) Network of protein interactions within the Drosophila inner kinetochore. Open Biol 6:150238. https://doi.org/10.1098/rsob.150238
169. Kiyomitsu T, Iwasaki O, Obuse C, Yanagida M (2010) Inner centromere formation requires hMis14, a trident kinetochore protein that specifically recruits HP1 to human chromosomes. J Cell Biol 188:791–807. https://doi.org/10.1083/jcb.200908096
170. Obuse C, Iwasaki O, Kiyomitsu T, Goshima G, Toyoda Y, Yanagida M (2004) A conserved Mis12 centromere complex is linked to heterochromatic HP1 and outer kinetochore protein Zwint-1. Nat Cell Biol 6:1135–1141. https://doi.org/10.1038/ncb1187
171. Pereira C, Reis RM, Gama JB, Celestino R, Cheerambathur DK, Carvalho AX, Gassmann R (2018) Self-assembly of the RZZ complex into filaments drives kinetochore expansion in the absence of microtubule attachment. Curr Biol 28(21):3408–3421.e3408. https://doi.org/10.1016/j.cub.2018.08.056
172. Zhang G, Lischetti T, Hayward DG, Nilsson J (2015) Distinct domains in Bub1 localize RZZ and BubR1 to kinetochores to regulate the checkpoint. Nat Commun 6:7162. https://doi.org/10.1038/ncomms8162
173. Griffis ER, Stuurman N, Vale RD (2007) Spindly, a novel protein essential for silencing the spindle assembly checkpoint, recruits dynein to the kinetochore. J Cell Biol 177:1005–1015. https://doi.org/10.1083/jcb.200702062
174. Tavosanis G, Llamazares S, Goulielmos G, Gonzalez C (1997) Essential role for gamma-tubulin in the acentriolar female meiotic spindle of Drosophila. EMBO J 16:1809–1819. https://doi.org/10.1093/emboj/16.8.1809
175. Wilson PG, Borisy GG (1998) Maternally expressed gamma Tub37CD in Drosophila is differentially required for female meiosis and embryonic mitosis. Dev Biol 199:273–290. https://doi.org/10.1006/dbio.1998.8900
176. Llamazares S, Tavosanis G, Gonzalez C (1999) Cytological characterisation of the mutant phenotypes produced during early embryogenesis by null and loss-of-function alleles of the gammaTub37C gene in Drosophila. J Cell Sci 112(Pt 5):659–667. https://doi.org/10.1242/jcs.112.5.659
177. Ehrhardt DW, Shaw SL (2006) Microtubule dynamics and organization in the plant cortical array. Annu Rev Plant Biol 57:859–875. https://doi.org/10.1146/annurev.arplant.57.032905.105329
178. Ju JQ, Zhang HL, Wang Y, Hu LL, Sun SC (2024) Kinesin KIFC3 is essential for microtubule stability and cytokinesis in oocyte meiosis. Cell Commun Signal 22(1):199. https://doi.org/10.1186/s12964-024-01589-8
179. Fu W, Chen H, Wang G, Luo J, Deng Z, Xin G, Xu N, Guo X, Lei J, Jiang Q, Zhang C (2013) Self-assembly and sorting of acentrosomal microtubules by TACC3 facilitate kinetochore capture during the mitotic spindle assembly. Proc Natl Acad Sci U S A 110:15295–15300. https://doi.org/10.1073/pnas.1312382110
180. Zhang Y, Tan L, Yang Q, Li C, Liou YC (2018) The microtubule-associated protein HURP recruits the centrosomal protein TACC3 to regulate K-fiber formation and support chromosome congression. J Biol Chem 293:15733–15747. https://doi.org/10.1074/jbc.RA118.003676

181. Wakefield JG, Bonaccorsi S, Gatti M (2001) The Drosophila protein asp is involved in microtubule organization during spindle formation and cytokinesis. J Cell Biol 153:637–648
182. Sun L, Gao J, Dong X, Liu M, Li D, Shi X, Dong JT, Lu X, Liu C, Zhou J (2008) EB1 promotes Aurora-B kinase activity through blocking its inactivation by protein phosphatase 2A. Proc Natl Acad Sci U S A 105:7153–7158. https://doi.org/10.1073/pnas.0710018105
183. Banerjee B, Kestner CA, Stukenberg PT (2014) EB1 enables spindle microtubules to regulate centromeric recruitment of Aurora B. J Cell Biol 204:947–963. https://doi.org/10.1083/jcb.201307119
184. Rogers SL, Rogers GC, Sharp DJ, Vale RD (2002) Drosophila EB1 is important for proper assembly, dynamics, and positioning of the mitotic spindle. J Cell Biol 158:873–884
185. Głuszek AA, Cullen CF, Li W, Battaglia RA, Radford SJ, Costa MF, McKim KS, Goshima G, Ohkura H (2015) The microtubule catastrophe promoter Sentin delays stable kinetochore-microtubule attachment in oocytes. J Cell Biol 211:1113–1120. https://doi.org/10.1083/jcb.201507006
186. Li W, Miki T, Watanabe T, Kakeno M, Sugiyama I, Kaibuchi K, Goshima G (2011) EB1 promotes microtubule dynamics by recruiting sentin in Drosophila cells. J Cell Biol 193:973–983. https://doi.org/10.1083/jcb.201101108
187. Wu C, Singaram V, McKim KS (2008) mei-38 is required for chromosome segregation during meiosis in Drosophila females. Genetics 180:61–72
188. Popova JV, Pavlova GA, Razuvaeva AV, Yarinich LA, Andreyeva EN, Anders AF, Galimova YA, Renda F, Somma MP, Pindyurin AV, Gatti M (2022) Genetic control of kinetochore-driven microtubule growth in Drosophila mitosis. Cells 11(14):2127. https://doi.org/10.3390/cells11142127
189. Hayward D, Wakefield JG (2014) Chromatin-mediated microtubule nucleation in Drosophila syncytial embryos. Commun Integr Biol 7:e28512. https://doi.org/10.4161/cib.28512
190. Ye AA, Deretic J, Hoel CM, Hinman AW, Cimini D, Welburn JP, Maresca TJ (2015) Aurora a kinase contributes to a pole-based error correction pathway. Curr Biol 25:1842–1851. https://doi.org/10.1016/j.cub.2015.06.021
191. Leca N, Barbosa F, Rodriguez-Calado S, Esposito Verza A, Moura M, Pedroso PD, Pinto I, Artes E, Bange T, Sunkel CE, et al (2025) Proximity-based activation of AURORA A by MPS1 potentiates error correction. Curr Biol 35:1935–1947. https://doi.org/10.1016/j.cub.2025.03.018
192. So C, Menelaou K, Uraji J, Harasimov K, Steyer AM, Seres KB, Bucevičius J, Lukinavičius G, Möbius W, Sibold C et al (2022) Mechanism of spindle pole organization and instability in human oocytes. Science 375:eabj3944. https://doi.org/10.1126/science.abj3944
193. Yu JX, Guan Z, Nash HA (2006) The mushroom body defect gene product is an essential component of the meiosis II spindle apparatus in Drosophila oocytes. Genetics 173:243–253. https://doi.org/10.1534/genetics.105.051557
194. Divekar NS, Davis-Roca AC, Zhang L, Dernburg AF, Wignall SM (2021) A degron-based strategy reveals new insights into Aurora B function in C. Elegans. PLoS Genet 17:e1009567. https://doi.org/10.1371/journal.pgen.1009567
195. Parton RM, Valles AM, Dobbie IM, Davis I (2010) Live cell imaging in Drosophila melanogaster. Cold Spring Harb Protoc 2010(4):pdbtop75. https://doi.org/10.1101/pdb.top75
196. Barbosa P, Zhaunova L, Debilio S, Steccanella V, Kelly V, Ly T, Ohkura H (2021) SCF-Fbxo42 promotes synaptonemal complex assembly by downregulating PP2A-B56. J Cell Biol 220(2):e202009167. https://doi.org/10.1083/jcb.202009167

Index

B. Mogessie (ed.), *Cellular Architecture and Dynamics in Female Meiosis*, https://doi.org/10.1007/978-3-031-97173-0

The manufacturer's authorised representative in the EU is Springer Nature Customer Service Centre GmbH, Europaplatz 3, 69115 Heidelberg, Germany. If you have any concerns regarding our products, please contact ProductSafety@springernature.com

Printed and bound by CPI Group (UK) Ltd, Croydon, CR0 4YY
30/09/2026
02225559-0009